MW01324345

2

REGULUS
The Forgotten Weapon

A Complete Guide to Chance Vought's Regulus I and II

David K. Stumpf

Turner Publishing Company
412 Broadway
P.O. Box 3101
Paducah, KY 42002-3101
(502) 443-0121

Turner Publishing Company Staff:
Editor: Amy Cloud
Designer: Lora Lauder

Library of Congress
Catalog Card Number: 1-56311-277-9
ISBN: 96-60266
Printed in the United States of America

Additional copies may be purchased directly from Turner Publishing Company.

Table of Contents

FOREWORD

In this unique document Dr. David Stumpf has put together a complete history of the Regulus Cruise Missile from its conception to its retirement from the Fleet. He has presented it from both the technical and the human side—which in itself is most unusual!

It is generally accepted that the most important part of any complex, successful and effective weapon system are the people involved. It is easy to have technical support in a matter of hours for anything except the "People System." That required more time and more people that could work the problem. This was certainly my experience when I was head of the first Regulus Guided Missile Training Unit, skipper of a nuclear Fleet Ballistic Missile Submarine, as well as Commander of a FBM submarine squadron.

This book covers the complete history of every unit that was involved with the missile program at Chance Vought and the Navy, as well as all key individuals identified with the background and previous involvement with the missile program. It would be impossible to fully cover the tremendous role of Nevin Palley or Sam Perry—outstanding engineers and problem-solving people at Chance Vought.

The Regulus missile was successful, cost effective, reliable and easy to support. It was a role model for cooperation and success between the contractor and the Navy, as was shown by the ease of the transfer of operational control to the Navy of a missile flying at 35000 feet on a mission. The program had to face tremendous technical problems—it started with conventional submarines dating from World War II and ended up on a nuclear submarine built specifically for Regulus missiles. It broke ground on the needed development of inertial navigation, missile guidance, impact accuracy and many others.

Putting this history of the Regulus Cruise Missile together took tremendous effort, dedication, sacrifice, and understanding. Dr. David Stumpf has provided anyone interested an insight into the complexities and difficulties encountered in order to achieve success. This will be a marvelous source of information to anyone involved in a complex program development.

W.E. Sims
Captain, USN (Ret.)

Foreword

This book, of course, is interesting to people like myself who had a part in the operational life of the Regulus program as a part of the nation's Strategic Deterrent System or who were involved in the research, development, testing and construction of the system components. At the same time, this subject will be interesting and instructive to anyone who wonders why it takes so long to bring a weapon system to an operational status. It is rewarding to note that the effort and resource expended to create this system provided much of the learning experience for the current and highly effective Tomahawk system.

The operational period of the Regulus system was really a story of the men who made up the crews struggling to meet their commitments. This book is a vivid reminder to me of those years and the wonderful and dedicated men who, because of their pride and professionalism, made the system work against every adversity.

The sea inself was an adversary, with our operating area being a region of the roughest weather in the Pacific Ocean. The Regulus submarines lost parts of their superstructure, encountered, damaging ice flows, experienced severe depth control problems and a rough ride for periods of over two months submerged.

Time was an adversary. Every ship in the system was working against the clock to repair the ship in port, train the crew in the local operating area, perform torpedo and missile training operations, load the ship for deployment, and make the sailing date so that you relieved your buddy on time! And then to do it all over again and handle the inevitable equipment casualities so as not to delay sailing. And, oh yes, for most of us there was a family to fit in there somewhere, but it doesn't take much imagination to figure out what priority they got in the scheme of things - that would simply have to wait for a shore duty tour!

From my perspective as the Executive Officer of USS *Growler* (SSG 577) for three patrols and Commanding Officer of USS *Grayback* (SSG 574) for four, the ships themselves proved to be the greatest challenge. With unreliable main engines, aluminum superstructures that could not sustain the pounding they got and the poor hull design that made depth control marginal in high sea states, these ships were badlly beaten up by every deployment. The supply system and the logistics support bases and shipyards were hard pressed each cycle to get them back on line and keep them there. I can not say enough about the people who manned these ships. With few exceptions, they were magnificent. They took the hardships in stride and took pride making the system work. They knew they were the best and that nothing was too tough for them. They held the attitude that no other group of people could do this job - and as far as I am concerned, they were probably right!

Upon reflection, perhaps the real heroes of the operational period were the wives. They raised our kids, paid the bills and fixed the car, and then met us at the pier with the family looking like a million bucks as if it were a walk in the park!

Where did all these great people come from?

John J. Ekelund
Rear Admiral, USN (Ret.)

Acknowledgements

After five years of research I obviously have many people to thank. While all of the interviews are listed in the appendices, I want to thank many individuals more completely. Jean Nott, the widow of Captain Hugh G. Nott, USN (Ret.), the first commanding officer of USS *Grayback* (SSG 574) provided names and addresses of several Regulus submarine program officers and this in a real sense was the start of the project . My initial contact and interviews with Captain Donald Henderson, USN (Ret.) and Commander Samuel T. Bussey, USN (Ret.), and Captain William Sims, USN (Ret.) in August 1989, led to a short contact list that eventually blossomed into 149 people. Captain Rex Rader, USN (Ret.), and his wife Betty, extended both their hospitality and enthusiasm to a perfect stranger during the very early stages of the project. While the book later turned in a different direction, Captain Rader's assistance is greatly appreciated. Rear Admiral Robert Blount, USN (Ret.), took time from his active work schedule, on short notice, to respond to my questions about the first deterrent patrols conducted by USS *Barbero* (SSG 317). His continued full support with copious quantities of memorabilia and photographs made piecing together the early deterrent patrol history of *Barbero* much easier. David Alan Rosenberg, Ph.D., provided me with the unbelievable information that USS *Growler* (SSG 577) had just been added to the INTREPID Air-Sea-Space Museum in New York City. Dr. Rosenberg also informed me of the Ph.D. thesis of Berend D. Bruins. This thesis more then ever convinced me that the time had come to write the complete history of the Regulus program since Bruins' work was being quoted so often and represents only one side of the program's history. Captain Marvin S. Blair, USN (Ret.), was the first correspondent from my mass mailing. His enthusiasm and support over the past six years, including critical but helpful suggestions on the manuscript, and a wonderful scrapbook of photographs, were especially important. Captain Charles Priest, Jr., USN (Ret.), the first commanding officer of *Growler* and his wife Susan, are perfect examples of the kind of camaraderie that developed during the interview process. Captain Jack O'Connell, USN (Ret.) and Commander Eugene Lindsey, USN (Ret.), fleshed out the initial material on *Barbero* provided by Rear Admiral Blount. Considering my first contacts with both was during the blizzard of December 1989, their interest and support were all the more appreciated. Captain Myron "Max" Eckhart, USN (Ret.) was perhaps the most patient of all my interviews as he explained the story of bringing the BPQ-2 version of the Trounce guidance system into operational status. Commander Morris A. Christensen, USN (Ret.) and his wife Patty, provided memorable moments during my visits and interviews. Commander Christensen's tenacity in his insistence that USS *Tunny* (SSG 282) carried out the first scheduled Regulus I deterrent patrol was the beginning of a six-month effort to locate and copy the pertinent unclassified ship's deck logs for all five submarines. Rear Admiral Daniel Richardson, USN (Ret.), provided the first records of submarine patrol dates. This compilation was the starting point of a two month effort by Doreen German, Ship's History Branch, Naval Historical Center, to confirm not only patrol dates but also several changes of command. Tom Stebbins, CWO, USN (Ret.), provided outstanding support for four years as well as reviewing the submarine chapters. He also supplied a large number of *Growler* crew member names and addresses.

Research on the heavy cruiser Regulus operations was particularly difficult. My first contact, Captain Robert Munroe, USN (Ret.), clarified the early history of the scope of the program and provided me with sufficient information to ask intelligent follow-on questions. Commander Joseph Zelibor, USN (Ret.) and Lieutenant Commander Kenneth Brisco, USN (Ret.) provided a wealth of photographs as well as operational deployment data.

Bob Lawson, formerly Editor-in-Chief of "The Hook," the journal of the Tailhook Association, made me an offer I could not refuse concerning data on Guided Missile Groups ONE and TWO. Bob introduced me to Commander Ernie Mares, USN (Ret.), a pilot with Guided Missile Group ONE and we published an article in the Spring 1991 issue of "The Hook". Captain George Monthan, USN (Ret.), one of the first Navy chase pilots of the program, provided patient review of more then one draft of the manuscript. His frank comments and attention to detail are greatly appreciated. Lieutenant Commander Robert Blount, USN (Ret.), Captain David Leue, USN (Ret.), Rear Admiral Dewitt Freeman, USN (Ret.), Commander Roy Mock, USN (Ret.), Lieutenant Commander Len Plog, USN (Ret.), Rear Admiral Jack Monger, USN (Ret.) and Commander Al Thayer, USN (Ret.) all provided extensive interviews and memorabilia.

Rear Admiral Russell Gorman, USN (Ret.) and Rear Admiral Jerry Miller, USN (Ret.), were instrumental in starting the process of obtaining my limited-access security clearance by speaking on my behalf to Dr. Dean Allard, Director of the Naval Historical Center. Vice Admiral Roger F. Bacon, USN, made the clearance possible. Mary Anderson from Naval Investigative Services was most helpful in clearing several paperwork log jams during this process. Bernard Cavalcante, Head of Operational Archives, Naval Historical Center, and his staff members Kathleen Lloyd and Ella Nargele were most patient and cooperative as I learned my way around their document listings. Their assistance and generosity made much of the more tedious document review more enjoyable and productive. Cary Conn, Freedom of Information Act Officer, Suitland Reference Branch, National Archives and Record Administration, was outstanding in helping with the voluminous records from the Bureau of Aeronautics' Guided Missile Division. Johnny Howard, Freedom of Information Act Officer, Pacific Missile Test Center, Pt. Mugu, was instrumental in arranging access to the Center's archives. Max White volunteered his time to locate documents and assist in my viewing the unclassified material. Debbie Franz-Anderson at the Puget Sound Naval Shipyard confirmed that USS *Halibut* (SSGN 587) was still in the inactive fleet at Puget Sound. Wendy Gully of the Nautilus Submarine Force Library and Museum was also very helpful as was Susan Lemmon of Mare Island Naval Shipyard.

Obviously Chance Vought Aircraft archives and personnel were critical to the successful research on this book. In 1989, Jan Dixson of Public Relations Department at LTV, was able to locate Bill Micchelli, one of the early field managers in the program. Bill graciously invited me to Dallas to meet retired Regulus program personnel as well as several that were still at LTV (now Loral Vought Systems) at the time. With this effort on his part, Bill made the book possible. My first interview was with Leroy Pearson, the Chance Vought Chief Test Pilot for Regulus I. Considering that Roy had just returned home from major surgery, I was hoping to simply meet him, introduce myself and begin corresponding. Roy was full of energy and we talked for hours, far longer then we should have. I thank Roy's wife, Lois, for not kicking me out of the house! Roy's collection of photographs and documents, including a copy of the 1961 Regulus I Flight Test Record, provided a critical framework on which to build the Regulus I story. Later in the same visit I met with Fred Randell and Robert Stewart, managers of Regulus I and II programs; and Sam Lynne, one of the first missile operations engineers. In each case, the interview started with "I don't have much to say" and ended one or two hours later. All three were of considerable help in piecing the story together. I especially thank Sam Lynne for the gift of one of the Chance Vought promotional Regulus I lighters. C.O. Miller, Ace Yeager, Joe Engle, Glenn Paulk, Bud Holcomb and Bill Coleman, all Chance Vought test pilots for either Regulus I and/or II, provided me with wonderful flight stories as well as patiently explaining how the flight testing was conducted. Palmer Ransdell and Bill Albrecht, field engineers during the program described many of the individual early flight operations in great detail, and provided many key photographs. Bill Cannon, a missile operations engineer for both Regulus I and II, provided not only a wealth of documents and photographs; in addition, he expedited access to the document and photographic archives of LTV and provided generous photocopy privileges. Carol Newton, his secretary, was more then patient in making sure my phone calls got through. My interviews with George Sutherland, an engineer in Regulus I and a manager during Regulus II, were invaluable. His field notes provided a basis for the Regulus II flight history. Morgan Wilkes provided many of the early research program photographs used in the book.

My first contact in the LTV Records Department was MaryAnne Lloyd. Over a three year period, Ms Lloyd located many more documents then I had any right to hope for. Whatever success this book achieves is in no small part due to her efforts. Tommy L. Wilson, Director of Public Relations for LTV and Michael Drake, Manager for Public Relations and Advertising, Missiles Division, provided access to Jim Blick, Manager of Cinematic Services at LTV. Without their cooperation and generous contribution to the costs incurred, the film-to-video transfer of Regulus I and II flights would not have been possible. Jerry David patiently spent three hours with me reviewing the film and at least that many preparing it for the copy process.

Lieutenant Julian "Joe" Morrison III, USN (Ret.) and his wife, Helen, helped tremendously as I put together the second rough draft of the manuscript. His attention to consistency and detail, particularly in the submarine chapters was invaluable. Thank you, Helen, for putting off the many projects while Joe worked with me on this book. Thank you, Joe, for all the input. The visit to Tennessee after the Great Storm of 1993 was indeed memorable. Captain Vance Morrison, USN (Ret.) and his wife, Libby, extended warm hospitality and advice to a stranger that his brother, Joe, knew only from letters and phone calls. Captain Morrison's critical review of the rough draft is greatly appreciated. Pat Acton, Joe's daughter, provided long distance research in Hawaii, locating several documents that helped complete the book. Thank you, Pat, for responding to my "Just one more request," messages.

My collection of photographs would have not been possible without the help of Wally Knepp and Ray Allbright. The quality of the work was superb and in many instances they were able to generate copy negatives from difficult source photographs.

Captain Wayne Hughes, USN (Ret.), encouraged me to begin my research in May 1989. Over the ensuing 6 years he provided support and constructive criticsm that kept me on track.

The deck plans included in Appendix V were generously provided by Jim Christley. Jim also provided timely encouragement along the way.

Bob Martin, Chief Editor, and Amy Cloud, my editor, proved to have high levels of patience and plenty of enthusiasm for this project. Their support is greatly appreciated. The layout was designed by Lora Lauder. Her skills in working around my Wordperfect manuscript and multitude of photographs was amazing.

Last and most important of all has been the incredible patience and support of my wife Susan. As I located hard-to-find people, photographs, and documents, Susan realized that this was a project that was meant to be. Her constant vigil to keep me on track was a wonderful expression of love and friendship, even if I didn't agree with her all the time. Innumerable phone calls, substantial phone bills, to say nothing of travel and photography costs; computer problems...the list could go on and on. The last 20 months of effort were while I fulfilled my role as househusband.

David K. Stumpf
April, 1996

PREFACE

The Regulus Missile program is a nearly forgotten part of the early evolution of strategic nuclear weapons. Dismissed by many leading Navy weapons system historians as a stop gap system, inferior to the later submarine launched ballistic missiles, Regulus was in fact an alternative path on our strategic nuclear weapons map, a path that eventually led the Navy to the extremely successful and versatile Tomahawk cruise missile program.

In the fall of 1947, the United States Navy, with Chance Vought Aircraft, Incorporated, as the prime contractor, began development of the SSM-N-8 missile, Regulus I. By 1955, when Regulus I was deployed to the Western Pacific theater of operations aboard the USS *Los Angeles* (CA 135), the Navy had its first truly rapid response, intermediate range nuclear weapon delivery capability. In 1955 and 1956, the aircraft carriers USS *Hancock* (CV 19) and USS *Randolph* (CV 15), armed with regulus I, made single deployments. Cruiser Regulus deployments on the West Coast became routine after 1955 with the addition of USS *Helena* (CA 75) and USS *Toledo* (CA 133) and the addition of USS *Macon* (CA 132) on the East Coast. By mid-1956 the Pacific and Atlantic Fleet each had a Regulus launch submarine available, the USS *Tunny* (SSG 282) on the West Coast and USS *Barbero* (SSG 317) on the East Coast. The first submarine Regulus I deployments were made in 1958 when both *Tunny* and *Barbero* were deployed during the Lebanon Crisis. When consecutive Regulus I submarine strategic deterrent patrols began in 1959 in the Pacific, the United States had a covert capability to launch missiles with a 500 nautical mile range carrying a thermonuclear warhead. The last Regulus I submarine deterrent patrol took place in the Summer of 1964. The last flight of Regulus I was on 6 June 1966 at the end of the fleet target drone program.

A second generation missile, Regulus II, capable of cruising at Mach 2 speeds with a range of 500 to 1,200 nautical miles, was designed for both submarine and cruiser launch platforms. A successful test program, beginning in May 1956, led to shipboard launches in the Fall and Winter of 1958 but Regulus II was canceled prior to reaching operational status with the fleet.

This book is divided into four parts. Part I is a short history of how the Navy became interested in guided missiles. Part II is a detailed history of the flight test and development of Regulus I and II, covering examples of the day-to-day operations as well as the yearly milestones for the program as it reached operational status. Part III represents the deployment histories of Regulus I, from the contribution of naval aviation and deployment on aircraft carriers, to heavy cruisers and finally the five submarines that patrolled the North Pacific. Part IV, the appendices and glossary, presents details of the missile guidance systems; missile support unit histories; guidance submarines; nuclear warhead; selected submarine construction information, deck plans, and a summary of Regulus I and II flight operations and missile production.

My resources for the information presented comes from a variety of official and unofficial sources. Navy and contractor documents were obtained from the Naval Historical Center, the National Archives or Chance Vought contract reports held in the Loral Vought Systems archives. One hundred and forty-nine interviews were conducted in person or via correspondence. While the accuracy of such recollections may be questionable after nearly 30-40 years, I was struck by the clarity of recall and the abundance of personal papers many had retained. Personal recollections were corroborated by official documents when possible and/or with additional interviews.

DEDICATION

This book is dedicated to those men who gave their lives during the development and operational periods of the Regulus program:

Lieutenant A. Rice, USN
Ensign R. Pekkanen, USN
Seaman J. R. Jensen, USN
Ensign A. Wagner, USN
Lieutenant (jg) K. West, USN
Lieutenant (jg) C. Niederlander, USN
Lieutenant (jg) W. Thompson
Lieutenant E. Schuler
Lieutenant (jg) A. Gunn
Lieutenant M. Grouch
Lieutenant (jg) Philips

Publisher's Message

Our nation's reputation as being the most powerful country in the world has a lot to do with our defense program. Long before we discovered our highly effective Tomahawk system there was a group of men knee deep in research and testing — giving birth to the Regulus Missile. The Navy, along with Chance Vought Aircraft, Inc. felt the need to develop an alternate plan to the nuclear weapons development and the Regulus missile was their answer.

Author Dr. David K. Stumpf has spent the last six years researching every aspect of the Regulus I and Regulus II Missiles. *Regulus: The Forgotten Weapon* not only contains the history of this complex system but also includes the people who made it happen. Information from the widow of the first commanding officer of the USS *Grayback* , interviews with Captain William Sims, USN (Ret.) and Captain Donald Henderson, USN (Ret.) plus flight stories from the greatest test pilots — C.O. Miller, Ace Yeager, Joe Engle, Glenn Paulk, Bud Holcomb and Bill Coleman. This book contains flight deck plans never before published and one of a kind photographs that have been held captive in personal libraries.

I would like to thank those who made this book a reality. To Dr. Stumpf for his relentless research about this nearly forgotten stage of America's defense program. I would also like to thank the U.S. Navy photo archives for their use of photographs. Any photograph that is not attributed to a specific person in this publication is courtesy of the U.S. Navy.

Time was an adversary, money was an adversary, even the sea was an adversary. But the hardships were endured and the Regulus missile took its place in history.

Sincerely,

Dave Turner

Dave P. Turner
President

The first successful flight of the Navy's LTV-N-2 Loon took place on 12 April 1946 from Pt. Mugu, California. (Courtesy of Dzikowski Collection)

Part I

A New Era In Weapons Systems

Chapter One: The Guided Missile - A New Weapon System

In the closing months of World War II, guided missile technology blossomed as a technological breakthrough for the delivery of offensive weapons. The German V-1 "buzz bomb" cruise missile proved to be a crude but potentially devastating weapon. Even more frightening was the technologically advanced V-2 ballistic missile. Given an equally advanced guidance system, the V-2 could well have changed the course of history. Submarine-fired unguided rocket weapons were used near the end of World War II when USS *Barb* (SS 220) and USS *Seahorse* (SS 304), using experimental 5-inch rocket launchers, bombarded both military and industrial targets on the Japanese home islands from ranges of approximately 4,500 yards.[1]

At the end of hostilities with Japan, military planners in the U.S. armed forces, as well as around the world, believed with varying degrees of foresight that they had witnessed the twilight of conventional warfare. For the next fifteen years, the struggle within the United States armed forces for supremacy in the delivery of strategic as well as tactical nuclear weapons was waged primarily between the Navy and the Air Force.

From 1947 to 1953 the field of guided missile technology in the United States grew at a tremendous pace, with over 114 separate missile weapons systems contracted as development projects during this period. At the end of the Korean War in 1953, the list had been pared down considerably with only 25 guided missile projects reaching flight hardware stage.[2] The Navy had only one surface-to-surface guided missile (SSM) program, Regulus, which was well into its third year of flight testing and development.

With the multitude of nuclear weapon delivery systems potentially available, which one, or ones, would be most feasible, efficient or economical? In the Fall of 1954, President Eisenhower created the Technological Capabilities Panel, led by James R. Killian, a highly respected scientist, to review the many possible options and prioritized them for consideration by Congress and the Administration. The Killian Committee report was submitted in February 1955. The committee felt that Soviet ballistic missiles would soon be a nearly overwhelming threat to the security of the United States. Foremost in its suggestions was the recommendation to proceed with both intercontinental and intermediate range ballistic missiles armed with thermonuclear warheads to counter the perceived near-term Soviet threat.

With the detonation of the first "deliverable" Soviet hydrogen bomb on 23 November 1955, one of the greatest fears of the Eisenhower Administration was realized. The Killian Committee's findings assumed even more importance as both the military and political ramifications of U.S. strategic deterrence were debated by Congress, the intelligence community, the military and the Administration. The intermediate range ballistic missile would be easier to develop and build since its guidance system needed to be accurate over only a 1,500 nautical mile range rather than the 6,000 to 7,000 nautical mile range of an intercontinental ballistic missile. Utilizing already established Allied military bases in Europe and the Mediterranean on the perimeter of the Soviet Union or the Iron Curtain countries, these shorter range ballistic missiles would be able to reach the Soviet Union quickly but would be much more vulnerable to attack in their forward positions. Once developed and operational, intercontinental ballistic missiles would be safely based in the United States and augment the intermediate range missiles.

The Killian Committee had strongly recommended the development of a sea-launched intermediate range ballistic missile. In November 1955 the Department of Defense created a joint Army-Navy project to direct development of a sea-launched version of the Army's liquid-fueled Jupiter intermediate range ballistic missile. The Navy preferred the use of a solid-fuel missile but had been unable to demonstrate that the timetable for its development would be faster for deployment then that of the Jupiter joint project. Nonetheless, the Navy realized that in order to keep in the mainstream of strategic nuclear weapon delivery, they would have to develop a fleet ballistic missile, liquid-fueled or not. With Regulus I already well into the development stage, the Navy had plans in place for building a longer range, Mach 2, follow-on to Regulus I, Regulus II. The Regulus missile family would thus be employed as an ancillary component to the Navy's strategic weapons during ballistic missile development as well as provide a viable alternative should unforeseen problems arise in the fleet ballistic missile program's development or basing phases.

The Navy in Transition

At the end of World War II, the United States Navy was the most powerful naval force in the world. The naval forces of Germany and Japan were destroyed or being dismantled; Russia, not yet perceived as a peacetime threat, was considered, due to her vast land mass, to be relatively immune to attack by sea and without need for a strong offensive naval force. Within this setting the Navy had to accomplish two fundamental tasks in order to retain its strong position within the military establishment. First, the Navy had to demonstrate that contrary to popular perception, the atomic bomb had not rendered naval tactics or basic naval strategic functions obsolete. Second, the Navy had to make a strong case for attaining an essential role in the delivery of atomic weapons, or face the function fostered by the Air Force of being merely an escort and support service.[3]

The Atomic Bomb vs. the U.S. Navy

The tremendous devastation wrought by atomic weapons had been conclusively demonstrated on two densely populated land targets. Inconclusive attempts to extrapolate the blast effects of the atomic bomb on naval forces led the Navy in September, 1945, to propose an actual test of the effects of atomic weapons on an array of ships. After considerable debate and infighting as to which service would be responsible for overall control of the test, President Truman settled the arguments on 10 January 1946 by approving a joint task force of Army, Air Force, Navy and civilian personnel.[4]

"Operation Crossroads", the official name for the test, was conducted by Joint Task Force ONE, commanded by Vice Admiral William Blandy, an experienced and widely respected ordnance officer. The original operations plan included three atomic weapons tests; Test ABLE, an air burst from a tethered balloon and Tests BAKER and CHARLIE, underwater bursts with the weapon suspended under an "LCU"-type amphibious landing craft for BAKER and a weapon placed much deeper, in a bathysphere, for CHARLIE. Once the Air Force was included, they made a strong case to change Test ABLE to bomber delivery since only three atomic bombs had been detonated to date with just two dropped from an aircraft. The aircrews and weapon assembly teams needed the practice. Practice, in one sense, was true, since the Nagasaki-type "Fat Man" implosion weapon, with a yield of 23 kilotons (KT, equivalent to 1,000 tons of TNT), had to be assembled by a 38 man team working over a 48 hour period.[5]

Test ABLE was designed to evaluate the effect of air pressure shock waves and intense radiation on an anchored fleet. Radioactive contamination of the ships was not expected to be of major concern since the fireball would not touch the surface (an airburst is defined as a nuclear explosion where the fireball at its maximum expansion does not touch the ground or water surface). The 93 ships of the test array included the German cruiser *Prince Eugen*, the Japanese battleship *Nagato* and light cruiser *Sakawa* as well as U.S. ships including the aircraft carrier USS *Saratoga* (CV 3), the battleships USS *Nevada* (BB 36), USS *Pennsylvania* (BB 38), USS *New York* (BB 34) and USS *Arkansas* (BB 33), and a combination of destroyers, submarines, attack transports, landing ships of several types, barges and a floating dry dock. The ships were made as "war ready" as possible, including loading fuel and ammunition on board as necessary. By the time Operation Crossroads was prepared for the first detonation, Joint Task Force ONE was composed of 42,000 personnel, 156 aircraft, 10,000 cameras and 251 ships (both in the target array and the support fleet).[6]

Several classes of ships, notably destroyers, attack transports and landing craft, were arrayed in a radial manner from the intended ground zero target, the battleship *Nevada*. This disposition was designed to allow evaluation of the gradation of the blast effect using the same type of ship in each arm of the radial array. Twenty of the 65 ships were located within one square mile of ground zero. Normal naval doctrine at the time for anchorage of this many ships would have had only 4 to 8 ships in a one square mile area. Similarly, a tactical array for a carrier task force in the open ocean called for only one capitol ship per square mile.[7]

The Test ABLE took place on 1 July 1946. The bomb was dropped from 28,000 feet by the B-29 "Dave's Dream" of the 394th Strategic Bomb Wing, Rosewell, New Mexico. The weapon detonated at an altitude of 520 feet and between 1,500 and 2,000 feet west of the target.

The only capital ships within one-half mile of the actual ground zero were the battleships *Nevada* and *Arkansas* and the heavy cruiser *Pensacola*. Little damage was done to the hulls and turrets of these ships, but the superstructures suffered considerably. At three quarters of a mile from ground zero there was little physical damage to the anchored vessels. Trucks and planes placed on the decks of the various ships were still recognizable, even on the *Nevada*. Five ships sank, six ships were seriously damaged, eight ships suffered seriously impaired efficiency and nine ships were moderately damaged. Several of the ships beyond 750 yards were reboarded on 1 July 1946 and used for crew quarters by 2 July.[8]

Test BAKER, detonated on 25 July 1946, at a depth of 90 feet, was designed to test the effect of the underwater shock wave on ship hulls as well as the ability to decontaminate the ships that were doused with radioactive water and other debris. Ninety-two ships were used, most of which were survivors of Test ABLE and arrayed in a similar manner. The atomic blast generated a column of water 2,200 feet in diameter and 6,000 feet in height, containing an estimated 10 million gallons of highly radioactive water. The largest ship near ground zero, the battleship *Arkansas*, was lifted vertically in the water column and sank moments later. Several smaller ships nearby also sank immediately. The stern of the *Saratoga* was lifted 130 feet into the air and she sank eight hours later. The battleship *Nagato* sank after four days. The second underwater test, Test CHARLIE, was canceled due to engineering difficulties with the bathysphere hardware.[9]

Evaluation of the results from these two tests indicated that no ship within a mile of either an underwater or air burst of an atomic weapon with a yield of 23 KT would escape serious structural damage or serious radiation injury to the crew. The Air Force and other critics pointed out that only 10% of the ships used in the test were left unscathed, indicating that massed surface fleets should be considered highly vulnerable to atomic weapon attacks and hence were obsolete. The Navy countered that the ships were massed, anchored and defenseless. Furthermore, the artificial arrangement of the ships in both tests yielded inconclusive results since tactical deployment formations were completely different. The Navy, however, did delay and eventually cancel the construction of the large battlecruiser USS *Hawaii* (CB 3) and battleship USS *Kentucky* (BB 66). Much to the surprise of the Navy and the consternation of the Air Force, the public and political opinions from the test results were that the Navy had dramatically proven that it still had a strong role to play in the atomic age.

Just what was this role to be? Determination of possible strategic directions had begun well before Operation Crossroads and were reaching their culmination as it became evident that the probable major menace to the United States was to be the Soviet Union. The Soviet Navy did not yet have a major naval presence of capital ships, most notably aircraft carriers, but was feared to have a significant and growing fleet of modern submarines due to the capture of advanced German U-Boat designs. The U.S. Navy focused its attention on naval aviation and anti-submarine warfare based on "killer" submarines. The aviators desired large "super carriers" for projection of naval tactical and possible strategic nuclear capabilities. Anti-submarine warfare advocates requested improved submarine technology and anti-submarine weapons.

USS Cusk *(SS 348) prepares to launch LTV-N-2 Loon missile on 7 July 1948. The missile container could hold one missile with wings detached. This long launch ramp was soon replaced by a shorter launcher which greatly improved* Cusk's *underwater performance. (Courtesy of Christensen Collection).* *Sequence continued on pages 14 and 15.*

Submarines and the Projection of Undersea Power

At the conclusion of World War II, submariners understood that the days of convoy busting, anti-shipping attacks were over. The perceived enemy, the Soviet Union, had only a rudimentary need for a merchant marine and had no major naval forces to protect shipping or to project power. Indeed, the initial major emphasis in construction for the Soviet Navy was that of submarines for use in protecting the Soviet mainland by forcing American aircraft carriers far from the Soviet coast. Naval strategists, both American and Allied, realized that anti-submarine warfare using submarines was going to be an arduous task.

Aside from the defensive anti-submarine warfare and submarine radar picket roles, a group of submariners led by Captain Thomas Klakring, a World War II submarine hero, saw an opportunity to combine the covert capabilities of a submerged submarine with the over-the-horizon projection possibilities of guided missiles. This idea moved from a conceptual stage in 1946 to the contract stage in 1947. Two options were possible for such a system; a cruise missile and a ballistic missile. The ideal submarine cruise missile was considered to be one that could be launched from a torpedo tube while submerged but the submariners realized that such a capability was considerably in the future, so surfaced launch was selected. The ballistic missile option was shelved temporarily in anticipation of development of necessary missile propulsion, decrease in atomic warhead size and improvement in long-range guidance capability.

The Navy's Guided Missile Program Before Regulus

Project DERBY

In April 1945, the Office of the Chief of Naval Operations (OPNAV) established Project DERBY at Pt. Mugu, California. Its mission was to train personnel in the assembly, operation and launch of the Air Force JB-2 cruise missile, an "Americanized" version of the German V-1. The Navy version was named "Loon" and had both KUW-1 and cruise missile LTV-N-2 designators. Modifications made in the missile from the German production version included radio command control to permit left, right and dive commands as well as the use of a radar beacon to assist in tracking the missile. Since mid-1945 the Navy had been observing the Air Force program in Florida with the JB-2 and had plans to utilize the missile as a bombardment weapon during the invasion of Japan. Fifty-one missiles had been purchased from the Air Force with another 100 to be built and delivered by Republic Aviation.[10] With the end of the war in the Pacific, the program was reoriented towards an intriguing concept, the development of submarine cruise missile launch capability.

Project DERBY launch activity began with the first Loon launch on 7 January 1946 from the beach at Pt. Mugu, California.[11] The pulse-jet engine died during the take-off run up the launch ramp and the missile splashed into the ocean one mile offshore. On 12 April 1946, the first successful launch took place and the Navy to authorized conversion of two submarines, one as a combination launch and guidance submarine and the other as guidance only. Slightly more than one year later, on 7 March 1947, USS *Cusk* (SS 348), launched and successfully guided a Loon missile.[12] Three months later USS *Carbonero* (SS 337) joined Project DERBY as a Loon guidance submarine.

Loon flight operations rapidly became routine but not without a steep learning curve. During the early launches the missile would fre-

Seconds after booster ignition, the booster rockets exploded, bursting the missile fuel tank. (Courtesy of Berry Collection)

Parts of the launch ramp shoot by the missile as it tumbles off the launcher onto the deck. (Courtesy of Berry Collection)

quently pitch up violently upon leaving the launcher guide rails. Dr. Wilhelm "Willie" Fiedler, a German V-1 scientist who was working at Pt. Mugu on other projects, pointed out the solution to the concerned project officers. The alignment of the multiple jet-assisted take-off bottles (JATO bottles) thrust had to be through the center of gravity of the combined launch sled and missile assembly, not just through the center of gravity of the missile.

The original 40 foot long launcher rails installed on *Cusk* were cumbersome and caused considerable drag while submerged. A much shorter 12 foot rail launcher was installed on *Carbonero* and proved very successful. *Cusk* was also modified by the addition of an on-deck missile hanger aft of the conning tower to permit transport of one Loon missile.

Project POUNCE

In May 1949, OPNAV directed the Commander, Submarine Force, Pacific Fleet, to determine the operational capabilities of the Loon missile. This task was given to Commander, Submarine Division FIFTY-ONE and code named Project POUNCE.[13] While the technical assistance of the NAMTC staff was still available, this program was to be self-sustaining if at all possible in order to make it a valid test. Project POUNCE would evaluate equipment used for launching, tracking and guiding missiles launched from submarines as well as evaluate the current maintenance and launching techniques for use in locations remote from major support facilities.[14] Project POUNCE program efforts culminated six months later in a combined fleet exercise, Operation Miki, conducted in the Hawaiian Islands fleet operating areas in November 1949. The submarine-tender USS *Sperry* (AS 12) deployed to Pearl Harbor with four LTV-N-2 missiles and associated support equipment. *Cusk* and *Carbonero* deployed from Port Hueneme under simulated wartime conditions.

The weather on the day of the test was far from optimal with 20-30 knot winds. The first launch by *Cusk* was successful but the missile splashed 25 miles downrange. *Carbonero* launched a second Loon that flew untouched over the fleet as they tried to shoot it down with anti-aircraft fire. With these attempts unsuccessful, aircraft were directed to shoot the missile down as it cruised at 10,000 feet and 300 knots. The missile was not shot down, finally splashing into the ocean due to fuel exhaustion.

The final report on Project POUNCE was cautiously optimistic. On the one hand, the submarine force had achieved a remarkable 83% launch success, 90% guidance control and 100% tracking record. Forward deployment had been demonstrated to be feasible. The short rail launcher concept had proven to be reliable and could easily replace the 40 foot launcher on *Cusk*. On the other hand, the Loon was found not to be acceptable as a tactical missile. If absolutely necessary, they could be made into tactical weapons but their best use would be as test vehicles for further research and development programs in support of newer, more up-to-date missiles.

The report finished with a ringing endorsement of the submarine launched guided missile concept by Commander John S. McCain, Jr.:

"...operations conducted during this period definitely have proven that the use of guided missiles from submarines as an offensive weapon is a progressive step in the expanding study of naval warfare."[15]

The report further recommended that an interim submarine guided missile program be continued. This would serve to retain the nucleus of technical support personnel in the submarine force and permit continued evaluation of newer guidance and launching equipment as it became available. Specific improvements were also listed in the report. These included: improved guidance, especially for submerged submarines; evaluation of a single booster rather then the cluster of boosters currently used; missile launch operations in rough weather to understand the difficulties that might develop; and, finally, inclusion of even this interim system in fleet exercises to further test the ability of the surface navy to combat such a new weapon system. After submission of the report, Loon launch operations continued on a reduced scale of approximately one launch per month.

Project TROUNCE

On 17 May 1950 OPNAV established Project TROUNCE within Project DERBY.[16] Project TROUNCE was tasked with the preparation submarine personnel for operational evaluation of the new Regulus missile which had just begun its test and development program. Manned drone aircraft were used to evaluate and improve submerged guidance techniques for use in the tracking of Loon. Project TROUNCE was also the beginning of a new guidance system that used paired-pulse radar signals for transmitting guidance commands to the missile. The primitive radio command control guidance system of Loon, while a distinct improvement over the original German guidance system used in the V-1, was considered too vulnerable to electronic countermeasures. The new radar command control system was given the name Trounce.[17]

Loon launches continued, albeit at a slower pace, and the Phase I Trounce guidance system was ready for testing in June 1951. Phase II was the Phase I system modified to prevent interference by electronic countermeasures and was to be ready in time for use in the Regulus program by the middle of 1952. Phase I tests indicated, however, that the Trounce system was sufficiently secure from electronic countermeasures so that the Phase II program could be reduced. Of all the new developments proposed in the Phase II portion of the program, the radar bearing and range accuracy as well as the strength of the radar transponder beacon were the most critical. The guidance computer could be no more accurate than its inputs and the range of the radar beacon signal return would be the deciding factor in the missile's maximum tracking range.

Commander Walter P. Murphy was Officer-in-Charge of Project DERBY during the beginning of the Trounce development program. The Trounce guidance system was more or less designed in the field by Paul Fiske and the civilian engineering staff of the Naval Electronics Laboratory (NEL), San Diego, working in conjunction with the officers and men of *Cusk* and *Carbonero*. The technical hurdle that had to be

Flames engulf Cusk *as the missile fuel is ignited. Lieutenant Commander Fred T. Berry, CO of* Cusk, *submerged the boat to extinguish the fire. As the smoke cleared, everyone thought* Cusk *had sunk. (Courtesy of Berry Collection)*

Cusk *surfaces much to the relief of all concerned. This photo sequence is probably the best remembered of the Loon program. (Courtesy of Berry Collection)*

USS Carbonero *(SS 337) preparing to launch a Loon LTV-N-2 missile on 19 May 1949. Note the short rail launcher compared to the earlier ramp launcher.* Carbonero *was not fitted with a missile hangar. (Courtesy of G. Peed Collection)*

overcome was the reliable and reproducible generation of paired-pulse radar signals as well as reception and decoding by the guidance system on board the missile. This required electronic time delays of reliable and consistent length. The time delays used were O.15, 0.2 or 0.25 seconds. Thus, a pulse pair with a time delay of 0.15 seconds was the command to turn 2.5 degrees to the right, a time delay of 0.2 seconds commanded a 2.5 degree turn to the left and the time delay of 0.25 seconds commanded the terminal dive to target. Murphy recalls that the encoder and decoders were designed to tolerate no more than plus or minus 0.025 seconds in the spacing between signals. This was pushing the state-of-the-art with the balky vacuum tubes available at the time.[18]

The submarine SV-1 air-search radar antenna was a source of many of the early problems with range and bearing accuracy using Trounce guidance. The SV-1 had not been designed with this kind of tracking in mind. The modifications were not trivial and further complicated the maintenance of the system. First, the radar had been modified to permit sector scanning of a 10-20 degree arc. Second, magnetron radar signal sources had to be carefully hand-matched to permit effective control by more than one guidance control station. This proved to be a continuing problem until the beginning of the Regulus program. The early solution was to "burn-in" a large number of magnetrons to eliminate the unreliable ones. Tubes were then selected whose characteristics matched closely enough to permit dual-station control of the missile.[19]

Two major problems with Trounce were soon evident. The transponder beacon signal was present in only the center seven to eight degrees of the twenty-degree sector scan. Since the pulsed radar command would only be properly received if it was sent when the beacon was essentially centered within the radar beam, coordination and training on the part of the Trounce operators was crucial, almost an art. Clearly this feature or deficiency would have to be rectified before the Trounce guidance system could become operational. The other major difficulty was the requirement that the radar system's magnetron be modified such that the system, which was not intended to send paired-pulses, could both transmit the guidance pulses and also work as a normal radar when required.[20]

On 4 May 1951 *Cusk* and *Carbonero* participated in Operation REX, a simulated Loon missile attack on San Diego. One escort aircraft carrier, three destroyers, five destroyer escorts and all aircraft available in the Southern California Sector of the Western Sea Frontier conducted anti-submarine operations to locate the two submarines but were unsuccessful. The simulated attack included surfacing, readying the missile for launch and having a manned Lockheed P-80 drone aircraft fly the simulated missile flight path while under Trounce control from the submarines. *Cusk* simulated the launch, controlled the aircraft for the first 55 miles and then transferred control to *Carbonero*. Neither the aircraft nor either submarine were detected even though *Cusk* had spent 18 minutes on the surface. Yet again, the covert potency of such a weapon system was demonstrated and, yet again, much to the chagrin of the surface naval forces, a submarine-launched guided missile proved to be extremely elusive.[21]

After one year of evaluation, the Trounce I guidance system was installed in Loon #622 and launched from *Cusk* on 28 June 1951. The launch and initial guidance was successful but 15 miles down-range the engine failed and the missile splashed. Twenty-two days later, Loon #1071 was launched from NAMTC and successfully guided by both *Cusk* and *Carbonero* as well as the flight test control center at NAMTC. While the dive command from *Carbonero* was not received by the missile, the NAMTC Trounce system successfully commanded a dive to impact.[22]

The Trounce I guidance system continued to evolve as refinements were made in the field to make the system more reliable. By 1952 all Loon flights were controlled by the Trounce I guidance system and the twenty-five war reserve Loon missiles were configured for Trounce I. Electronic countermeasures tests conducted at the end of the Project TROUNCE indicated a low susceptibility.[23] The Phase II test and evaluation process resulted in recommendations for modifications to the SV-1 radar. The new system would be designated as the SV-5 radar. This included the associated CP-98 guidance computer for automatic calculation of course corrections to the missile. The SV-5 radar was fabricated by Stavid Engineering of Plainfield, New Jersey, and permitted more accurate range and bearing information for input into the computer. Field tests began in late December 1952. The submarine community felt that the Trounce system was now co-equal to the rival, bipolar navigation system which required a launch submarine and two picket signal boats near the target. Combining the high performance of the Regulus I missile with Trounce guidance would result in a much more formidable weapon than Loon.[24]

Project SLAM

With the completion of the Phase II objectives of the Project TROUNCE program, OPNAV replaced Project TROUNCE with Project SLAM (Submarine Launched Attack Missile) on 10 September 1952. Project SLAM was to continue readying fleet personnel for Regulus missile operations, while launching Loon missiles to maintain proficiency and testing operational tactics.[25]

On 1 January 1953, Project DERBY was formally disestablished and Guided Missile Unit FIFTY (GMU-50) formed in its place under Commander Submarine Squadron FIVE at the Naval Air Missile Test Center.[26] GMU-50 was further expanded in late 1953 with the addition of the personnel from Guided Missile Training Unit FIVE who were training in Regulus flight operations at Pt. Mugu and Edwards Air Force Base, California.

Project SLAM was organized to maintain Loon assets in such a state that they could be used as an interim tactical weapon prior to Regulus becoming operational. The remaining Loon missiles would be launched at a rate projected to expend them by June 1954, approximately five missiles per quarter. With the introduction of Regulus into the Operational Development Force program, projected for the 1954 time frame, Loon operations were curtailed nine months early. The last launch of a Loon missile took place 11 September 1953 when LTV-N-2 #121 was lost due to the misalignment of the single booster rocket.[27]

Cusk and *Carbonero* were converted to Regulus guidance submarines, operating in the West Coast Regulus program as well as deploying to Pearl Harbor during the Regulus strategic deterrent program of 1959-1964.

Endnotes

1 *Sink'em All, Vice Admiral Charles A. Lockwood, 1984, Bantam Books New York, page 327.*
2 *How We Fell Behind in Guided Missiles, Trevor Gardner, 1958, Air Force Historical Foundation 51:3-13.*
3 *While technically still the Army Air Force at this time, Air Force is used throughout the book for consistency.*
4 *Operation Crossroads-1946. Defense Nuclear Agency Report Number 6032F. 1984, page 18.*
5 *US Nuclear Stockpile, 1945-1950, David Alan Rosenberg, Bulletin of Atomic Scientists May 1982, page 29.*
6 *Operation Crossroads-1946. Defense Nuclear Agency Report Number 6032F. 1984, page 94.*
7 *Ibid., page 90*
8 *Ibid. pages 191-192.*
9 *Ibid., page 19.*
10 *The History of Pilotless Aircraft and Guided Missiles, Rear Admiral Delmar S. Fahrney, 1958, Naval Historical Center, Operational Archives; pages 586-596; 798-818.*
11 *The Pt. Mugu facilities were part of the field test program from the Naval Air Modification Unit, Johnsville, Pennsylvania. Pt. Mugu did not become the Naval Air Missile Test Center until 1 October 1946. Days of Challenge, Years of Excellence: A Technical History of the Pacific Missile Test Center, 1989, page 5. United States Printing Office.*
12 *The first successful launch of a Loon from Cusk had taken place on 18 February 1947 but crashed three miles down range (History of Pilotless Aircraft and Guided Missiles, Rear Admiral Delmar Fahey, pages 586-596).*
13 *Project POUNCE, CNO sec ltr OP3420/nr(sg)s78; ser 0092P34 of 9 May 1949. Naval Historical Center, Operational Archives, COMSUBDIV 51.*
14 *Ibid.*
15 *Project POUNCE: Final Report, FB-51/A1(POUNCE) ser 0017GM, page 2; 16 January 1950. Naval Historical Center, Operational Archives, COMSUBDIV 51.*
16 *TROUNCE was considered to be the acronym of Terrain Radar Omnidirectional Underwater Navigational Computing Equipment. There is no historical evidence for this. TROUNCE was simply the name of the project and it was used as the name of the guidance system, Trounce.*
17 *CNO sec ltr ser 00258P34 of 17 May 1950*
18 *Personal communication with Captain W. "Pat" Murphy, USN (Ret.), November 1990.*
19 *Ibid.*
20 *Ibid.*
21 *Submarine Guided Missile Program (Phase II - Project TROUNCE). Semi-Annual Report for April - September 1951. FB4-11/51:AARF:jms A1, Ser 004, 16 October 1951, page 2. Naval Historical Center, Operational Archives, COMSUBDIV 51.*
22 *Submarine Guided Missile Program (Phase II-Project TROUNCE) Semi-Annual Report for April-September 1951. FB4-11/51:ARF:jus A1, ser 004, 16 Oct 1951, page 3, enclosure (1). Naval Historical Center, Operational Archives, COMSUBDIV 51.*
23 *Semi-annual report on Submarine-Launched Attack Missile Program (Project SLAM) - period 1 Oct 1952 to 31 March 1953; page 4, (enclosure 2). Naval Historical Center, Operational Archives, COMSUBDIV 51.*
24 *Semi-annual report on Submarine-Launched Attack Missile Program (Project SLAM) - period 1 Oct 1952 to 31 March 1953; page 1, (enclosure 2). Naval Historical Center, Operational Archives, COMSUBDIV 51.*
25 *CNO sec ltr OP511/jek ser 0036P37 of 10 September 1952. Naval Historical Center, Operational Archives, COMSUBDIV 51.*
26 *Ibid.*
27 *"Semi-annual report on Submarine Launched Attack Missile Program (Project-SLAM) - period 1 Oct 1952 to 31 March 1953; page III-C; Naval Historical Center, Operational Archives COMSUBDIV 51.*
The History of Pilotless Aircraft and Guided Missiles by Delmer S. Fahrney, 1958, Naval Historical Center, Operational Archives, page 1126.

TM-1146 is launched on 11 October 1956 from Norton Sound. *This was the first long range flight, and guidance was successfully passed between* Tunny *and* Carbonero *during the 47 minute flight. (Courtesy of Wilkes Collection)*

Part II

Regulus I and II Flight Test and Development

Chapter Two: Chance Vought Enters The Missile Business

As World War II came to a close, the management of the Chance Vought Aircraft Division (Chance Vought), of United Aircraft Corporation, realized that production of the F4U "Corsair", the mainstay of the division for the past several years, would soon be nearing an end. Advanced projects would be needed to keep the company at the forefront of the newly emerging technologies of jet aircraft and guided missiles. Three such projects were already in progress, the XF5U, XF6U and XF7U. The XF5U or "Flying Pancake," while propeller driven, could take off and land on 50 feet of runway. Part of the success of this project was the extensive use of a sandwiched composite material, Metalite, a thin skin of aluminum covering a balsa wood core. The XF5U was not continued to the production stage. The XF6U "Pirate" was the company's first jet aircraft. The Navy ordered 30 F6U-1's in 1948, making it the first production naval aircraft equipped with an afterburner. The XF7U "Cutlass" was in the design stage in 1945. Two unique features of the F7U illustrated that Chance Vought was on the cutting edge of aircraft design. The extensive use of Metalite in the large wing surface areas provided significant weight reduction and a concept that Chance Vought called the "ailavator" was used for control since there were no horizontal tail surfaces.

In the Fall of 1945, research and development money for new technology weapons systems was still readily available within the Bureau of Aeronautics (BuAer). While intra-service political infighting continued between the naval aviators and the surface Navy, BuAer was able to expand its guided missile programs with increased attention to the surface-to-surface missile program. A study committee was formed, headed by Commander Grayson Merrill, to review the status of American, English, and German guided missile technology and recommend future directions for the BuAer programs.[1] The committee submitted a 69 page report on 15 December 1945 entitled "Study of Requirements for Pilotless Aircraft for Fleet Use in 1950".[2] This report resulted in eighteen BuAer study contracts dispersed amongst twelve of the leading aircraft manufacturers. These contracts were part of a concerted BuAer effort to get the aircraft companies seriously involved in the new technology of guided missiles. Guided missiles were seen as a weapon of the future and the government did not want to wait until the aircraft companies reached this conclusion themselves.[3]

Chance Vought had become a part of this process prior to the release of the committee's final report. Based on the preliminary findings of the committee, in October 1945 BuAer invited Chance Vought to submit proposals for several P/A (pilotless aircraft) designs with ranges from 25 to 300 nautical miles. Several months later, Chance Vought learned that it was to receive just the short range surface-to-surface P/A VI study contract. This contract was signed in June, 1946 and by the middle of 1947, the P/A VI supersonic ramjet powered missile was the focus of most of Chance Vought's attention in the guided missile field.[4] At this point BuAer decided to develop an "interim" subsonic P/A that could be integrated into the fleet aboard fleet-type submarines as early as possible. By not relying on the development of the ramjet engine for this interim missile, the Navy was opting for an early entry into guided missiles utilizing proven power plant technology. On 21 May 1947, Chance Vought submitted an initial proposal for this missile. Guidance concepts for the P/A VI project were found to be directly applicable to the new interim missile with operational concepts centered around the readily available fleet-type submarine launch platform.[5]

In June 1947, Chance Vought submitted a complete proposal regarding the subsonic missile program. Two features of this proposal made it interesting to the Navy. First, Chance Vought proposed developing a recoverable flight test vehicle utilizing landing gear and a parabrake. The recoverability feature would greatly reduce the cost of flight test and development. Second, the flight test vehicle and production tactical

Full-scale mockup of the Chance Vought PA-6 supersonic surface-to-surface missile. The study contract was signed in June 1946 for this short range, ramjet-powered missile, and scale model testing was carried out for one year. The program was canceled when Chance Vought received the Regulus contract in November 1947. (Courtesy of Loral Vought Systems Archives, Author's Collection)

Table 2.1 General Characteristics for Interim Guided Missile for Adaptation to Present Fleet-Type Submarine

Range:	500 nautical miles maximum.
Speed:	High subsonic, Mach Number 0.85-0.95.
Warhead:	At least 3,000 lbs.
Accuracy:	0.5% of range from nearest terminal guidance station point of impact.
Targets:	Fixed.
Guidance:	Basic guidance to be inertial, programmable; provision for superimposed electromagnetic wave command correction in course. A minimum of one aircraft or submarine will exercise command guidance within a minimum of 75 miles of the target. SHORAN is the preliminary guidance system.
Altitude:	Up to 40,000 feet; variable and controllable.
Configuration:	Length - 30 feet maximum. Body Diameter - about four feet. It is desired to carry two missiles in a ready hangar that will replace the conning tower of the submarine. Wings - wings of about 60 square feet area each; 10 feet in length and six feet chord; detachable or fold back. Wings to be rigged out by two men in 40 seconds. Weight, Stowage and Handling: Weight of 10,000 to 12,000 lbs in accordance with preliminary estimates acceptable. Missile to be stowed and handled in cradle; may be removed from cradle when passing through reloading hatch between missile compartment and ready hangar.
Fuel:	Gasoline or kerosene; fuel to be carried in a tank external to submarine pressure hull.
Launching:	Missile to be launched from short launcher on deck of submarine. To be loaded directly on launcher from ready hangar. It is desirable that arrangement for loading and firing be such that the missile can be fired within 60 seconds of opening hangar door. Solid fuel booster desired for launching, to be put in place after missile is in ready hangar.
Time Scale:	Missile to be ready for NAMTC evaluation tests in January 1951.[13]

missile would differ as little as possible. Chance Vought suggested that by doubling the design safety factor, structural testing could be eliminated for both the flight test vehicle and tactical missile. BuAer concurred.[6,7]

Two Submariner Officers Conferences were held in early August 1947 to resolve the outstanding design questions of this submarine launched missile to ensure compatibility with submarine operations. The Submarine Officers Conference concept had begun in 1926 as a method of ensuring adequate discussion and interchange of ideas leading to the improvement of the submarine service.[8] I. Nevin Palley, the Program Engineer for Guided Missiles at Chance Vought, his assistant, R.H. Carter, and J.M. Shoemaker, the Chief Engineer for Chance Vought, attended the 4 August 1947 meeting where representatives from Office of Chief of Naval Operations, Bureau of Ships, BuAer, Office of Naval Research, and the submarine personnel from the West Coast finalized the design constraints for the new missile (See Table 2.1).[9] It is interesting to note that during these meetings there was a great deal of discussion concerning the advantages and disadvantages of submerged versus surfaced launch of the missile.[10] Both methods were endorsed by the conference representatives. None of the submarine officers voiced any concern about surfacing and firing a missile within four or five nautical miles of an enemy coast, let alone 100 nautical miles. Much more concern was expressed that money would be wasted trying to enable this interim missile to be launched from a submerged submarine. The general design criteria for the subsonic missile were summarized in a memo circulated at the end of the meetings as well as selection of a site for the initial research and development program. Safety was paramount and with much of Southern California within the flight range of the missile, a "destruct" system was requested. Chance Vought's suggestion of the recoverable flight test vehicle was readily accepted. Provided the recovery capability was successfully implemented, the cost of the flight test program would be reduced significantly when compared to a very similar Air Force missile program currently underway by the Glenn Martin company, the MX-771 "Matador".

Armed with this information, Palley, Carter and Shoemaker returned to Stratford, Connecticut, and preparation of the detailed formal proposal began. The original supersonic P/A VI proposal was shelved at this time since BuAer's emphasis was now on the subsonic version. On 19 August 1947, BuAer formally requested Chance Vought to submit a proposal for the initial phases of research and development for Project Regulus, now designated XSSM-N-8. Contained in this request were the general characteristics for the missile and the proposed changes necessary to adapt the system to fleet-type submarines.[11] On 3 October 1947 Chance Vought submitted a preliminary proposal to BuAer covering the cost of fabricating one flight test vehicle. The estimated cost was $4,997,309. On 17 November 1947 BuAer issued a Letter of Intent to Chance Vought for Contract NOa(s) 9450.[12] On 28 November 1947 the initial engineering design work was started and on 23 December 1947 Chance Vought accepted the contract.

With the authorization to design the first flight test vehicle came the detailed specifications for the construction of the airframe. The powerplant was to be the Allison J-33 jet engine, then currently used in the Lockheed F-80 "Shooting Star". The J-33 was a proven engine; by choosing it, BuAer and Chance Vought limited the performance of the missile, but at the same time eliminated concerns over possible delays in engine development. Above all, the goal was to produce as quickly as possible a flight test vehicle that differed as little as possible from the tactical vehicle.

Wind tunnel testing began utilizing 0.07 and 0.32 scale models. Since the effects of conditions in the transonic speed flight regime were not well known, the design of the missile originated with dorsal and ventral rudders but no horizontal tail surfaces. This cruciform symmetry, with an internally mounted engine, permitted a symmetrical cross-section which in theory would permit balancing of buffeting forces during the transition to the supersonic terminal dive maneuver. Wind tunnel tests verified the initial design concepts, but, as will be seen later, only two test flights with the ventral rudder surface installed were flown.

In April 1948, BuAer requested a proposal from Chance Vought that would cover work towards: a) fabrication of one Regulus Flight Test Article, complete in all details and ready for flight testing upon delivery to the Navy; b) instrumentation of the Flight Test Article; c) fabrication of a launcher; d) fabrication of a mockup; e) fabrication of the Beacon Guidance System; f) fabrication of ten test missiles and airborne beacon guidance units. On 1 June 1948 Chance Vought responded with a proposal for 30 missiles, citing the economy of scale in purchasing materials for the larger order.[14] On 20 September 1948 Chance Vought was awarded contract NOa(s) 9450 that superseded the original Letter of Intent issued 10 months earlier. Simultaneously, BuAer gave the go-ahead for the 30 flight test vehicle production. On 30 November 1948, BuAer informed Chance Vought that due to budgetary constraints a new proposal was necessary reflecting the costs for the fabrication and testing of 10 complete air frames instead of the agreed upon 30. By 17 December 1948 the production plan for 10 airframes was completed.[15]

Early in 1949 the Air Force Matador project began to enter into Regulus funding discussions as the Navy and Air Force started competing for limited guided missile funding. Matador was quite similar in scope to Regulus. The same Allison J-33 engine was used, but unlike Regulus, the missile did not have cruciform symme-

Preliminary one-quarter scale wind tunnel test model for Project Regulus, known initially ***as*** *Chance Vought at Project VF-357. Initially designed with a cruciform shape for stability during the terminal dive to target, the ventral vertical fin was removed relatively early in the program, greatly simplifying launch and recovery operations. (Courtesy of Loral Vought Systems Archives, Author's Collection)*

try. Instead, Matador had a "T" tail with standard control surfaces and a single booster was used launch. More importantly, Matador was not recoverable. With each of the early Regulus flight test vehicles costing upwards of $400,000, the ability to recover the missile was a tremendous cost saving. Nonetheless, with the first Matador flight scheduled for January, the Office of Chief of Naval Operations was concerned that not enough was being done to get Regulus ready for flight. Submarine missile stowage, servicing, and launch problems did not appear to be high priorities within Chance Vought at this time, according to the Office of Chief of Naval Operations staff. Had decisions been made on use of a removable wing? If so, how would it be reattached quickly? The wing position of the booster rockets on Regulus could cause stowage problems and seemed much more cumbersome than the single booster of the Matador. Had the size of the tank envisioned for missile storage been adequately considered in terms of handling gear for the missile? These questions were quickly addressed within Chance Vought and work started in completing the necessary analyses. [16,17]

On 10 May 1949 the Research and Development Board, Department of Defense, directed that the Navy Regulus and Air Force Matador Programs be combined under the jurisdiction of the Navy.[18] The Navy and Air Force discussed the directive and recommended that the Matador test program for 1949 be continued while the Navy reviewed the adaptability of the Matador missile to submarine operations.[19] As a result of directives by the Research and Development Board, on 20 May 1949 Chance Vought was authorized to begin construction of the Lot I missiles, a total of five airframes. One missile was to be completed by December 1949 and one each in February, March, April and June of 1950. A flight test date of 30 June 1950 was selected to demonstrate the ability to conduct radio command control landings. With completion of the first flight test vehicle in November 1949, qualification tests were initiated at Dallas. The autopilot and radio command control had not yet been installed; this would be done in the field as the equipment became available.

In addition to flight test vehicle fabrication, Chance Vought had orchestrated the conversion of one TV-1 and one TV-2D (Navy training versions of the Lockheed F-80 "Shooting Star"; one and two seats respectively) for use in training of airborne and ground radio command control operators.[20] The first ground control operator for Chance Vought, William W. Sunday, began training at Bell Aircraft in November 1949. Sunday, a former naval landing signals officer, trained using a Bell Aircraft P-63 "Kingcobra" propeller-driven aircraft that was modified to respond like the TV-1 aircraft.

Chance Vought had decided to pattern TV-1 and TV-2D training flights and missile guidance work along the lines of operation at Utility Squadron FOUR (VU-4), based at Naval Air Station, Chincoteague, Virginia. VU-4 provided target drone services to the Atlantic Fleet using radio controlled aircraft. Familiarization flights were made using propeller-driven F6F-5K "Hellcat", F7F "Tigercat", and F8F "Bearcat" aircraft, the latter two being the control or chase aircraft and the former the drone. The autopilot and radio command control equipment were not the same as those to be used in the Regulus program, but the experience gained in setting up a standard operating procedure later proved to be invaluable. Training proceeded smoothly from airborne control of flight to airborne control of takeoffs as well as ground controller operations. Difficulties encountered integrating the radio command control equipment into the TV-1 prevented use of the jet aircraft during this training program. Final discussions of the Regulus training program with the VU-4 staff centered around electronic interference with the radio control equipment. Some of the interference problems at Chincoteague were felt to be due to nearby television transmitters and the point was raised that the proximity of Muroc Air Force Base to the *Los Angeles* basin might invite similar problems. Chance Vought's Chief Test Pilot for the Regulus Program, Leroy (Roy) Pearson, and Sunday, the lead ground control operator, completed the training program in late December 1949 and returned to Dallas to prepare for the beginning of field operations.[21]

Endnotes

[1] *The History of Pilotless Aircraft and Guided Missiles by Delmer S. Fahrney, 1958, Naval Historical Center, Operational Archives, page 1126.*
[2] *Ibid.*
[3] *Personal interview with Captain Robert Freitag, USN (Ret.). Freitag was an early and important advocate within BuAer for the guided missile[s] in genel interview with Captain Robert Freitag, USN (Ret.). Freitag was an early and important advocate within BuAer for the guided missiles in general and Regulus in particular.*
[4] *Fahrney, pages 1157-1158.*
[5] *Fahrney, page 1175.*
[6] *As a result, there were no instances of Regulus I missiles breakingup in flight.*
[7] *Fahrney, page 1176.*
[8] *The Fleet Submarine in the U.S. Navy by J.D. Alden, 1979, Naval Institute Press, pa[g]e 16.*
[9] *Submariner's Conference, 4 August 1947, Chaired by Captain Thomas Klakring. Operational Archives, Naval Historical Center, pages 1-11.*
[10] *An earlier meeting on 1 August 1947, not attended by Chance Vought personnel, had already discussed many of the submarine operational issues.*
[11] *Regulus I Program Synopsis, page 3. This document was kindly provided by Clell Blackwell of LTV Aerospace. It will be abbreviated as RPS for reference purposes.*
[12] *RPS page 3.*
[13] *Submariners Conference, August 1947, pages 2-3. Naval Historical Center, Operational Archives.*
[14] *RPS, page 3.1*
[15] *RPS, page 7*
[16] *Report of Visit to the Navy Department on 4 and 5 January 1949. Chance Vought departmental memo of 17 January 1949. Loral Vought Systems Corporation Archives, A50-18 Box 1. This source will be abbreviated as LVSCA for reference purposes.*
[17] *Personal letter to I. Nevin Palley from E.J. Greenwood. 17 January 1949. Personal papers of Roy Pearson.*
[18] *The date of this decision is listed in a memo to the chairman of the Research and Development Board, "Approval of Matador as Technically Feasible as a Surface-to-Surface Guided Missile." National Archives, Bureau of Aeronautics, GM/93 Project Regulus, Record Group 156, Box 5.*
[19] *The Navy and Air Force response is referred to on page 1 of "Report on Evaluation of Regulus and Matador Projects" 7 March 1950. National Archives, Bureau of Aeronautics, GM/93, Project Regulus, Record Group 156, Box 5.*
[20] *These aircraft were originally designated TO-1 and TO-2. In 1952 the designation was changed to TV-1 and TV-2D which are used throughout remainder of the book.*
[21] *Pilot Training Reports, 14 December 1949 to 21 December 1949. LVSCA A50-18, Box 1.*

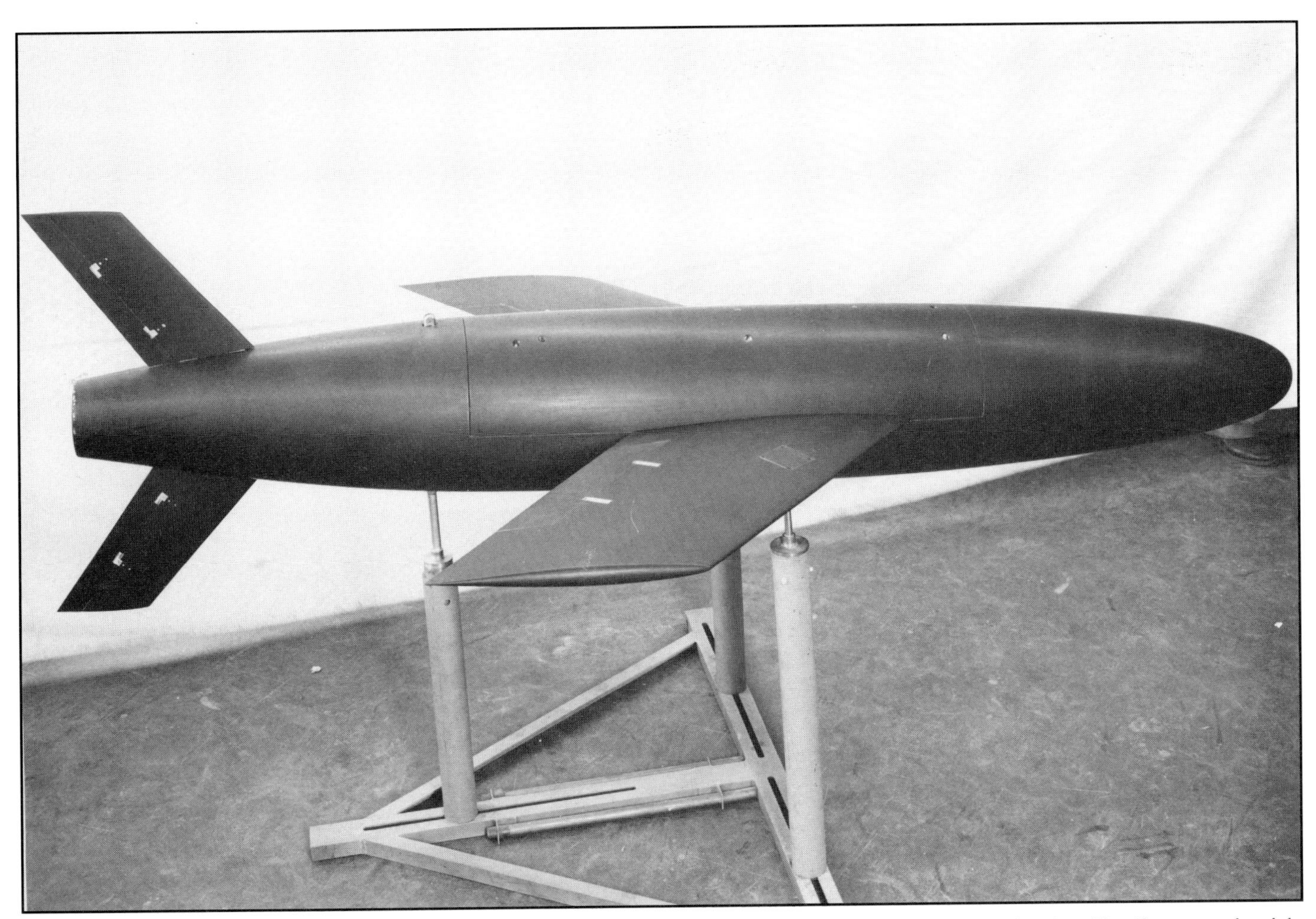

Three-quarter rear view of a 1947 wind tunnel model of Regulus I, clearly showing the ventral fin and ailavator controls on the wing. The ailavators replaced the horizontal stabilizer; a concept that had been successfully demonstrated with the Chance Vought XF7U "Cutlass." (Courtesy of Loral Vought Systems Archives, Author's Collection)

Chapter Three: Regulus I Flight Test and Development 1950-1952

Chance Vought realized that a key factor in the Regulus missile design was the time it would take to develop the tactical missile. Consequently, the program borrowed heavily from the company's standard aircraft construction and development experiences. Regulus was built in three variants. The flight test vehicle (FTV), later renamed the flight training or fleet training missile (FTM was the official Chance Vought designator, SSM-N-8 the official Navy designator) was equipped with tricycle landing gear. In the tactical missile (TM within Chance Vought, SSM-N-8a within the Navy) the landing gear was replaced with additional fuel and a nuclear warhead. The target drone (TD designation within Chance Vought, KDU-1 within the Navy) was nearly identical with the FTM, differing in that telemetry for target miss indication and greater fuel capacity was included.[1]

In each version, Regulus I was 34.5 feet long, not including the noseboom, and 4.75 feet in diameter. Launch gross weight was 13,685 pounds. The wingspan was 21 feet with the wings unfolded and, depending on missile production lot number, either 12.33 feet or 10.83 feet with wings folded. When resting on its landing gear with the tail unfolded, the FTM version stood Regulus I stood 8.83 feet tall; the tactical version was 8.19 feet from launch slippers to top of the tail. All three variants were launched from a short rail launcher using two JATO (Jet Assisted Take-Off) solid fuel boosters rated at 2.3 seconds firing time with 33,000 pounds of thrust each. This permitted launches from the small deck spaces of submarines and cruisers as well as the more spacious flight decks of aircraft carriers. The range of Regulus I was 500 nautical miles with a high altitude cruise and terminal dive to target. Late in the program external fuel tanks were used to extend range to 550 nautical miles but this feature was not incorporated in the tactical missile.[2]

Flight Test

1950

The flight test program was conducted jointly by the Navy and Chance Vought. The early phases of the program were more concerned with the development of the missile than demonstration of the complete weapon system. Chance Vought was given the responsibility for initial overall planning and control of the flight test program; maintaining the flight test vehicles and repairing them as required; collect and analyze flight test data and work as much as possible with naval personnel to ensure adequate instruction and training as the program progressed. BuAer had final approval of all plans and operations. The Navy would provide test range and base facilities.[3] This arrangement concept evolved into the contractor Phase A and Navy Phase B flight test programs. Phase A was contractor-directed, with assistance by the Navy as needed. Phase B was just the reverse, with the Navy directing the program and the contractor providing assistance as required. Virtually all direct contact with the Navy took place through the Regulus Project Officer at the Naval Air Missile Test Center (NAMTC), Pt. Mugu, California. The Regulus test program at Pt. Mugu had been established one year earlier, on 4 January 1949, by direction of BuAer. Lieutenant Commander Paul Goldbeck was the first NAMTC Regulus Project Officer.

After two years of preparation and planning, the Regulus flight test program began with the arrival of Chance Vought personnel at Edwards Air Force Base, California, on 12 February 1950.[4] The large dry lakebeds at Edwards af-

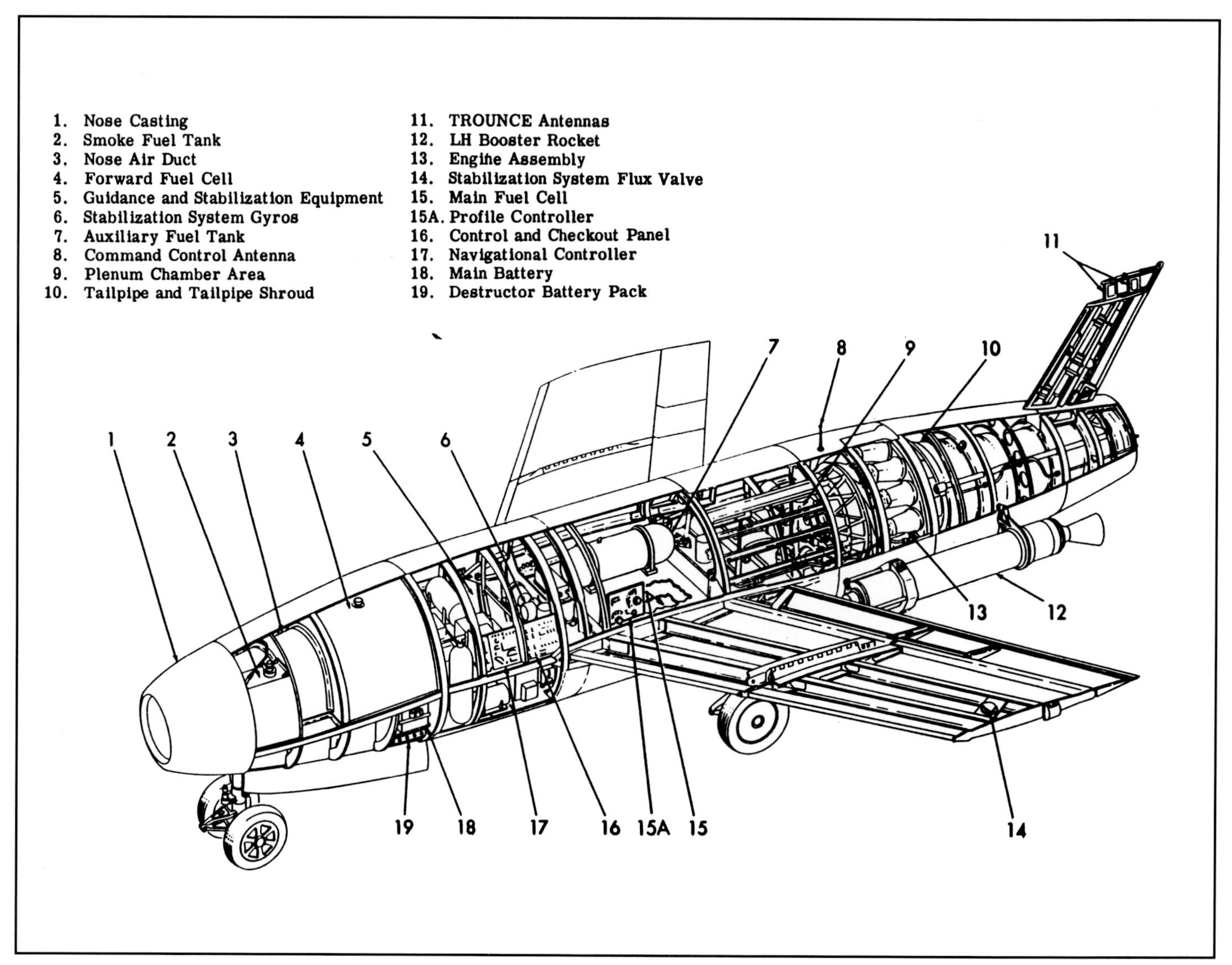

Figure 3-1. Regulus I Fleet training missile (SSM-N-8) general interior arrangement.

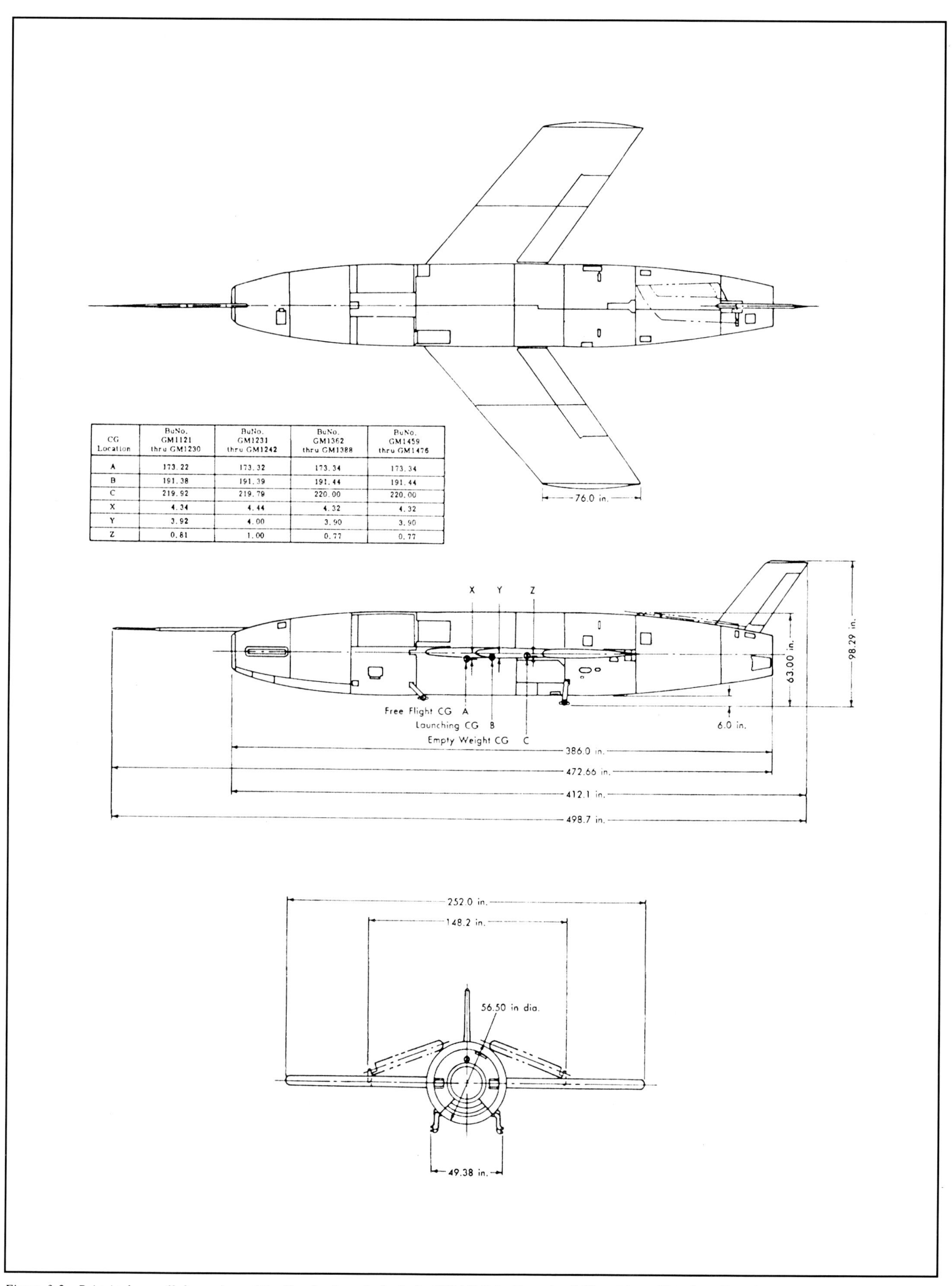

CG Location	BuNo. GM1121 thru GM1230	BuNo. GM1231 thru GM1242	BuNo. GM1362 thru GM1388	BuNo. GM1459 thru GM1476
A	173.22	173.32	173.34	173.34
B	191.38	191.39	191.44	191.44
C	219.92	219.79	220.00	220.00
X	4.34	4.44	4.32	4.32
Y	3.92	4.00	3.90	3.90
Z	0.81	1.00	0.77	0.77

Figure 3-2. Principal overall dimensions of the Regulus I tactical missile (SSM-N-8a). Two wing-fold positions, 54-inch or 70-inch were available on Regulus I to accommodate the different hangar dimensions of Tunny *and* Barbero *(70-inch) versus* Grayback, Growler *and* Halibut *(54-inch).*

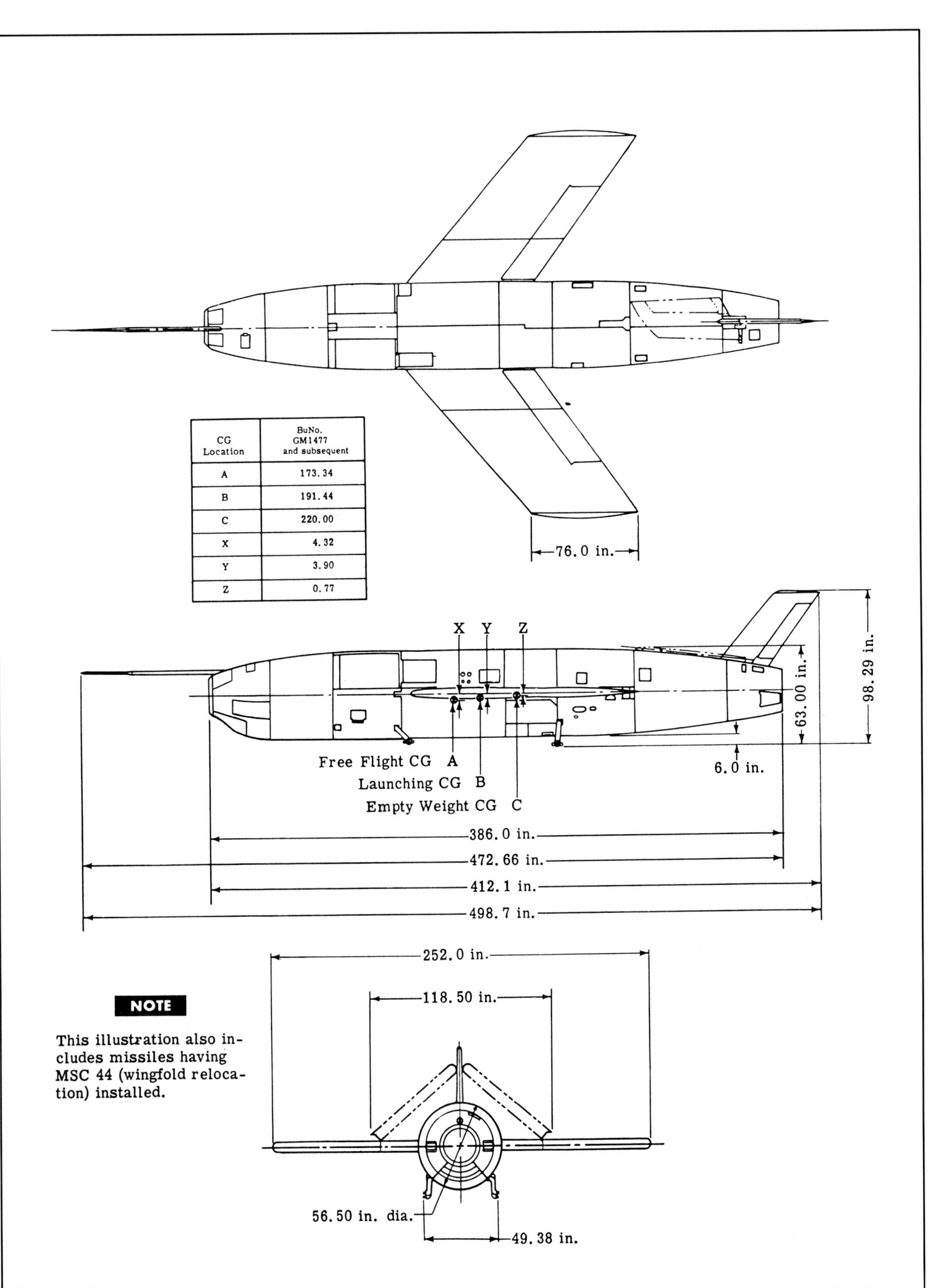

CG Location	BuNo. GM1477 and subsequent
A	173.34
B	191.44
C	220.00
X	4.32
Y	3.90
Z	0.77

Figure 3-3. Principal overall dimensions for "bulged-chin" version of the Regulus I tactical missile. (Courtesy of Loral Vought Systems Archives, Author's Collection)

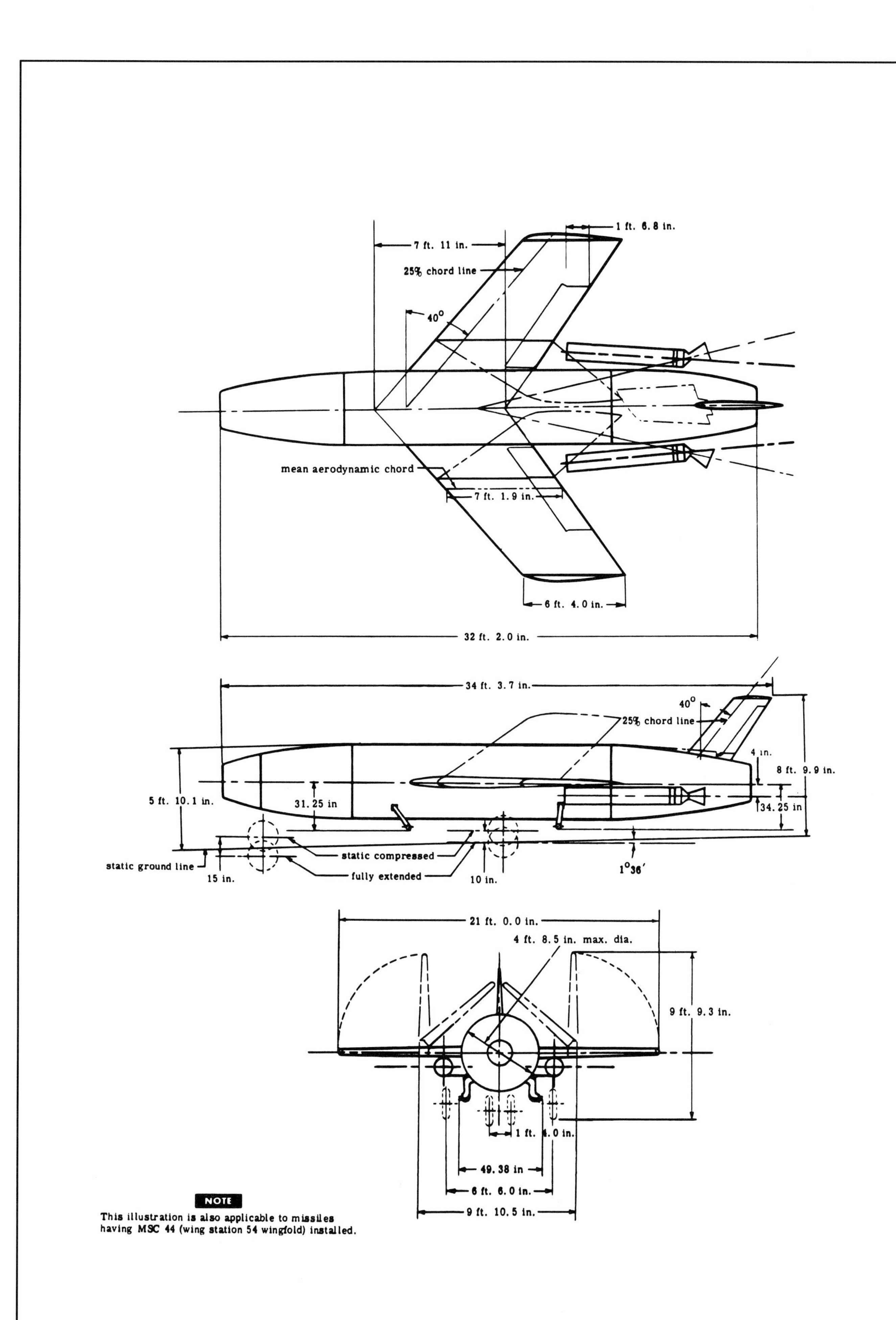

Figure 3-4. Principal overall dimensions for the fleet training (SSM-N-8) and target drone (KDU-1) versions of Regulus I. (Courtesy of Loral Vought Systems Archives, Author's Collection)

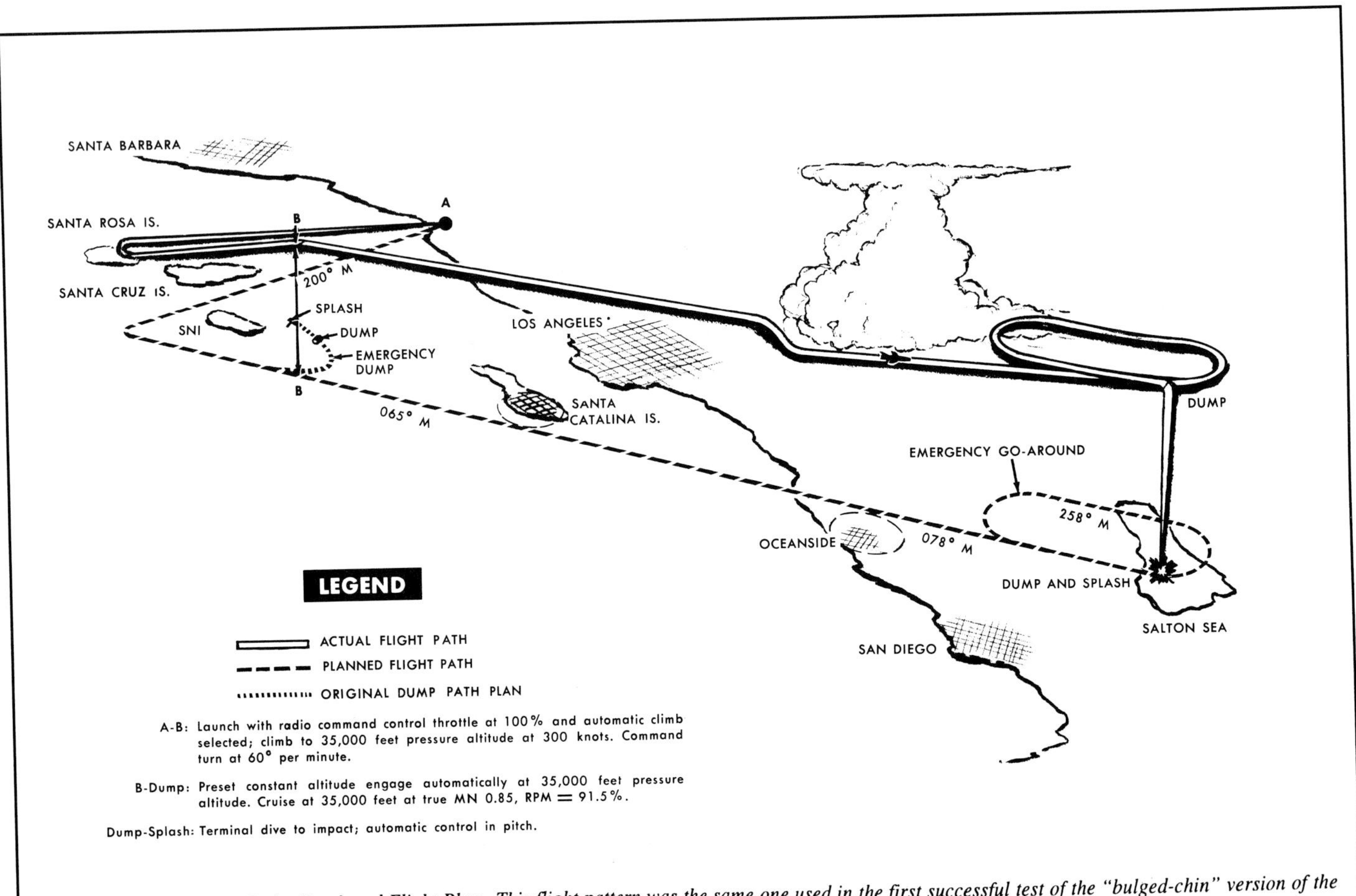

Figure 3-5. Flight #400 Radar Track and Flight Plan. This flight pattern was the same one used in the first successful test of the "bulged-chin" version of the tactical missile. (Courtesy of Loral Vought Systems Archives, Author's Collection)

Several of the key personnel in the early days of the Regulus program. L to R: Paul Baker, Chief of Aerodynamics and Flight Testing, CVA; Lt. Dewitt Freeman, USN, Navy Regulus program pilot; Leroy Pearson, Chief Test Pilot, Regulus Program, CVA; C.O. Miller, Regulus control pilot, CVA; Bill Sunday, Regulus control pilot for recovery, CVA; and I. Nevin Palley, Chief Engineer, Missiles, CVA. (Courtesy of Loral Vought Systems Archives, Micchelli Collection)

L to R: Bill Micchelli, Field Operations Manager, CVA and Sam Perry, Assistant Chief Engineer, Missiles, CVA. (Courtesy of Loral Cought Systems Archives, Pearson Collection)

forded Chance Vought the necessary room for evaluating the suitability of recoverable missile operations. Once the missile's flight characteristics were proven, launches would then be made at Pt. Mugu with recovery offshore at the Navy's landing strip on San Nicolas Island, one of the many islands off the Southern California coast.

Chance Vought's operating facilities were housed in a dilapidated World War II hangar at the Main Base (presently known as the old South Base). I. Nevin Palley was Guided Missile Division Manager and Chief Missile Engineer. Samuel O. Perry was Assistant Chief Engineer and Director of the Flight Testing Program. William S. Micchelli was Manager, Airframe section; Robert Stewart was Manager, Electronics Section and Roy Pearson was Chief Test Pilot.[5] With the arrival of the first flight test vehicle, scheduled for 21 February 1950, work began in earnest to ready the shop areas for the missile, electronic equipment and chase aircraft.

Radio command control of propeller-driven aircraft had become routine during World War II. The success of such a system with jet-powered pilotless aircraft was a challenge that Chance Vought would be one of the first to face. The Regulus radio command control system had two modes of operation. The first was known as "beep" control where only "on" or "off" commands could be sent. A selector switch with 20 positions worked together with a switch set at

Flight Test Vehicle 1 (FTV-1) in the Main Base, EAFB, hangar facilities on 2 March 1950, one month before the beginning of taxi tests. Wind driven sand and dust complicated repair work on engine and electronics. (Courtesy of Loral Vought Systems Archives, Micchelli Collection)

1. **Radio Command Control Console**
2. **Bank-Turn Control**
3. **Throttle Control**
4. **Power Switch**
5. **Carrier Switch**
6. **Channel Select Switches**
7. **Execute Switch**
8. **Proportional Pitch Control**

Typical radio command control installation for TV-2D aircraft. The equipment was modularized and could be rapidly removed if necessary.

The first and only "piloted" test of Regulus I took place on 12 April 1950. Missile checkout and support vehicles on the dry lakebed. Top left: The tow-truck was connected to FTV-1 by a 250 foot tow rope. Note the TV-2D chase aircraft parked behind the tow truck. Top right: Leroy Pearson poses for the camera before climbing onto the wing to serve as emergency control. Bottom left: Pearson laying on the wing as the slack is taken out and the test begins. I. Nevin Palley was on the opposite wing to counter balance Pearson's weight. (Courtesy of United States Air Force, Pearson Collection)

"decrease" or "increase" to command such things as parabrake deployment, landing gear up or down, brakes on or off, or any of the automatic flight control system functions such as altitude or climb settings. The second method of control was the proportional system. This provided variable continuous control for pitch, rate of turn and throttle settings. Unlike propeller-driven aircraft, where changes in throttle settings resulted in nearly instantaneous changes in speed, with the turbojets of this time frame there was a short lag before the engine responded. This made the challenge of radio command control flight with the TV-1 training drone and the missile all the more interesting.

The cockpit positions in the TV-2D director aircraft were referred to as ABLE and BAKER. The ABLE position was in the front cockpit where the radio control panel was located. While the ABLE pilot would use occasional eye contact to observe the missile or drone control surfaces during maneuvers, he primarily kept his vision straight ahead. The BAKER pilot flew in tight formation with whatever maneuvers the drone or missile executed through command control by ABLE. Thus, the perception of the ABLE pilot was as if he was in the drone or missile since his commands were immediately reflected by changes in the flight path of the TV-2D. Considerable training was required to insure that ABLE refrained from turning and watching the drone or missile too often or for too long.

The TV-1 pilot position was known as CHARLIE. During initial flight testing, the TV-1 could serve as both a missile surrogate or a backup control aircraft by simply switching control panels prior to flight. In the surrogate missile function the CHARLIE pilot was basically along for the ride as the ABLE pilot in the TV-2D controlled the TV-1, including landing maneuvers right up to the point of touch-down. Needless to say, this made for some interesting rides for the CHARLIE pilot. Early flight operations with the TV-1 served a critical function in addition to that of training TV-2D pilots. Since it was equipped with the same autopilot as the missile, deficiencies in the angle-of-attack stabilization and the autopilot yaw-bank control function were observed and rectified prior to the first missile operation.[6]

The two ground-based radio command control operators were known as FOX-1 and FOX-2. FOX-1 checked out the missile before launch to insure proper control surface response. As the airborne TV-2D made its pre-launch checkouts with the missile, FOX-1 would visually confirm each response. FOX-2 was stationed at the end of the taxi test area or runway to control the missile during its initial rolling takeoff and then turn control over to the ABLE pilot as the TV-2D swooped down for control pickup.

With the Air Force reluctant to permit the missile taxi testing to tie up the marked runways, Pearson and Sunday spent a full day evaluating areas between runways to the north and south of the hangar facilities, finally settling on the area to the south. This would be somewhat inconvenient in that a small convoy of support vehicles would have to be driven approximately six miles to and from the test site, towing the missile. At the same time, this put Regulus I testing out of the main operations area, hopefully preventing frequent calls for temporarily halts while other programs flew overhead.

During the first week of April 1950, three FTV's were in various stages of flight readiness at Edwards. The first two flight test vehicles were completely assembled and were in use as maintenance and troubleshooting procedures were ironed out; FTV-3 was assembled but the engine was not installed. Radio command control reception from the TV-2D on the ground as well as in the air was also evaluated. A nagging problem with the Main Base facilities was the excessive wind driven dust and sand. More than one ramp test failed due to these materials contaminating the hydraulic fuel or fouling relays. Another unforeseen problem was the amount of electrical "noise" generated by the missile's electronics as well as from shop and other Edwards flight operations. Since spurious signals of sufficient strength might create erroneous radio commands, these problems were certainly not trivial. Field modifications were made to much of the electronics to mitigate these problems.

Regulus I on a Weber dolly. Notice that the landing gear are not touching the hangar floor. This dolly permitted the full simulation of a complete flight operation including lowering of landing gear and control surface movement. (Courtesy of Loral Vought Systems Archives, Ransdell Collection)

Flight operations began 10 March 1950 with Pearson's and Sunday's checkout flights in the TV-2D with Lieutenants Dewitt Freeman and Ed Schuler, the first Navy pilots in the Regulus program. The Navy provided chase pilots in order to facilitate training for the operational groups. Freeman and Schuler flew out to Edwards as necessary at this point since one or both of them were also needed for flight operations at Pt. Mugu.[7] The first missile operations on the Edwards dry lakebed took place 10 April 1950. Prior to any self-propelled missile taxi tests the brakes had to be aligned to insure that there would be positive control of the missile on the taxi and landing roll out portions of flight operations. This was a simple matter in piloted aircraft, almost mundane. With an unmanned missile, however, a little ingenuity was needed. A semi-tractor truck towed the missile on a 40 foot rope. The brake servomotor was locked in the neutral position so that when the rope was detached from the missile via a single-point bomb shackle quick release mechanism, the missile would glide to a stop, allowing wheel alignment to be checked. Once the alignment was established, the brakes shoes could be burned in and brake actuation forces equalized. The tests proceeded smoothly and the missile was ready for the ground operation using control as needed from either the FOX-1 ground console or the TV-2D flying a pattern overhead.[8]

Roy Pearson clearly remembered the first and only "piloted" test of Regulus I on 12 April 1950. Concerned over the possible lack of braking control, if either the truck mounted console or the TV-2D was unable to maneuver the missile, the decision was made to have Pearson lie on the left wing with a brake control switch in his hand. In order to balance his weight on the left wing, Palley would ride the right wing. Pearson and Palley donned crash helmets and leather jackets and climbed onto the missile wings. The tow began and as luck would have it, control from FOX-1 could not be established and the TV-2D's signal faded soon after the missile brakes were energized. Pearson now had the unique opportunity to drive the missile for several minutes as he brought it to a controlled stop. During the second run, the FOX-1 control worked perfectly and the missile rolled to a straight ahead stop.

As the towed taxi tests progressed, the TV-1 and TV-2D flight operations began in earnest. Simulations of missile operations commenced with rolling takeoffs of the TV-1 controlled from the TV-2D and landings with waveoffs controlled by both the TV-2D and FOX-1. Pearson, Sunday, Freeman and Schuler interchanged ABLE, BAKER and CHARLIE positions as well as FOX-1 in order to understand the various viewpoints of each control operator's duties. On 10 May 1950, after two weeks of bad weather, the first low-speed taxi run took place. After a complete pre-flight check of all missile and control systems, Pearson commanded missile release from the TV-2D as they approached from two miles behind. FTV-1 was released from its holdbacks and immediately began a left hand turn. Pearson commanded "beep" right brakes, but the turn continued. He then signaled "beep" brakes again and held the control for two seconds. The missile straightened out and rolled down the runway holding a straight course. Several seconds later FTV-1 began to turn to the right. Pearson realized that the missile was headed towards some lakebed obstructions approximately a mile away. After unsuccessfully trying to correct the turn, he shut the engine down, whereupon the missile straightened out and came to a stop. Differential braking was going to be more difficult then anticipated.

On 23 June 1950 welcome news from two fronts reached Chance Vought operations at Edwards. First was the word that after almost a year of review, the Navy had concluded that combining the Air Force Matador program and Regulus into one program was a bad idea. The Navy therefore recommended to the Research and Development Board that Regulus proceed on its own. The Board concurred and the Navy retained full control over Project Regulus.[9] The second piece of heartening news was the signing of Contract NOa(s) 12035 which ordered 10 FTMs, designated Lot II.[10,11,12]

Aircraft flight training operations proceeded simultaneously with the taxi tests as control procedures and power settings for the TV-1 were refined to more closely simulate the low-speed operation of the missile. By 26 June the TV-1 was being routinely controlled from take-off, through patterned flight, to hands-off touchdown, guided by both the TV-2D and FOX-1 controllers. On 9 August 1950, FTV-1 reached 190 knots ground speed during the first high-speed taxi test, but tire tread separation occurred with subsequent tire failure. This was completely unexpected as standard aircraft tires had been used. For the next test a camera was mounted to record the details of the problem. This second test on 24 August ended with the missile doing a ground loop as the tires again failed. Review of the film revealed that while the tires operated normally through the beginning of the test, soon a standing wave could be seen in tread as the tires overheated and the rubber became extremely flexible. Subsequent tread separation led to complete failure and the resulting disintegration. The tire manufacturer, B.F. Goodrich Tire Company, was consulted and suggested that tires made of the rubber used on the Indianapolis 500 race cars and reinforced with nylon cord be used. This was agreed upon and a set of tires made for testing. The third high-speed taxi test with the new tire designed was a complete success. Though the tires were extremely hot after roll out, no evidence of tread separation or im-

Right-hand main landing gear wheel from the first high-speed taxi-test on 9 August 1950. The tires failed due to excessive heat build up during prolonged taxi runs. (Courtesy of Loral Vought Systems Archives, Micchelli Collection)

FTV-1 being readied for the first low-speed taxi test 10 May 1950. The white cord at the rear of the missile is the parabrake line. The missile checkout equipment stand can be seen just forward of the wing. (Courtesy of United States Air Force, Pearson Collection)

minent tire failure was seen. As an added safety precaution, wire tire cages were used to protect the ground crew working on the missile immediately after recovery.[13]

On 11 September 1950, Lieutenant Ed Schuler was killed in an aircraft accident at Point Mugu. The replacement pilot for Schuler was C.O. Miller, a civilian pilot newly hired by Chance Vought. Between 3 October 1950 and 9 November 1950, three high-speed taxi tests were conducted during which the missile became airborne for brief distances. The purpose of these tests was to evaluate the sensitivity of the flight controls during taxi operations. Pearson noted that FTV-1 responded to nose-up pitch commands smoothly. In contrast, the missile seemed extremely sensitive to nose-down pitch commands, a problem to be aware of during the first full-fledged flight.

On 22 November 1950, thirty-six months and five days after the first BuAer Letter of Intent for Project Regulus was issued, the first flight of Regulus took place. The standard dress rehearsal, which consisted of the TV-2D controlling the TV-1 from takeoff through the flight pattern to landing, was completed by 0700 hours. After aircraft refueling and last minute equipment checks, the chase aircraft took off at 0800. Nine minutes later FTV-1's engine was started and the pre-flight countdown began using the FOX-1 controls. Pearson and Miller were flying as ABLE and BAKER pilot respectively in the TV-2D, with Lieutenant Freeman flying CHARLIE as back up in the TV-1. As they approached from three miles out at 170 knots, Pearson centered the missile rudder, increased the missile engine speed to 100% rpm and when he was three-quarters of a mile away, commanded missile release. FTV-1 headed straight down the runway. At 120 knots missile airspeed Pearson signaled for 8 degrees pitch increase and the missile responded immediately. At 170 knots the missile began to drift left but Pearson easily corrected it back to the centerline. Miller concentrated on maintaining his slightly faster speed, since once the missile left the ground, even with the landing gear down, it would rapidly catch up to him. Ten feet above the lakebed, Miller manuevered close in on the missile's left wing, as they had practiced many times before. On schedule, the missile rotated and became airborne at 190 knots with 11 degrees pitch. After a few seconds at this angle, and remembering the experiences of the previous high-speed taxi tests, Pearson commanded a slight downward pitch. As the missile responded, he began a right turn to avoid flying over buildings near the main runway. When the change of heading had been accomplished, he returned the missile to zero bank and FTV-1 straightened out on its new course.

Pearson realized that with the decreased pitch attitude the missile was beginning to pick up too much speed and commanded a slight increase in pitch. No pitch response was noted and the missile began a bank to the right. At this point the missile ceased responding to commands and started slight roll oscillations which then increased greatly in magnitude. As Miller maneuvered to keep the missile in sight, it complete two rolls and passed under the TV-2D, missing by only 10-15 feet. Seconds later the FTV-1 slammed into the lakebed surface at a speed of 220 knots from an altitude of 3,500 feet.[14,15] The ground crew rushed to the crash site which was directly off the end of the main runway. They established a reference grid and mapped the location of FTV-1 debris prior to collecting it for analysis. Lieutenant Freeman remembers quickly landing the TV-1 near the crash site and walking up to the missile engine fan disk lying on the ground. Nearby was the cruciform imprint of where the missile had hit. He recalls being amazed at how small an indentation it had made on the lakebed and then wondering what the base commander was going to say about the mess just off his main runway.[16]

Fortunately a complete telemetry record of the flight was available as well as most of the photo-observer film. Review of both sources began simultaneously at Edwards and Chance Vought in Dallas. Palmer Ransdell, a flight test

FTV-2 being checked out after the first successful flight and recovery of a Regulus I missile on 29 March 1951. The cages encasing the main landing gear were used to protect personnel from the hot tire and brake assemblies. (Courtesy of Loral Vought Systems Archives, Author's collection)

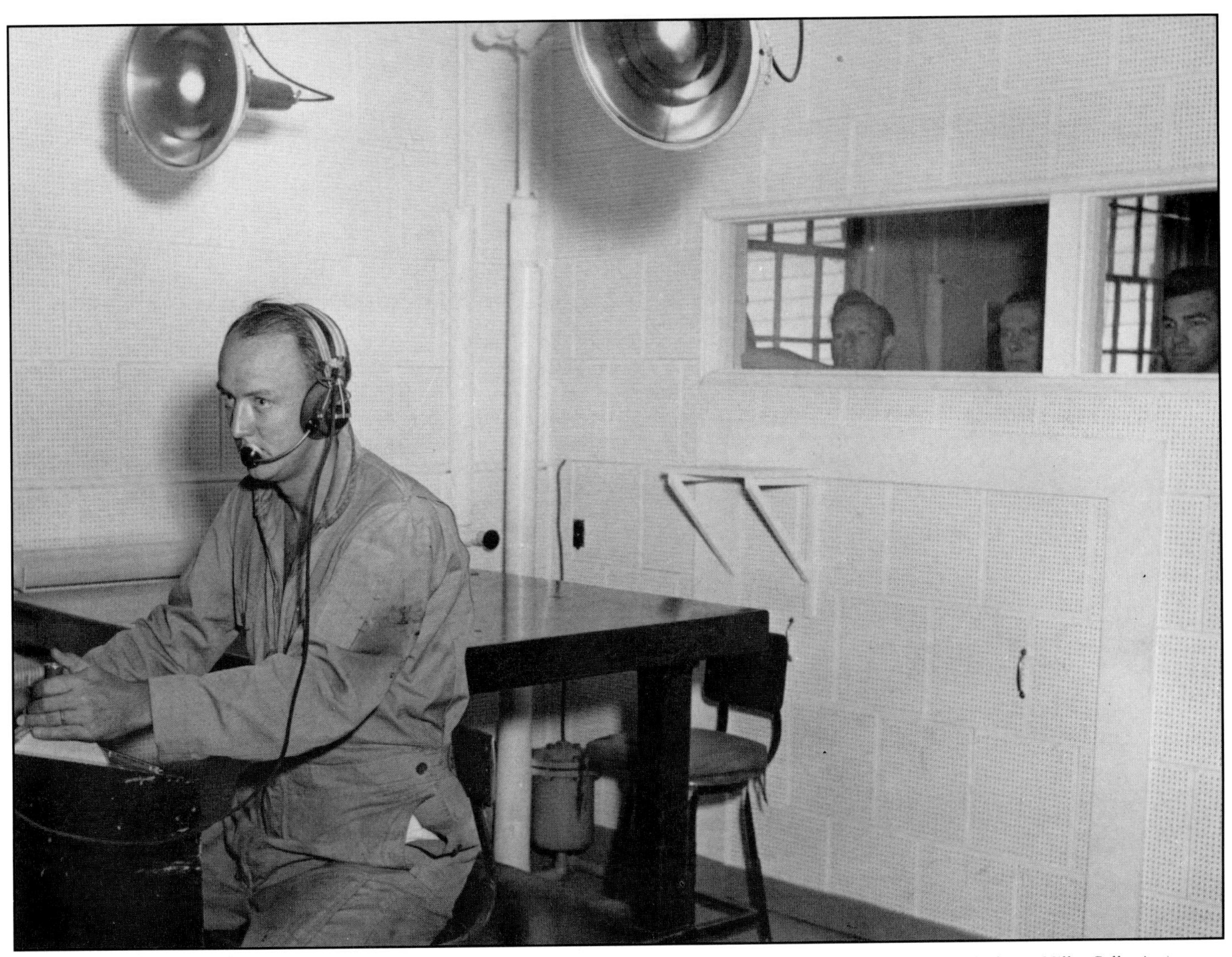

C.O. Miller at the flight control console for the Out-Of-Sight-Control station at EAFB. (Courtesy of Loral Vought Systems Archives, Miller Collection)

engineer on the program in Dallas and part of the crash investigation team, recalls that the telemetry information had narrowed down probable cause of the failure to a hydraulic pump and motor combination. Review of the telemetry data indicated that the left ailavator hydraulic system had experienced a two-step downward transient that resulted in a full-down positioning of the control surface. This information indicated that the ailavator had lost its radio command control response, indicating a control failure.

Attention was focused on reproducing the same telemetry recorder trace utilizing a spare pump and motor set in Dallas. Ransdell rigged a pump and motor to simulate the failure by replacing the bolts holding the motor-pump assembly with long pins that allowed separation of the hydraulic pump and the motor while the system was running. The telemetry system was started, the motor turned on and then the motor slowly separated from the pump. The telemetry trace soon became an exact match to the flight test data as the motor ceased to turn the pump's spline. With this information a search was made of the collected crash debris and a fractured spline pin was found in the left ailavator pump assembly. This hollow pin, made of brass, had failed from metal fatigue.

Hindsight now clarified the cause of the crash. FTV-1 had been the only airframe used in all of the testing, from initial control check-out to the low and high-speed taxi tests on the lakebed. The autopilot system was a modified A-12 *Sperry* autopilot and a fail-safe feature used in piloted aircraft had been overlooked. If necessary, the pilot could overpower the hydraulic system due to a frangible link which was the hollow brass spline shaft. Repeated testing of the system had fatigued the spline to nearly the point of failure, but the inspection program in effect at that time did not check this particular component. The accident investigation committee recommended that all splines in the hydraulic pump-motor combination be replaced with solid steel pins. No further flight failures were caused by this combination of events.[17]

Continued project funding was still very much in question at this point and a crash on the first attempted flight didn't help matters at all. Palley proposed to the NAMTC project office that a formalized program of exhaustive reliability runs be started. The missile and the ground crew would carry out 20 completely successful simulated missions before the missile could be designated as "flight qualified". This concept would insure that not only was the missile ready but that the ground crews became thoroughly familiar with all of the flight systems and their telemetry responses. The project office and BuAer concurred with the plan and the reliability program began immediately with preparations for the first flights of FTV-2 and FTV-3.

What might at first appear to be a relatively simple task was in reality quite complicated with an unmanned missile. Bill Micchelli, the Chance Vought engineer managing the airframe section at Edwards, recalls that one of the major efforts concerned the flight profile reliability tests. The missile was mounted on a test stand, permitting its engine to be run at full power for the duration required by the flight plan. Commands to move the control surfaces were made through the autopilot. This required the removal of the gyroscopes to a three axis table so that they could be tilted to actuate ailavator movement. Since all commands were by radio signal during these reliability tests, the transmitter was removed from the TV-2D and placed on a 20 foot tower next to the missile. Each deflection of the control surfaces was noted to insure that all commands were responded to correctly. This process was quickly developed into a pre-flight routine that enormously enhanced the success of the program.[18]

1951

On 14 February 1951, BuAer ordered 14 additional missiles, designated Lot III, FTM-1016 through FTM-1029. In early February

1951, Chance Vought Regulus operations were moved to a larger hangar at the North Base area of Edwards, allowing more space for missile testing and repair. By early March 1951 both FTV-2 and FTV-3 were ready to begin the new reliability program test sequence prior to reaching flight status. A reoccurring problem with faulty ailavator and rudder hydraulic servopumps that caused a three week delay in the completion of the reliability runs was traced to dirt contamination of the hydraulic fluid. After a visit to the hydraulic system manufacturer's El Segundo factory, a new hydraulic fluid cleaning apparatus was devised and quickly solved what had been a persistent problem.

Reliability runs using FTV-2 began on 20 March 1951, with 14 test runs successfully completed before the autopilot gyroscopes were removed and placed on the tilt table to complete the testing. The remaining six runs were completed without incident, proving that the reliability program's attention to every detail was working as intended. The ground equipment had run for 27 hours without failure while the missile had run for 20 hours successfully. The FTV-3 reliability test was equally successful and both vehicles were put into flight status. These results impressed all concerned, bringing a new spirit of success to the entire program staff. Palley's leadership and drive had been successfully transmitted to the entire Chance Vought Edwards project team.

On 29 March 1951, 18 weeks after the crash of FTV-1, FTV-2 was towed to the south lakebed for launch. Pearson and Miller were again in the TV-2D ABLE and BAKER positions, respectively, and Lieutenant Freeman was once again flying CHARLIE in the TV-1. Visibility was 10 miles, with overcast at 15,000 feet, and a slight wind from the south. The dress rehearsal flight utilizing the TV-1 was completed successfully and FTV-2 was readied for launch down the seven mile main runway on Rogers Dry Lake. After completion of the final control checks, the aircraft entered the approach pattern at four miles out. As Miller passed over the three-quarter mile point, Pearson signaled for release of the missile. FTV-2 started the takeoff roll with a slight left hand turn which Pearson quickly adjusted. The missile continued to roll parallel to the runway center stripe and at 170 knots airspeed lifted into the air as Pearson signaled the nose up pitch command.

At 100 feet altitude Pearson signaled a right turn to head the missile away from the Main Base facilities. The turn was completed smoothly as Pearson returned the control to the zero position and the missile easily resumed level flight. Pearson and Miller both recall heaving a sigh of relief in reaching and passing the ill-fated part of the pattern from the previous flight. Throttle was reduced to 95% RPM and a climb to 2300 feet completed at 250 knots airspeed. Pearson then signaled the beginning of the cross-wind leg of the flight pattern. This was the first large turn commanded to the missile and Pearson remembers that he did it in small increments to evaluate the missile's stability which remained excellent. The turn was completed at 2,500 feet and 240 knots airspeed. The turn into the down-wind leg of the pattern was executed equally smoothly.

Towards the end of the down-wind leg, the missile was hit by several wind gusts causing some roll instability that was corrected by the autopilot. At this point Pearson had an opportunity to visually inspect the control surfaces of the missile. He detected a low amplitude, high-frequency jitter that was similar to that seen on the ground runs but noted that it did not seem to interfere with the flight response of the missile. The turn to the base leg was equally uneventful and at 220 knots Pearson turned the missile on to the final approach. As FTV-2 passed over the edge of the lakebed, Pearson commanded a small left turn to land the missile off the left side of the runway, so that if a crash occurred the runway would not be damaged. Touchdown was smooth and only one-quarter mile past the selected position. When the nose gear contacted the runway, the drag chute released, blossomed and then collapsed. Differential braking was selected at 170 knots for a few seconds and then released since the brakes were not designed for use at this high a ground speed. Once the mis-

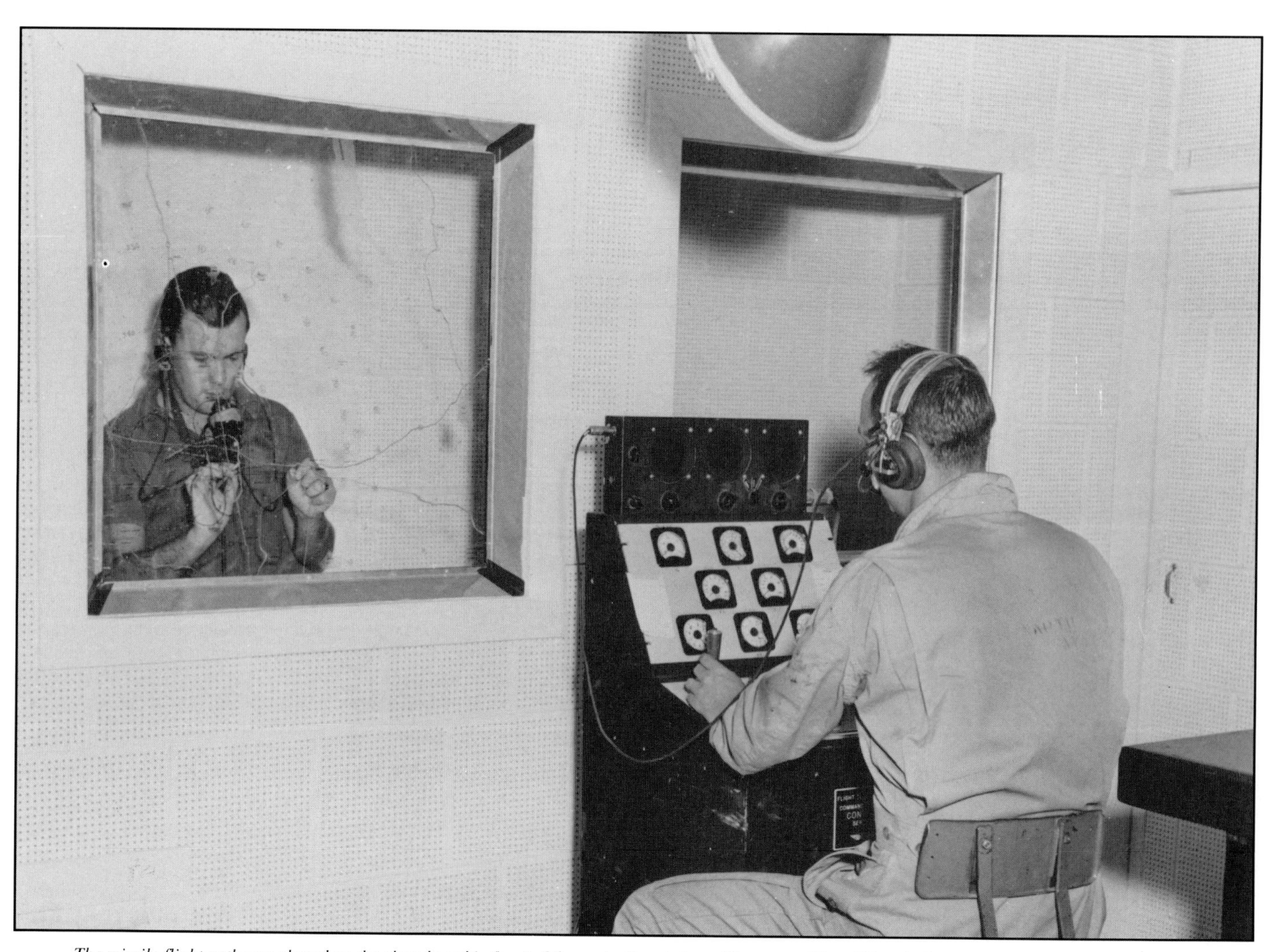

The missile flight path was plotted on the clear board in front of the controller station. (Courtesy of Loral Vought Systems Archives, Miller Collection)

sile had slowed to 50 knots, brakes were again selected and the missile brought to a halt at the three and one-quarter mile mark. In his post-flight report, Pearson noted that the missile actually flew better then the TV-1 in terms of responsiveness to his commands. He had not attempted to make a precision landing and had intentionally landed off the main runway in case the missile crashed. Pearson was quite convinced that precision landings could be routinely executed in the future. The drag chute failure was the only malfunction on an otherwise perfect operation.[19] The first flight of FTV-2 was the beginning of a string of 13 successful flights over the next eight months.

The first flight with a Navy pilot controlling a Regulus missile took place 12 June 1951. Lieutenant Freeman flew as the ABLE pilot during a high-gain control setting evaluation for yaw and pitch. The flight was completely successful. By June 1951, the ground and maintenance crews was averaging 10-15 days between flights. On 21 June 1951, the eighth flight of the program and the fifth for FTV-3, the first tests of the missile's automatic flight stabilization and automatic airspeed control systems were conducted. The automatic stabilization system received angle of attack information from a boom mounted vane positioned at the tip of each wing. The boom vibrated and as a result the missile would wander all over the sky. Flying 10-15 feet from the missile's wing was indeed an exacting task during these tests. Shortening the boom helped somewhat but the final solution was to place the boom on the nose of the missile.[20]

The next missile to fly was FTV-5, the last of the Lot I production run. FTV-4 had been designated to receive an alternate autopilot from Honeywell and was still at the Dallas plant. The delay in FTV-5's use was caused by the need to install retractable landing gear on the missile so that the high-altitude, high-speed tests could begin. The initial design of the Lot I missiles had required only extensible landing gear since most of the launches were to be JATO boosted with the gear retracted. The landing gear would then be dropped and locked on landing approach using aerodynamic forces. Delays in JATO booster development resulted in a requirement for retractable landing gear installation which, in turn, required moving the landing gear aft on the airframe. FTV-5 was the first missile with these changes. On 7 July 1951, FTV-5 made its first flight with the new landing gear system extended and successfully landed. FTV-5 made the first test of landing gear retraction during a medium altitude, 10,500 feet, and a medium speed, 360 knots, flight on 27 September 1951. All test objectives were met except that the main landing gear doors would not close completely.

Reliability test run stand for remote control of missile via radio signals. C.O. Miller is behind the console, I. Nevin Palley to his left. (U.S. Air Force Photograph, Miller Collection)

The main landing gear door closure problem had to be resolved before the flight program could expand to higher speed, higher altitude and longer duration flights. A quick fix was simply to remove the main gear doors to reduce drag. This temporary fix permitted testing at 400 knots and 10,000 feet on 11 October 1951. This flight included partially successful test of the ground based out-of-sight-controller as well as the automatic altitude controller. The out-of-sight-controller station was a room at the Chance Vought North Base hangar with a control station containing readouts of airspeed, engine rpm, altitude and heading. A technician was in communication with a Navy radar van and he plotted the missile position on a glass panel for reference by the controller as he monitored his instrument panel. C.O. Miller was the first out-of-sight-controller operator and successfully controlled FTV-5 until the first turn was commanded. At this point the missile's radar transponder failed and radar track was lost, not to be regained until the missile was on approach to landing.

Recovery of Regulus I on the lakebed. The TV-2D is on the left, controlling the missile through to touchdown and initial taxiing. The TV-1 is on the right, serving as back up in case of primary control failure in the TV-2D. While the missile cannot be identified, it has to be FTV-2, -3 or -4 since no landing gear doors can be seen. (Courtesy of Loral Vought Systems Archives, Ransdell Collection)

Paul Wandell, a Chance Vought airframe engineer at Edwards, came up with a novel solution to the gear door closure problem that was also almost his undoing. The need to have some manner of additional power-assist was apparent. Engineering support in Dallas, however, insisted that the design in place was sufficient. Finally Palley decided to fix the problem in the field, regardless of what Dallas thought. Wandell had suggested earlier, somewhat humorously, that if the gear doors could not be made to come up to the missile, the missile should be pulled down to the doors. Wandell located, ironically enough at a war surplus equipment store called "Palley's Surplus", a B-36 bomb-bay door pneumatic piston that fit into the air duct of the missile. In the compressed position the gear doors were down. With application of bottled compressed air to the cylinder piston, the cylinder strut could be extended and through the use of several pulleys, the main gear doors could be pulled shut with sufficient force to overcome the aerodynamic forces. Great in theory and very workable in fact.

Palmer Ransdell recalls working with Wandell on one of the reliability tests prior to the next flight of FTV-5. They were cycling the new gear door closure system. The landing gear retraction mechanism was sequenced through a series of cams and microswitches so that when the retract command was given, the gear would fold up into the wheel well. As the gear reached the full-up position, a microswitch was activated which then allowed compressed air to push the B-36 piston to the extended position and pull the gear door shut. After a test cycle, a system reset was required to rest all of the switches. Ransdell forgot to perform the system reset. The next gear retraction cycle was started and the door began to close at the same time the gear retracted. The sequence was automatic and could not be interrupted so Ransdell and Wandell watched in fascination as the landing gear doors and wheel assemblies met at an intermediate position, blocking any further movement of the doors or the gear. As gear door closure system pressure built up, part of the linkage broke. With a loud bang, the piston shot out of the nose of the missile, continued across the hangar, impacting the wooden wall like an arrow. Luckily, no one was hurt. This part of the hangar quickly became known as the Wandell Shooting Range with a large bull's eye target painted at the point of piston impact.[21]

On 17 November 1951, five days short of one year from the ill-fated first flight attempt, FTV-5 made the first high speed, high altitude flight, climbing to 25,000 feet under out-of-sight-controller direction. Upon turning into the high speed test leg of the pattern and assuming a level flight attitude, the missile quickly pulled away from the TV-2D which was at full throttle and could not go any faster. The test plan had anticipated this event and an Air Force North American F-86E "Sabre" high-speed chase aircraft, piloted by Captain John W. Konrad, was positioned 2,000 feet above the FTV. As soon as the missile's throttle was increased to 100%, Konrad pushed over in a shallow dive, enabling him to keep pace with the missile for a short period. When the F-86E reached the same altitude as the missile, it too began to drop behind. Meanwhile, the TV-2D had flown at maximum speed across the corner of the pattern and successfully picked up the missile once it reached the end of the high-speed run. This first high-speed, high altitude test was a complete success with the missile setting a new record for the program in both altitude, 26,500

One of the more amusing features of the Regulus program were the various attempts to conceal the fact that it was a missile and not a piloted aircraft. After the crash of FTV-2 on Mirage Lake, a fake aircraft canopy such as that pictured here with Leroy Pearson, was taken to the crash site in an effort to conceal the missile operation. (Courtesy of Loral Vought Systems Archives, Pearson Collection)

Vertical center of gravity test on FTV-3, 24 January 1952, at Pt. Mugu. After the failure of a single crane during a preliminary test, two cranes were used for this operation. FTV-3 has been fitted with the lower vertical tail. (Courtesy of Loral Vought Systems Archives, Author's Collection)

feet, and speed, Mach 0.93. The out-of-sight-control system had succcessfully commanded the missile in the transonic cruise speed range without any apparent problems, confirming that the concept of a remote control, high subsonic speed missile was feasible.

The third flight of FTV-2 took place on 29 November 1951. This flight was to evaluate control in a shallow dive as well as achieve maximum Mach numbers in two dives and pull outs. The flight test portion went according to plan with the first dive from 34,100 feet pulling 2.2 "g's" and the second, from 31,000 feet pulling 2.8 "g's". The missile reached a new record speed of Mach 1.01 during the dives and control did not appear to be a problem at this speed. The flight ended in one of the more amusing stories of the early Regulus program and one that is often retold. After the completion of the 18 minute test program, Roy Pearson, flying the ABLE position in the TV-2D, turned the missile into the approach for landing. A failure in the autopilot control circuit caused the missile to perform a snap roll to the right, followed by oscillations about all three axes. It then went into a spin, crashing into Mirage Lake right next to the highway. After Pearson landed the chase aircraft at the North Base hangar, he and Palley rushed over to the crash site, Pearson still is his flight gear. Sam Perry brought over the fake cockpit canopy which had been used on several occasions to disguise the missile as a manned aircraft. He placed it amongst the wreckage. The missile's parabrake was collected and laid out next to the wreckage. At this point a rancher came over to investigate and was jubilantly told by Perry how lucky it was that the pilot had been able to parachute to safety. Pearson just smiled and agreed.[22,23]

Operation SPLASH

By the end of September 1951, Regulus I had flown successfully in 11 of 12 flight tests at Edwards. Several months earlier, in consultation

Preparting FTV-3 for the first booster rocket shock test 12 January 1952. Originally planned for the wide open spaces of the dry lake at EAFB, the test was moved to Pt. Mugu due to standing water on the lakebed surface. (Courtesy of Loral Vought Systems Archives, Author's Collection)

FTV-3 is eased onto the Short Rail Mark I launcher at Pt. Mugu on 30 January 1952. The large cylinder beneath the missile is part of the hydraulic system for elevating the launcher. This launcher was the prototype for the launchers later used on USS Tunny *(SSG 282) and USS* Barbero *(SSG 317). (Courtesy of Loral Vought Systems Archives, Author's Collection)*

with BuAer representatives at Pt. Mugu, I. Nevin Palley saw an opportunity to make a bold move to illustrate the progress of Project Regulus. He proposed to BuAer a demonstration of the basic Regulus I tactical mission, complete with the first boosted launch of the missile. The boosters were now ready to be shipped from Aerojet General Corporation; the high speed, high altitude capability of the missile would be demonstrated in the next series of flights; and, counting on their successful completion, he suggested scheduling the tactical demonstration for late January 1952. The specific objectives agreed to for this demonstration flight were: first, evaluation of the boosted launch and terminal dive portions of the flight profile; and second, the evaluation of operational problems that might be encountered in conducting follow-on flight testing at the NAMTC Sea Test Range at Pt. Mugu.[24]

Just as important as the risk of the first boosted launch going haywire was the decision to install the out-of-sight-controller equipment on board the USS *Cusk* (SS-348). *Cusk* was one of two launch and guidance submarines currently used in Project DERBY, the Navy's first guided missile program utilizing the Loon guided missile. Using *Cusk* would further demonstrate the feasibility of the tactical concept.

Booster operations were not new to Pt. Mugu personnel since the Loon guided missile, which was part of an ongoing flight test program, was launched with JATO bottles. The Regulus I boosters were much larger, however, providing 33,000 pounds of thrust over a 2.2 second period. Chance Vought's and the Navy's experience with this size booster was limited to contractor tests completed eighteen months earlier in April and May 1950. Chance Vought had constructed a true "zero" length launcher and successfully test-launched concrete and steel "dummy" Regulus missiles.[25] The tests had been successful in that the dummy missiles had been easily boosted. The telemetry data told another story. High shock loads generated by the rupture of the nozzle closure diaphragm indicated that further research was needed before these types of boosters could be used with a missile.[26]

Seventeen months later, in late October 1951, the booster tests resumed using a dead load "sled" as a dummy missile. Two Aerojet JATO 2.2KS-33,000 booster rockets were used.[27] A short rail launcher designated as SR MK I and built as a prototype for installation on board submarines was used. Booster and launch rail slipper ejection systems were a critical part of these tests because if either of the two boosters or any of the four launch slippers did not eject, the excessive drag would either cause the missile to crash or else severely limit its flight performance. In the first two tests all systems operated as planned and the "g" shock force was well within tolerances. The third test in early December was partially successful but the boosters failed to eject. The decision to continue with Operation SPLASH as planned meant that the booster ejection system would have to be carefully checked and rechecked prior to launch.

FTV-3 had been chosen for Operation SPLASH several months earlier. It represented, in every sense of the word, the success of Project Regulus to date with seven successful flight operations. FTV-3 was modified by removing the landing gear and placing an additional fuel saddle tank above the intake duct. The parabrake was also removed and an APN/33 radar beacon added to aid in tracking the missile. Terminal dive controller electronics were added; and, for the first time, the lower vertical tail surface was attached (taxied take-offs had precluded the use of the fin prior to this point).

In late December the newly configured missile was fully fueled and ready for determination of the missile's center of gravity (c.g.) for use in booster alignment. While calculation of the c.g. from blueprints would give an approximate value, the earlier booster launch experience with the dummy missiles indicated that a more accurate method was needed. The longitudinal c.g. was the least critical and determined by weighing the missile at the wheel points. The deceptively simple method of suspending the missile from a bar passing through the nose was decided upon for determining the lateral and vertical c.g.'s, with the vector sum of all three determining the overall c.g. This required the use of a crane that could lift the 13,000 pound, 35 foot long missile, 40 feet into the air. Bill Micchelli, the Chance Vought engineer in charge of the field work for Operation SPLASH, insisted that the candidate crane, which looked pretty much the worse-for-wear, be tested with a dummy load. Pt. Mugu personnel agreed and part way through the test, the crane failed. Needless to say, two cranes were used with the missile. Once the c.g. was located an optical boresight system was attached to an insert in each JATO bottle nozzle and the thrust vector aligned to pass within one-quarter inch of the c.g.

On 1 December 1951, Chance Vought personnel occupied the newly established Project Regulus facilities at Pt. Mugu and immediately begin preparations for the launch. On 10 December 1951, Chance Vought received the official go ahead for Detailed Firing Test Plan #7: Operation SPLASH. Installation of the out-of-sight-controller equipment in the *Cusk* began in late December.

By early January 1952, only two aspects of Operation SPLASH remained to be evaluated; the booster rocket taxi test to confirm booster hardware attachment, and the control of a manned drone throughout the extended range flight pattern. Roy Pearson clearly recalled the first booster test he participated in. Originally planned for the wide open spaces of Edwards, rainy weather precluded the use of the lakebed, so a runway at Pt. Mugu was used instead. FTV-6 had been outfitted with numerous accelerometers to permit detailed recording of the "g" forces from booster ignition. Special JATO bottles were used that would fire for only one-half second but which possessed both the ignition and initial thrust characteristics of the normal 2.2 second duration boosters. To add to the realism, the missile would be restrained by a shear-pin system that would permit the engine to reach full power prior to JATO bottle ignition, simulating as nearly as possible the operational launch conditions. The boosters, coupled with the missile engine at full power, would accelerate the missile to 40 mph before an engine shut-off lanyard triggered engine shutdown. Pearson's role was to be stationed at the end of the 2,500 foot runway, using radio command control to keep the missile headed down the center of the runway. If the lanyard system didn't work, he could also control brake action and parabrake deployment.[28]

On 12 January 1952, as he stood in the slight drizzle covering Pt. Mugu, Pearson wondered if perhaps this was really such a hot idea. At T minus one minute the J-33 engine was brought to full thrust and after the engine reached and maintained constant power, the boosters ignited with a tremendous roar and billowing clouds of white smoke. The shear-pin gave way as planned and Pearson saw FTV-6 barreling down the runway straight at him. From his vantage point it seemed that the lanyard had not triggered en-

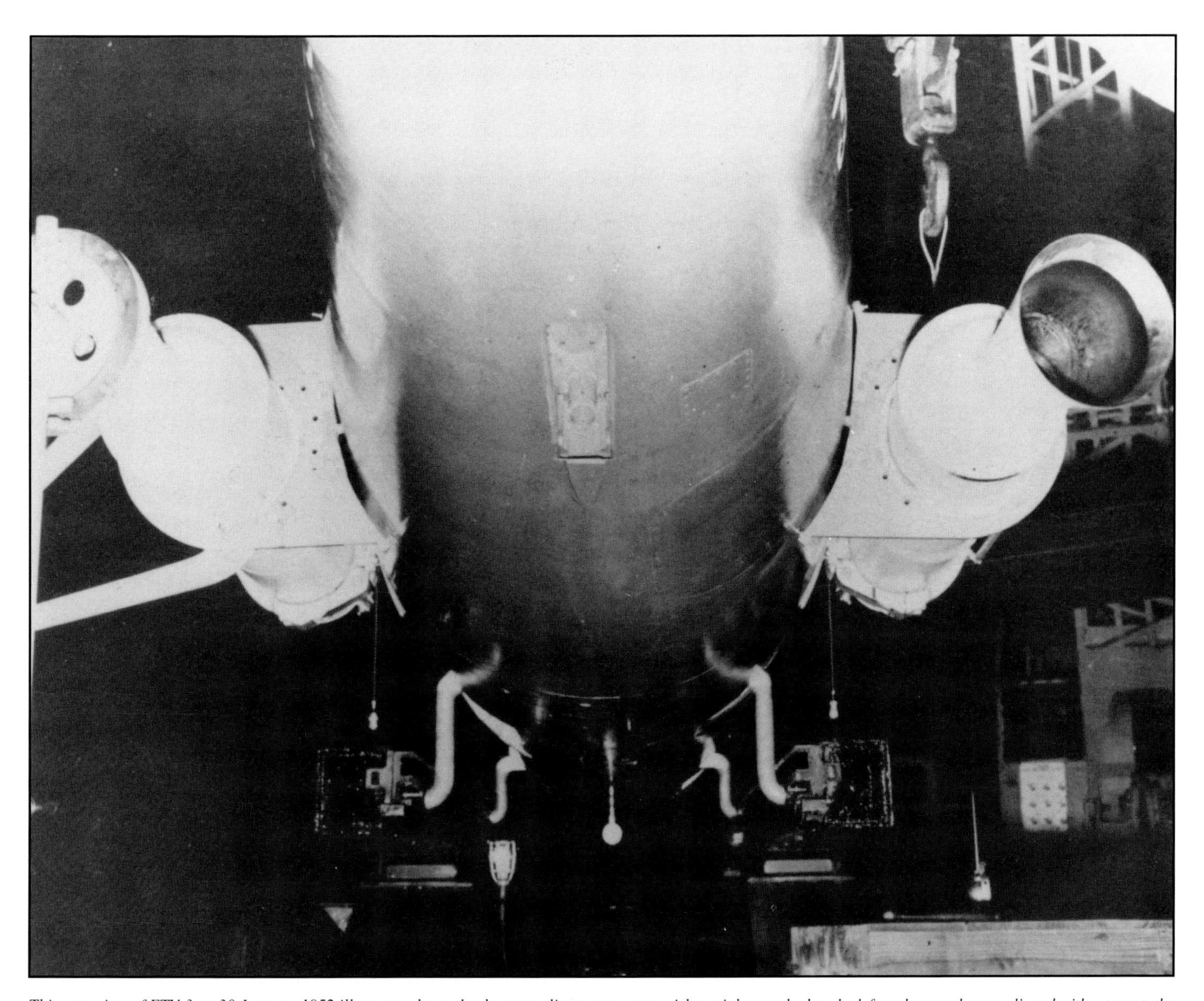

This rear view of FTV-3 on 30 January 1952 illustrates the rocket booster alignment system. A boresight attached to the left rocket nozzle was aligned with a target, the small white circle in the lower center of the picture, placing the booster thrust vector through the center of gravity of the fully fueled missile. The process was then repeated for the right rocket nozzle. (Courtesy of Loral Vought Systems Archives, Author's Collection)

Moments after booster ignition and launch of FTV-3 during Operation Splash. Although FTV-3 was not being recovered on this flight, the lower vertical tail could be blown off using compressed air, and the missile recovered using normal procedures. The trailing wire is an instrumentation umbilical. (Courtesy of Loral Vought Systems Archives, Author's Collection)

gine shutdown, so Pearson signaled the command and was relieved to see the missile immediately begin to loose speed. He signaled brakes after 1,000 feet and watched as the missile ended the roll out with a gradual right turn onto the shoulder of the runway. The missile was quickly towed back to the end of the runway. After a rapid visual inspection, the engine was again run up to full thrust and a complete flight system check quickly conducted. As had been hoped, all systems worked and the critical first booster test passed.[29]

Evaluation of the Sea Test Range for TV-1/TV-2D operations began in early January 1952 and was found to be satisfactory. The first check-out of the newly installed out-of-sight-controller on *Cusk* took place on 3 January 1952. Pearson flew the TV-1 to San Diego and served as the drone for C.O. Miller to control from the submerged *Cusk*. Pearson's signal strength meter showed a signal at 120 nautical miles but it was weak and fluctuating. The submarine's SV-1 radar did not trigger the transponder on the TV-1 so only intermittent skin tracking of the plane was available. Work to remedy these problems took three weeks as the *Cusk* had other operational duties. On 21 January 1952 a second out-of-sight-controller range check was run with the TV-1 as *Cusk* made the transit from San Diego to Port Hueneme. Control was effective out to 125 nautical miles but again the radar transponder did not work effectively with the submarine's radar.

The next day a full dress rehearsal of the flight pattern was run using the TV-1 as the missile and two TV-2D's as control and standby control respectively. Launch was simulated by having the TV-1 and primary TV-2D control aircraft fly a normal pickup pattern over the launcher at Pt. Mugu. After the initial climbout to 3,000 feet, control was transferred to Miller on the *Cusk*. Control for the first two legs was excellent as the drone responded to Miller's airspeed and altitude commands. Once the drone reached 25,000 feet the constant altitude controller was engaged and performed perfectly. At the beginning of the turn onto the return leg of the pattern, carrier signal from the out-of-sight-controller dropped to 20% of normal and the drone returned automatically to level flight. Carrier signal then came back to full strength and the drone finished the left turn. Confusion between Miller and Harry Brackett, the pilot of the drone, resulted in a transfer of control to Pearson in the TV-2D. Carrier from *Cusk* was lost and so Pearson transferred control to the out-of-sight-control operator on San Nicolas Island who guided the drone to the simulated dump point. *Cusk* regained carrier signal and Miller took over, signaling a simulated terminal dive to impact at Begg Rock. While the transfers of control indicated that the methodology was well rehearsed, the fluctuating carrier signal problem from *Cusk* had to be analyzed and quickly corrected if the upcoming launch was to take place as scheduled.[30] On 28 January 1952 full control by the *Cusk* was achieved throughout the entire flight. The first run-through using the missile on the launcher and then a TV-1 flight around the flight pattern and under the control of *Cusk* took place 29 January 1952. *Cusk* again maintained full control for the test and it appeared that all the bugs in her out-of-sight-controller installation had finally been eliminated.

With 15 miles visibility and light variable westerly winds, 31 January was a perfect day for Operation SPLASH, at least at Pt. Mugu. Winds at 35,000 feet were 120 knots from the northwest. The weather at San Nicolas Island, however, was low, scattered to broken clouds at 1,000 feet and Begg Rock, the target point, was completely obscured. The *Cusk* was in position five miles at sea from the launch pad, having participated in the final systems check at 0850. This had included a full dress rehearsal using the TV-1 and TV-2D. With all systems ready, the missile's engine was started at noon. Fifteen minutes later, the two control aircraft took off with Pearson and Lieutenant Billy May as ABLE and BAKER in the lead TV-2D; Howard Mabey in the backup TV-2D and Harry Brackett in the TV-1. Two USAF F-86 high-speed safety-chase and gunnery aircraft took off at the T minus 10 minute mark. They positioned themselves to rendezvous with FTV-3 as it accelerated away from the TV-1 and TV-2D's.

As May brought the aircraft around on the final approach to the launch pad, Pearson called out the T minus 2 minute, 1 minute and 20 second warnings. At T minus 20 seconds the automatic launching timer started and 20 seconds later the JATO bottles ignited. The perfect alignment of the boosters was a welcome sight as FTV-3 climbed quickly away from the pad without a discernable roll. Slipper and booster ejection was flawless but Pearson and the ground crew suddenly realized that the missile was in a nose high attitude that was increasing quickly. Pearson commanded 3-4 degrees decreased pitch to prevent any chance of a stall and the missile responded properly. Full engine power was maintained and FTV-3 accelerated rapidly away from Pearson as expected.

Lieutenant Commander Charles B. Momsen, Jr., captain of the *Cusk*, observed the launch from the periscope and gave Miller a continuous description. One minute after booster ignition Pearson transferred control to Miller. Immediately after this transfer FTV-3's airspeed began to increase beyond the planned value for the automatic climb system. At an airspeed of 425 knots, Miller turned off the autoclimb controller and increased pitch to re-establish the correct rate of climb and correct airspeed. By the time the missile reached its cruise altitude of 34,000 feet, the airspeed had decreased to 205 knots. Miller signalled level flight and the airspeed quickly built back up to 480 knots. He engaged the altitude controller and it maintained altitude for the cruise portion of the flight. When FTV-3 reached Mach 0.85, this speed was held while one of the F-86 chase aircraft dove from a position above the missile to verify the Mach reading and visually inspect the missile. Miller then advanced the throttle to 100% and at Mach 0.9 the missile out ran the F-86.

Pearson and May still had the missile in sight and were inside the flight pattern 12,000 feet below. Shortly after the missile reached Mach 0.9, a steady stream of smoke replaced the intermittent puffs that were used as a tracking aid. This was not an engine fire but rather a visual indicator that carrier signal had been lost. Only a few seconds were left in which Pearson could signal the emergency to Miller and coordinate the transfer of control before the missile destruct system was automatically activated. Miller followed the procedure practiced countless times before in the drone flights. He radioed to Pearson, "Stand by, take control." When Miller said the word "take," he switched off his control and on Miller's saying "control," Pearson switched on his transmitter. If signals from both operators reached the missile simultaneously, it would go into the destruct mode. The missile sped away from Pearson since his throttle setting was at 100% as he had left it during the climb out from launch. He quickly reduced throttle to 91% and watched as the missile disappeared from sight, headed out to the seaward end of the test range.

The *Cusk*'s radar plot was inoperative but Morgan Wilkes, a Chance Vought engineer at the San Nicolas Island radar station, had a solid return from the missile beacon. He relayed the missile's location to May who then flew inside the programmed flight pattern, keeping the missile within radio control range and allowing Pearson to initiate the pattern turns as called for by Wilkes' radar track. On the final turn towards the target area Wilkes lost contact. The F-86 chase aircraft still had sight of the missile and told Pearson that the missile was responding to the control signals properly. Wilkes re-established radar contact and through Pearson vectored the missile to the target. Three miles from the dump point Pearson engaged the dive controller which initiated the pre-programmed dive to impact. An F-86 aircraft orbiting near the target area observed the beginning of the terminal dive and reported that the pushover had been smooth with the missile's wings level. By 15,000 feet the missile was in a vertical dive and at 3,000 feet disappeared into the overcast, impacting 25 minutes and 33 seconds after launch. Wilkes had been able to maintain radar transponder contact and thus had a radar fix on the impact point. A dye marker had been placed in the nose of the missile as an indicator of impact position. An observation boat located the dye stain approximately one mile from target center.

Operation SPLASH was a complete success in its primary objectives; a boosted launch and the completion of a simulated tactical mission. FTV-3 had withstood "g" loads of -2g to +3g during oscillations in the vertical dive and remained intact until impact. The basic premise of a long-range tactical mission had been clearly demonstrated. The drop out of carrier signal from *Cusk* was considered a relatively minor technical glitch.[31] Accuracy was acceptable since the terminal dive controller was still at the developmental stage.

One week later the problems with the out-of-sight-controller equipment on the *Cusk* had been isolated and identified. Miller had tested the equipment prior to leaving the boat at the end of the operation and signal strength was excellent. Confirming the signal drop out, Brackett, the TV-1 pilot, reported that he too had experienced signal loss in the TV-1 at the same time as the missile. Miller noted that the radio antenna was mounted on the top of the search radar antenna and recommended that the coaxial connector from the transmitter to the rotating SV shaft be examined carefully for possible "dead" spots. This was found to be the case and was easily remedied.

1.3 seconds after ignition. (Courtesy of Loral Vought Systems Archives, Author's Collection)

1952

After a delay of two months due to rainy weather at Edwards, flight operations returned to normal in April. Chance Vought was now conducting two flight test programs, one at Edwards, the other at Pt. Mugu. Preparations began immediately for the next boosted launch using FTM-1006. In addition to continuing evaluation of boosted launch procedures and specifically trying to correct the errant pitch-up motion at launch that was seen with FTV-3, the launch site now had a simulated submarine hangar located behind the launcher to test for heat and blast effects from the JATO bottle exhaust as well as prolonged exposure to the missile's J33 engine exhaust. FTM-1006 was configured for recovery and would test the feasibility of using compressed air to blow off the lower fin during the landing approach. The last but equally important feature of the flight plan would be recovery of the missile at San Nicolas Island. San Nicolas Island had a long runway but it was perched on top of a 500 foot cliff with violently changing updrafts that could cause havoc with manned aircraft landings, not to mention unmanned guided missiles.

Flight test vehicle 1006 was successfully launched on 3 April 1952. The second boosted launch was as successful as the first but the same pitch-up transient was seen. The rest of the flight was uneventful as the first Lot II production missile passed its performance, stability and control tests. The lower fin was ejected clear of the missile as the landing gear was lowered on approach to San Nicolas Island. Pearson was kept busy during the landing as the unpredictable wind shifts over the cliffs made for some delicate maneuvering. FTV-1006 landed successfully and the drag chute deployed correctly, a fitting end to this acceptance flight.[32]

In April 1952, the Office of the Chief of Naval Operations directed Pt. Mugu to study the feasibility of the Regulus Assault Missile (RAM) concept. Earlier drone aircraft programs in World War II and those currently in development for deployment to the Korean conflict spurred efforts to evaluate the feasibility of the RAM program. Escorting fighters using radio command control would guide the missile to the target. The guidance aircraft would then break away to a safe distance and detonate the warhead. Jet fighters would be deployed from carriers or shore installations to guide missiles launched from carriers, cruisers or submarines. Such a system would permit rapid and early deployment to the fleet while the more sophisticated ship-based automatic guidance equipment continued development. The goal was

First shipboard launch of Regulus I. FTM-1015 lifts off from USS Norton Sound *(AVM-1). Flight was a complete success though upon landing the parabrake collapsed and FTM-1015 hit the chain arresting gear at 100 knots, flipping over on its back. (Courtesy of Kilpatrick Collection)*

FTM-1008 on the CVA desert launcher in position at starboard catapult aboard USS Princeton *(CV 37), on 16 December 1952. Note the metal sheathing positioned on the flight deck to deflect engine and booster exhaust. (Courtesy of Loral Vought Systems Archives, Pearson Collection)*

Just after ignition. FTM-1008 is beginning to move off the launcher and the deflectors have worked well, directing the exhaust off the flight deck. This was the first launch of a Regulus missile from a warship. (Courtesy of Loral Vought Systems Archives, Pearson Collection)

FTM-1008 just before booster exhaustion and ejection. (Courtesy of Loral Vought Systems Archives, Pearson Collection)

fleet operations capability by early 1954. Each semi-independent RAM Detachment (RAM Det) would be composed of four pilots, a missile checkout and launch team, four McDonnell F2H-2P "Banshee" control aircraft, a portable launcher, missile checkout and firing equipment and a supply of Regulus I missiles.[33]

Chance Vought realized that the RAM program provided an excellent avenue of continuity for the Regulus program. The submarine Regulus program was still limited to the single fleet-type boat then under conversion. Even with immediate authorization for additional submarines, the new construction or conversion lag time would mean at least two years before the new boat(s) were operational. Chance Vought quickly moved to accomodate these new flight operations while simultaneously continuing the original test plan objectives. On 12 May 1952, Chance Vought flight test pilots Roy Pearson, Harry Bracket, Bob Fissette and C.O. Miller flew the first simulated RAM missions at Edwards as part of the Regulus I flight test program. The single seat TV-1 manned drone served as a surrogate missile. The TV-2D airborne control aircraft served as the controller with a second TV-2D acting as a spotter. The flight plan called for RAM "attacks" on a mesa well away from the Edwards flight operations area. A typical RAM flight pattern had both the drone and the control aircraft approach the target at 2700 feet. The control aircraft ABLE pilot would align the drone in azimuth, allowing for drift, and when three to four miles away, the control aircraft BAKER pilot would execute a rapid turn 45 degrees to the right, leveling out one mile away. The ABLE pilot then commanded the pop-up maneuver to the drone which would pull up to 5,000 feet, the optimum height for an air burst nuclear detonation for the proposed missile warhead. The detonation signal would be sent as the drone was observed by the TV-2D, now five miles away, to pass over the target. Accuracy during these first RAM profile attempts was at first erratic but was soon narrowed to a consistent 500 feet. After twelve flights over a two day period Pearson was convinced that the concept was a feasible extension of the routine missile control flights. Film from a vertically mounted camera on the TV-1 confirmed Pearson's observations.[34]

As the assault missile program progressed, BuAer contracted with Chance Vought to prepare RAM attack profile performance comparisons with the three likely airborne control aircraft, the McDonnell Aircraft Corporation's F2H-2 "Banshee"; the F9F-6 "Cougar" of Grumman Aircraft and Chance Vought's F7U-3 "Cutlass". Although all three aircraft had both good and bad points, the F9F-6 was considered the best choice for modification into the first RAM control aircraft. While the F9F-6 was slower then Regulus I and had considerably less range, these drawbacks were overcome by the ready availability and reliability of the Cougar within the fleet.

While the details of the RAM flight operations were being investigated at Edwards, the Navy assault drone operation of Captain Robert Jones was in the midst of shipping out to Korea. On 2 July 1952, on the way out to Korea from the East Coast, Captain Jones and Lieutenant Commander Larry Kurtz visited the Chance Vought factory in Dallas, Texas, to demonstrate the rudimentary television guidance system they had developed for use against railroad tunnels on the coast of North Korea.[35] Chance Vought engineers were quick to realize the potential of television guidance for RAM operations.

In mid-July 1952, two McDonnell F2H-2P "Banshee" aircraft were ferried to Edwards for conversion into interim control aircraft for the RAM program. Practice flights commenced immediately to flight qualify the radio command control installation. The evaluation program for the Navy was carried out by Lieutenant Billy May and Lieutenant George Monthan. One of the original Navy "Blue Angel" flight team members, Lieutenant May was a veteran in the Regulus program. Lieutenant Monthan had just recently joined the program at Pt. Mugu.

On 8 August 1952, a full-scale RAM operation using FTV-5 was conducted by Chance Vought with Navy backup. Roy Pearson, flying the ABLE position, was the primary controller throughout the mission. Lieutenant May flew secondary chase in the newly converted F2H-2P. Four specific objectives were listed for this flight. First, could the missile be controlled at low-altitude and high-speed? Second, could the missile maintain a constant heading to target if lined up three to four miles away? Third, could the control pilot accurately command the missile to pull-up and push-over to the desired altitude after his break-away maneuver? Fourth, could the missile's position be accurately marked by the control aircraft three to four miles abeam of the target?

Following launch, FTV-5 was throttled back to 85% to permit the slower chase aircraft to keep in formation with the missile. The first run to the target was made at low-altitude with a pull-up and simulated detonation at 1,500 feet. A second run was made with pull up to 3,000 feet. Pearson misjudged the airspeed and almost stalled the missile. Luckily for the program and the military brass viewing the operation, he pitched the missile over in time to keep from stalling. In the post flight report, Pearson strongly endorsed the RAM mission profile that Chance Vought had developed and added that he felt the missile could be easily controlled even as low as 100 feet. The missile accurately held its commanded heading during the final run-in. Pearson recalls that the operation was much easier than ones carried out with the manned drone aircraft.[36] With the success of this demonstration, training of the Navy RAM teams began in earnest.

In and amongst the assault missile program experimentation, the flight test program continued. On 17 June 1952, the first north lakebed operation took place using FTV-5. The south lakebed areas were underwater from recent rains and the north lakebed test area was much nearer the Chance Vought hangar. This was the sixth flight for FTV-5, proving that the recoverable flight test vehicle concept was working much better than the originally estimated three flight average. This twentieth flight in the program was the first high altitude, high Mach Number test specifically designed to evaluate the effect of the absence of the lower vertical fin on stability and maneuverability. It was also the first to take place on the Aberdeen Bombing Range. The mission was a success as far as the flight of FTV-5 was concerned. One high Mach Number dive from 35,000 feet was accomplished, resulting in a speed of Mach 1.08, with a successful pull-out. Flight stability was unaffected by the absence of the lower vertical fin. Out-of-sight-controller control had been used throughout the flight and worked perfectly. The tests confirmed what was already suspected; the lower fin was not necessary for stability during the terminal dive maneuver and it was eliminated on all remaining missiles, greatly simplifying the design of the short rail launcher and permitting continued taxi-take offs from Edwards for the time being.

On 28 October 1952, the first warhead prototype flight operation was successfully conducted using FTM-1013B, the "B" designating a FTM modified with a prototype warhead nose section.[37] While tactical missiles were already in production, the first missiles were not scheduled to be delivered until the middle of 1953. Even though one launch slipper did not eject, the climb to 34,000 feet and terminal dive at Mach Number 1.08 was successful. With Lot II qualification flights completed, modifications were made to the nose section of additional Lot II missiles to accommodate inert warhead components for further testing.

The first shipboard launch of Regulus I took place on 3 November 1952 from the USS *Norton Sound* (AVM-1) with the successful flight of FTM-1015, also the first missile with folding wings. A mobile launcher designed by Chance Vought and built by the Freuhauf Trailer Company was utilized with great success.[38] The flight was a successful RAM mission and demonstrated the robustness of the missile when, on landing at San Nicolas Island, the parabrake blossomed and then collapsed. Despite the use of differential brakes, FTM-1015 hit the chain arresting gear at the end of the runway while still moving at 100 knots. It flipped over on its back but was recovered and refurbished for later use in warhead evaluation tests.

Several more "firsts" for the program were completed before the end of the year. On 18 November 1952, after five months of exhaustive and often frustrating testing, FTM-1009 was launched for the first missile guidance test using the bipolar navigation system (BPN). Launch was uneventful but when the guidance system was engaged by the control aircraft, FTM-1009 turned in the wrong direction. BPN guidance was disengaged and the missile recovered at San Nicolas Island.[39] On 16 December 1952 the first launch of Regulus I from a warship took place. Since RAM operations were to use aircraft carriers as the Regulus launch platform, FTM-1008 was taken on board USS *Princeton* (CV 37) along with the desert launcher.[40] Lieutenant Commander William "Pappy" Sims, the Officer-in-Charge of Guided Missile Training Unit FIVE (see Appendix II), the first submarine Regulus I support unit, remembers that Chance Vougeht had to bring aboard its own power supply in order to checkout and launch the missile. The large launcher and all of the support equipment and cabling made this a less then ideal demonstration in many ways but, in the end the missile won at least lukewarm support as it roared off the launcher and had a successful flight. While the launcher was cumbersome and took up too much space, more importantly, carrier aviation now had another atomic bombardment weapon, one with a longer range then the available escort aircraft.[41]

Endnotes

[1] *FTV's were painted either white with red lettering or high visibility red with white lettering. FTM's were painted high visibility red with white lettering. TM's were painted either Navy blue with white lettering or in several cases battleship gray for ventral surfaces and Navy blue for the dorsal surfaces.*
[2] *Regulus I Master Plan, 15 February 1955, pages 17-25. LVSCA, A50-24, Box 7.*
[3] *RPS, Flight Test, page 11.4.*
[4] *Regulus Flight Test Program Weekly Activity Memorandum (RFTPWAM), 24 February 1950, page 1, LVSCA A50-18, Box 1.*
[5] *Personal interview with Bill Micchelli.*
[6] *Ibid.*
[7] *Personal communications and interviews with Leroy Pearson, November 1989, February 1990.*
[8] *RFTPWAM, 27 April 1950, page 1. Also personal interview with Leroy Pearson.*
[9] *"Navy Decision with Respect to the Combination of the Regulus and Matador Projects", 4 April 1950. National Archives. BuAer GM-93/7, Project Regulus, Record Group 156, Box 5.*
[10] *RPS, Contracts, page 3.3.*
[11] *These missiles were designated FTM-1006 to -1015. Chance Vought changed to this missile numbering system with the Lot II missiles. FTM stood for Flight Test Missile or Fleet Training Missile.*
[12] *Regulus I missile production statistics are compiled in Appendix VII.*
[13] *RFTPWAM, 5 September 1950, page 4.*
[14] *Personal interview with Leroy Pearson.*
[15] *RFTPWAM, 8 December 1950. Flight Reports 159, 160, 168 and 169 attached.*
[16] *Personal communications and interview with RADM Dewitt Freeman, USN (Ret.).*
[17] *Personal interviews with Palmer Ransdell, Leroy Pearson and C.O. Miller.*
[18] *Personal interview with Bill Micchelli.*
[19] *RFTPWAM 24 March 1951, Flight Reports 208 and 228. Personal interview with Leroy Pearson.*
[20] *Personal interview with Fred Randall, a Chance Vought aeronautical engineer.*
[21] *Personal interview with Palmer Ransdell.*
[22] *Personal interview with Leroy Pearson.*
[23] *For a description of missile production, and flight test results from 1950-1961, see Appendix VII.*
[24] *Flight Test Report XSSM-N-8 (Regulus) Operation Splash, 19 June 1952. Pacific Missile Test Center Archives, NAMTC-MT-31-52.*
[25] *"Zero" length referred to the distance the missile traveled along the launch rails prior to becoming airborne. This launcher was a prototype and truly zero length, the missile lifted off, at an angle, directly from the launcher receptacles, no rails were used.*
[26] *Test of Regulus Zero-Length Launcher, PMTCA, NAMTC-TMR-57, 15 December 1951, pages 1-7. This report covers the 24 April and 16 May 1950 launcher tests.*
[27] *The 2.2KS indicates a burn time of 2.2 seconds; 33,000 is the individual booster rocket thrust in pounds.*
[28] *Flight Test Report XSSM-N-8 (Regulus) Operation Splash, 19 June 1952. Pacific Missile Test Center Archives, NAMTC-MT-31-52.*
[29] *Personal interview with Leroy Pearson.*
[30] *RFTPWAM, 11 February 1952, LVSCA A50-18, Box 1.*
[31] *Flight Test Report XSSM-N-8 (Regulus) Operation Splash, 19 June 1952. Pacific Missile Test Center Archives, NAMTC-MT-31-52.*
[32] *The solution to the pitch-up attitude transient problem that both FTV-3 and FTV-6 had shown at the end of the booster thrust phase of launch was to make the pitch gyroscope erection system in the autopilot inoperative until the end of the automatic climb control portion of the launch phase. Personal interview with George Sutherland, guidance system engineer.*
[33] *Missile Test Department Memorandum Report No. 7-53. Naval Air Missile Test Center Regulus Assault Missile Evaluation Program, 23 March 1953, p. 1, Naval Missile Test Center.*
[34] *RFTPWAM, 12 May 1952, LVSCA Archives, A50-18, Box 1.*
[35] *Personal communication with Captain Larry Kurtz, USN (Ret.).*
[36] *Personal interview with Leroy Pearson.*
[37] *This modification was internal and involved reshaping the engine intake duct up and over the warhead compartment. Drawing is on page 51.*
[38] *This launcher was known as the "desert launcher."*
[39] *Regulus Flight Test Index. Chance Vought Report 10863. LVSCA A50-24.*
[40] *Ibid.*
[41] *Personal interview with Captain W. "Pappy" Sims, USN (Ret.).*

Regulus field operations staff 25 June 1952 posing on the edge of the dry lake at Edwards Air Force Base, California. (Courtesy of Micchelli Collection)

Chapter Four: Regulus I Flight Testing 1953-56

The remaining Regulus I flight test history is organized as Chance Vought Phase A Operations, Navy Phase B Training and Fleet Operations, and Operational Development Force (OPDEVFOR) Phase C Operations sections. This provides a reference framework for missile operations leading up to the Regulus submarine deterrent strike patrols that began in 1959.

1953

Chance Vought Phase A Operations

Lot III missile qualification flights were successfully completed and the final changes in the radio command control equipment accepted. Bell Aircraft Corporation had undertaken the continuing improvement of the instrumentation reliability, culminating in the Lot III equipment with higher power transmitters and receivers that were contained in pressurized equipment bays. Several additional automatic features were added to the Regulus I flight control system in 1953. These included a dead reckoning system for heading control immediately after launch and automatic throttle control. The dead reckoning system brought the missile to a pre-selected heading using a magnetic compass. This heading remained in effect until the primary guidance signals were received. Evaluation of the bipolar navigation (BPN) guidance system continued, primarily with TV-1 and TV-2D operations. Only two missile flights took place using BPN with sporadic system operation on both flights.

On 29 May 1953 the first Regulus I Trounce guidance flight was combined with the seventh warhead evaluation launch. Trounce was the coded pulse radar guidance system that had been developed at Pt. Mugu during Project Derby (see Appendix I). Due to cloud cover, the photo-theodolites could not track the missile for determination of terminal dive accuracy, but the Trounce guidance system worked well.

On 15 July 1953 the first Regulus I launch from a submarine took place as FTM-1018 was successfully launched from USS *Tunny* (SSG 282) (see Chapter Ten). The launch was conducted by an all-Navy team assisted by Jack Welch, the Chance Vought field engineer assigned to this program milestone. The low-altitude flight was combined with a training flight for pilots of Utility Squadron THREE, Regulus Assault Missile Detachment, culminating in a pass over the San Nicolas Island airfield at 1100 feet and 420 knots (See Chapter Eight).

On 3 September 1953, Bob Fissette, a Chance Vought test pilot, controlled one of the shortest Regulus I flights. The Missile, FTV-4, which had a second source autopilot made by Honeywell, had been flown three times with various degrees of success.[1] After a normal launch with normal booster and launch slipper ejection, the missile began flying with a sinusoidal pitching motion. Fissette was flying the ABLE position in the TV-2D and immediately brought the missile around for recovery. He eased FTV-4 as close to the ground as he dared and as it made its next downward motion, he gave it a downward pitch command, effectively slamming it onto the ground for a successful recovery. Subsequent investigation revealed that a linkage in the autopilot pitch gyroscope had been sprung. This allowed slack in the control which had created a "dead" spot around the neutral position, causing the over-correction in upward and downward pitch.[2]

On 24 September 1953, FTM-1023B was launched from Edwards carrying a prototype W-8 warhead with dummy nuclear components installed. The W-8 was an earth penetration warhead version of the Mk 5 nuclear bomb for use against hardened targets such as command centers or submarine pens. The flight plan called for a high altitude cruise to the Naval Ordinance Test Station range at Inyokern, California, followed by a terminal dive to impact at the target.[3]

George Sutherland was the Chance Vought flight test director for this operation and monitored the flight from the out-of-sight-control room at Edwards. The launch was routine except that all four launch slippers failed to eject. This was aggravating to Sutherland for two reasons. The first was that a new slipper ejection system had not been installed due to lack of time. Since the missile was an FTM in the "B" tactical configuration it could not be recovered for repair of the problem and then re-launched. The second reason was that with all four slippers hung up, the increased drag prevented reaching the desired test altitude of 35,000 feet and also decrease the range significantly. With no option but to proceed with the test, Sutherland quickly began estimating the probable reduction in range and made an educated guess where to start the terminal dive sequence.[4]

Although only reaching 26,000 feet altitude, the missile became supersonic in its terminal dive, a critical flight objective. Instead of hitting the designated target area in the middle of Wilson Mesa, it hit near the edge, on hard granite. Bill Albrecht, a Chance Vought Regulus I field engineer, recalls driving up to the impact site in the Inyokern Test Range later the same day and finding the entry point for the warhead: a circle barely two feet in diameter. The inert warhead had actually penetrated the granite, leaving fine granite dust around the entry hole. For an area 300 feet around this point, the ground was strewn with pieces of the missile and desert plants had pieces hanging from their limbs like tinsel. The largest piece was the turbine disc of the J-33 engine. Most of the fragments were 4-5 inches in size down to one-half an inch, although some of the access panels on the missile had been blown free and were intact.[5]

On 16 October 1953 FTV-4 was launched from Edwards in the first of three flights that evaluated television guidance for Regulus I. The television components had been scrounged from the NAMTC warehouses as leftovers from World War II experiments. Television technology was to be used with the penetration warhead program to give the missile the necessary accuracy for use on hardened targets. When the television guidance system failed to meet expectations, the penetration warhead project was cancelled, and so were the remaining television guidance flights.[6]

Navy Phase B and Fleet Training Operations

Phase B Navy operations began in 1953 but the training program began in early April 1951 when Guided Missile Training Unit FIVE (GMTU-5) had begun working with Chance Vought personnel at Edwards (See Appendix II). Stationed at Pt. Mugu, GMTU-5 personnel were assigned specific areas of expertise and worked closely with Chance Vought engineers. The first Phase B flight operation conducted by an all-Navy team took place on 6 February 1953. The launch was successful but the launch slippers and both boosters failed to eject. The additional drag of the launch hardware made the missile stall and crash on landing approach. One month later GMTU-5 again prepared a missile for launch with Chance Vought engineers reviewing the process and providing assistance as necessary. The launch and hardware separation was successful and as the missile cruised at 11,000 feet and 370 knots, Sam Perry, the Chance Vought Regulus I Field Program Director, turned to Lieutenant Commander Sims and stated, "There, she's all yours".[7] GMTU-5 personnel completed the flight and recovery without incident. One week later, GMTU-5 successful launched and recovered FTM-1012 and Phase B operations had "officially" begun.

On 15 April 1953, the only boosted launch accident in the Regulus program on the dry lakebed took place. The right-hand booster fired normally but the left-hand booster ignited six-tenths of a second later. The thrust of the right-hand rocket was sufficient to move the missile along the launcher until the booster ejection system was activated, ejecting the left-hand rocket just as it ignited. The loose rocket flew into the left wing, completely severing it from the fuselage, then cartwheeled across the desert. Palmer Ransdell, a Chance Vought field engineer, was observing the launch from atop the roof of the Chance Vought hangar. He was amazed at the sight of this overgrown firecracker tumbling wildly across the desert floor. He realized that the program had been fortunate that this was the first such drastic failure in 46 launches.[8] This failure instigated a complete review of both the booster and launch slipper ejection system. Dr. Willie Fiedler designed a new ejection system based on the capture and storage of the high pressure exhaust gases from the booster. Six months later the first flight using the new integral ejection system took place. Both boosters and the front slippers that incorporated the newly designed system ejected cleanly.[9]

With the deployment of Submarine Weapons Assembly Teams FIVE and SIX to GMTU 5 in June 1953, and the stockpiling of Mk-5 bomb-to-warhead conversion kits, the Navy now possessed the capability to assemble and install W-5 atomic warheads on Regulus I missiles (See Appendix IV). While this could be called the beginning of the operational status of the program, this was more true in theory than in fact.

1954

Chance Vought Phase A Operations

By 1954 Project Regulus was moving

steadily forward. The two major guidance systems, BPN and Trounce, were well into their test phases. In late 1953 the Chief of Naval Operations had made the decision to have Project Regulus undergo OPDEVFOR evaluation. This was a thorough evaluation of the missile and all support functions pertaining to fleet use. This program was designated Project Op/S317/X11, "Evaluate the REGULUS Guided Missile for Service Use," and assigned to the Commander, OPDEVFOR, for execution with a priority "A". The project evaluation was assigned to Captain Francis D. Boyle, with operations to begin in August 1954. All OPDEVFOR missile operations were to be conducted by Navy personnel, from acceptance of the missile to launch, recovery and refurbishing.[10] There were seven operational concepts that BuAer and the Office of the Chief of Naval Operations had delineated for evaluation: 1) the submarine launched assault missile (SLAM) with Trounce guidance; 2) Regulus Assault Missile (RAM) with radio command control, and 3) SLAM with two-boat Trounce, all to begin evaluation in August 1954. To follow were: 4) RAM with cruiser Trounce in December 1954; 5) SLAM with bipolar navigation in June 1955; 6) the KDU-1 air-to-air missile target drone in April 1955; and 7) SLAM with a new terminal guidance system called "Beagle," to begin in January 1955.[11,12]

In late 1953, Sam Perry, Chance Vought's Assistant Chief Engineer, Missiles, and Bill Micchelli, Supervising Engineer for the Regulus I Missile Inspection and Test Unit at Chance Vought's plant in Dallas discussed the upcoming need for ongoing technical assistance to Regulus fleet operating units. Both Perry and Micchelli had been closely involved in the early field operations with Regulus I at Edwards and Pt. Mugu. Both knew the complexities of the missile system, from maintenance to flight planning. Perry felt strongly that the complete missile system was too complex to simply turn over to the Navy and expect the total system to work without any problems. Micchelli's in-depth experience with the factory inspection and test operations supported Perry's convictions. They agreed that the method of technical support used in Chance Vought's aircraft programs, that of a single specialized technician for engines or hydraulics or sheet metal, etc., would be inadequate during the fleet evaluation process as well as in continuing support of fleet units.

Perry and Micchelli listed all of the engineers in the Regulus program in a matrix that ranked them for knowledge, "people skills," ability to work with difficult situations, prior education, etc. They took the top ten and formed the first Missile Operations Engineer "class," bringing the selected engineers to Dallas for several weeks of training. The field program managers complained long and loud as they lost many of their best engineers. All of the candidates were already proficient in at least one aspect of the Regulus program. This program cross-trained them so that they were well versed in each of the major missile subsystems. The classes were taught as much by the pupils as the instructors since field experience was a crucial aspect in solving missile operation problems. In return, the graduates of the course gained detailed knowledge of missile production and assembly. Thus, they knew where to call and who to talk to in order to get an answer to practically any question.[13]

Upon graduation, missile operation engineers were assigned to each Regulus support unit. Funding for field technical support personnel was usually considered part of the BuAer contract overhead, but Chance Vought was able to convince Captain Robert Freitag in BuAer that the unique services rendered by these engineers should be funded as a separate line item in the contract.[14] Conferences were held twice a year in Dallas with the missile operation engineer staff from each unit and their Navy counterparts meeting for several days of program updates and exchange of information.

At the beginning of the OPDEVFOR program, Captain Boyle requested that an missile operations engineer be assigned to Guided Missile Evaluation Unit ONE so that questions concerning missile capabilities, test design and operation critiques could be answered quickly. Ray Hill was assigned the duty with some speculation as to how a company representative could be objective in such a situation. Since Hill stayed with the OPDEVOR team throughout the Regulus I evaluation, he must have satisfied all the critics.

Developmental flights of Regulus continued as preparations for OPDEVFOR evaluation began. The Phase A test program worked in conjunction with the accelerated training of the Navy support units. Chance Vought had to demonstrate through the Phase A contractor evaluation program that each of the missile systems, including all support and test equipment, met all contractual specifications. Through the Navy Phase B operations, OPDEVFOR evaluated the serviceability, reliability and accuracy of each tactical concepts.

Captain Boyle planned to begin the evaluation process in August 1954 with a schedule calling for 146 missile launches to be conducted through June 1956. The Air Force had informed BuAer that Edwards did not have room to accommodate Project Regulus after June 1954. A similar situation existed at Pt. Mugu since this many launch operations simply could not be interwoven into the already overburdened scheduling. The solution was to shift the Edwards Regulus operations to the little used Marine Corps Auxiliary Air Station (MCAAS), at Mojave, California, with flight operations to commence in August 1954. MCAAS was isolated but close to Naval Ordnance Test Station China Lake and the Inyokern test range for further warhead tests. Contractor development and lot qualification flights could take place at MCAAS as well as virtually all of the Navy Phase B Regulus assault mission training.

Understanding the importance of the success of the OPDEVFOR evaluation, Chance Vought established its own internal organization to simulate the OPDEVFOR process. Where were the weaknesses in support activities? What skill levels and facility requirements were necessary? The simulation evolved into a paper, and in some cases operational equivalent of OPDEVFOR. The Chance Vought evaluation group reviewed results from a simulated unit that duplicated the Navy's Receive, Assemble, Maintain, Inspect and Storage unit that received missiles from Chance Vought, and prepared them for use within the fleet; a Regulus assault mission detachment that would prepare and launch the missile as well as provide chase aircraft; and an missile operation engineering unit to provide technical support.[15]

A significant development prior to the beginning of the OPDEVFOR evaluation program was the Phase A flight on 4 May 1954. This was the first successful flight of a production tactical missile from Pt. Mugu and was controlled by Trounce guidance from both the Flight Test Center on land and the guidance submarine USS *Carbonero* (SS 337). In theory, at least, the Navy now had an emergency war capability with Regulus I.[16]

The successful launch and recovery of FTM-1031 on 11 June 1954 marked the 113th flight in the Regulus I program and the last from Edwards. After two months of renovating facilities and transferring equipment and personnel, flight operations resumed at MCAAS, Mojave, with the first launch taking place 4 August 1954, the tenth flight of FTM-1031. Twelve more launches were conducted from MCAAS during 1954. The majority were Chance Vought performance and modification qualifications of missile automatic flight controls for climbout, airspeed and letdown control.

Phase A flight operations from Pt. Mugu also continued during the OPDEVFOR program with Trounce and BPN guidance system development flights. The first successful long range BPN flight took place on 2 September 1954. Two additional successful BPN operations were conducted by the end of 1954.

Navy Phase B and Fleet Training Operations

Navy Phase B flight operations for 1954 consisted primarily of Regulus assault mission training launches and training sufficient personnel in GMU-50 (GMTU-5 had been disestablished and the personnel absorbed into GMU-50, see Appendix II) to permit establishment of two East Coast Regulus support units. By August 1954, the beginning of the OPDEVFOR program, 11 flight operations had been conducted during 1954 with one missile lost at launch and one on landing. The first Navy flight operation from MCAAS took place 16 November 1954, three months after the cessation of Regulus operations at Edwards. Guided Missile Units FIFTY-ONE (GMU-51) and FIFTY-THREE (GMU-53) were formed on the East Coast. GMU-51 was the submarine support unit stationed at the Naval Mine Depot, Yorktown, Virginia. GMU-53 was the Regulus surface ship support unit stationed at the Naval Air Station Chincoteague, Virginia (See Appendix II).

OPDEVFOR Phase C Operations

OPDEVFOR operations were conducted by missile teams from GMU-50, the West Coast submarine Regulus support unit and GMU-52, the West Coast surface fleet Regulus support unit (See Appendix II). The OPDEVFOR evaluation began with the successful launch of FTM-1033 from the *Norton Sound* on 20 August 1954. This was the second flight of FTM-1033, the first

FTM-1033 being loaded onto the SRMK-2 launcher aboard Norton Sound. *The successful launch and flight of FTM-1033 marked the beginning of the OPDEVFOR testing program for Regulus I. (Courtesy of Loral Vought Systems Archives, Wilkes Collection)*

Regulus I flight operations aboard Norton Sound *required delicate maneuvering to line up the Regulus I transportation dolly with the SR MK 2 launcher rails. (Courtesy of Loral Vought Systems Archives, Wilkes Collection)*

missile delivered to the Phase B program. The missile was recovered at San Nicolas Island after a Regulus assault mission flight over the island by pilots from the newly formed VU-3 RAM detachment (see Chapter Eight). With six years of experience launching Loon missiles and one year with Regulus and a five-for-six success rate, submariners were anxious to begin the OPDEVFOR evaluation program. Flight operations on board *Tunny* had indicated that the missile hardware was reliable. Booster alignment, while a problem in the past, had not caused any recent difficulties. The problem now was the Trounce guidance system. Without adequate manuals and a maintenance support system, the Trounce guidance system was more likely to be in a state of repair than operational, despite the best efforts of submarine electronic technicians.[17]

Tunny was utilized in six OPDEVFOR operations in 1954. The first launch took place on 26 August 1954. The parabrake deployed just after reaching 1700 feet, and the missile spun into the water. One week later a successful flight with Trounce guidance to 133 nautical miles was conducted. The missile was lost prior to recovery due to fuel exhaustion but the flight was rated a success. On 8 October 1954, the third OPDEVFOR launch was a failure when the right booster failed to fire. After one month of analysis and refinement of procedures, *Tunny* resumed flight operations. On 18 November 1954, *Tunny* successfully launched a fleet training missile and guided it using the Trounce system to the recovery area near San Nicolas Island. On 7 December 1954 a low-altitude Trounce guidance flight was successful in all respects. Two days later a successful two-boat Trounce flight was conducted and the guidance system problems seemed to be behind them.[18,19] The heavy cruiser USS *Los Angeles* (CA 135) conducted three OPDEVFOR operations in 1954 in support of the Regulus assault mission operation certification process. The first was on 28 October 1954. After a successful launch, pilots from VU-3 RAM detachment escorted the missile to San Nicolas Island and made a successful Regulus assault mission pass prior to recovering the missile. The second was lost at launch. The third was another successful Regulus assault mission operation.[20] Three OPDEVFOR operations in three days were conducted aboard the USS *Hancock* (CV 19) in the Fall of 1954, again in support of the Regulus assault mission program evaluationa (see Chapter Eight). On 18 October 1954, a GMU-52 missile team successfully launched FTM-1034 and while the parabrake deployed in flight and was lost, the missile was successfully recovered after a Regulus assault mission pass over San Nicolas Island. Two mobile launchers were evaluated for ease of use on board carriers and both were found to be too cumbersome and disruptive of flight deck operations (see Chapter Eight).[21]

1955

Chance Vought Phase A Operations

In late February 1955, Chance Vought began the final series of tests of the Bendix Corporation BPN guidance system (See Appendix I). This guidance system had been in test and development, as government furnished equipment, since July 1952 and had gone through three modifications. BPN (XN-3), the final version, was tested from 24 February 1955 to 29 April 1955 using manned drone aircraft, three fleet training missiles and three tactical missiles. Limitations of previous versions of BPN included insufficient range, complex service requirements and vulnerability to electronic countermeasures. With over 200 manned drone and eight missile flights to date, these were the final attempts to evaluate BPN as a possible guidance system for Regulus tactical operations from submarines.[22]

On 24 February 1955 the first BPN (XN-3) guidance flight was launched from Pt. Mugu. Guidance signals were provided from two mobile vans 100 miles apart on the California coast, presenting ideal signal sources for the tests. Signal acquisition was successfully completed and automatic guidance commenced as planned. The safety-chase aircraft reported intermittent roll oscillations of the missile as well as fuel venting. Guidance was discontinued and the missile successfully recovered. Review of the telemetry data indicated that an area defense radar had induced transient signals in the airborne signal decoder. On 27 April 1955, the first tactical missile equipped with BPN (XN-3) guidance equipment was successfully launched and controlled on the outbound leg of the pattern. Roll oscillations were again observed. With cooperation from Pt. Mugu range controllers, the interfering radar was turned off and the oscillations ceased. The missile was turned under radio command control guidance and inbound

Chains were hooked around the missile launch slippers and used to pull the missile onto the launcher. (Courtesy of Loral Vought Systems Archives, Wilkes Collection)

BPN guidance was successful.[23] The last BPN (XN-3) flight two days later was the second test of the *Sperry* radio controller terminal dive controller. Unlike the controller used in the Trounce guidance system, the *Sperry* system was controller by the BPN master station. A variety of problems developed in the master station at Pt. Vicente but these were quickly overcome by transferring to radio command control while these problems were resolved. The terminal dive was successfully completed.[24]

In addition to the limited missile test flight operations, forty-one manned drone flights were used to evaluate serviceability, effectiveness of electronic countermeasures, guidance range and accuracy. The conclusions from both the missile and drone aircraft flights were mostly negative. Effective range was difficult to determine since both land-based picket stations had been in ideal locations and submarines were not used. Electronic countermeasure susceptibility was still a significant problem and not readily resolved. On the other hand, accuracy to the release point for terminal dive was good. Chance Vought presented the results of the testing to the Navy for a decision on the future of the BPN guidance program.[25]

The only engine design change in the program took place on 1 March 1955 when the Allison J-33-A-18A was chosen to replace the J-33-A-14. With a higher rated thrust, and ability to withstand longer periods of "military power" operations, the J-33-A-18A was the standard engine for the remainder of the program.[26,27]

Four years earlier, in late 1951, the Guided Missile Division of BuAer had expressed interest in the use of Regulus I as a target drone. Chance Vought responded with an informal proposal to BuAer shortly thereafter, setting forth the lead-time necessary to begin production of the drones. The proposed drone program was postponed due to lack of funding but was revived a year later with the first of three BuAer orders for Regulus I KDU-1 target drones (designated as TD's within Chance Vought).[28] Chance Vought target drone development flight operations began on 23 February 1955 with the launch and recovery of a KDU-1 (TD-1097) from MCAAS. Target acquisition tests with Sidewinder missiles were successfully completed during the flight. The first "live" air-to-air missile launch against a KDU-1 took place on 26 May 1955 with a Sidewinder launched from an F7U-M. The Sidewinder missed, much to the scarcely concealed joy of the Chance Vought launch team, demonstrating that Regulus I was not as easily defended against as its detractors claimed. The first successful interception of a KDU-1 occurred one month later when a Sparrow air-to-air missile scored a direct hit on TD-1102. The target drones were too expensive to use as general anti-aircraft practice drones and were used primarily in the evaluation of the Sidewinder and Sparrow air-to-air missiles as well as the Terrier surface-to-air fleet air defense missile. Twenty-one additional flights took place in the Phase A testing of the KDU-1 during 1955.[29]

Two new missile launchers were tested in 1955. The first was the Short Rail Mark 3 (SR MK 3), the new 6' short rail launcher that was to be installed on board all of the cruisers, replacing the SR MK 2 which had proven to be over engineered and difficult to operate. The SR MK 3 was the first launcher to incorporate design changes made since the lower vertical tail surface had been removed from production missiles. The result was a much more compact and less complicated launcher.

The second launcher was the induced pitch launcher or IPL. Dr. Fiedler and his team of scientists at Pt. Mugu realized that many of the problems incurred with the Regulus launch equipment to date revolved around the hydraulics and support arms needed to elevate the SR MK 2 launcher. Fiedler's team designed and built a launcher that permitted the missile to be launched from a nearly horizontal position and then, due to the curved nature of the launch rails, pitch up as the missile moved forward and cleared the rails. The resulting equipment was very simple in design and had few moving parts. Although intended for use on cruisers, the IPL was used primarily on shore but did deploy once on board USS *Randolph* (CV 15) in 1956 (see Chapter Eight).

In early 1955 it was apparent to all involved that the use of mobile launchers on aircraft carriers was simply not going to be feasible. A final evaluation during OPDEVFOR testing in March 1955 had resulted in a successful launch but carrier officers were adamant in their objections to the concept. Chance Vought had anticipated this problem and with BuAer approval instigated an in-house design effort for a new launch concept. The new launcher was designed to utilize the new, high powered C-11 steam catapults on USS *Hancock* (CV 19) to launch Regulus missiles, eliminating the use of JATO bottles altogether.[30,31] In late June 1955, the first tests using dummy mass sleds demonstrated the utility of the new Steam Assisted Regulus (STAR) carts. The catapult was operated at higher pressures than had been previously utilized, much to the consternation of the *Hancock*'s commanding officer who was justifiably concerned about irreparable damage to the newly installed catapult equipment. The catapults worked perfectly and the first test with a live missile was ready to be conducted. On 19 July 1955, GMU-52 personnel launched FTM-1058. The missile flew straight off the catapult without so much as a dip below the flight deck. A second launch one day later was also successful and *Hancock* departed for deployment to the Western Pacific six weeks later carrying four Regulus I tactical missiles and four STAR cart launchers (see Chapter Eight).[32]

The Trounce IA improvement evaluation program began on 3 March 1955 with a successful launch from Pt. Mugu. Trounce IA was the first improvement program for the Trounce guidance system with attention directed to improved transponder and encoder reliability. Guidance to 35,000 feet and 100 nautical mile range was accomplished with the missile receiving all commands. Seven additional successful Trounce IA flights took place during 1955.

Navy Phase B Operations and Fleet Training

The five Phase B operations in 1955 were warhead evaluation tests. These flights used radio command control in typical assault missile flight profiles. Three of five launches were successful and the test objectives met satisfactorily. Most of the fleet training flights utilized target drone missiles as the launch teams trained in launch operations, while the pilots flew safety-chase missions and conducted missile passes near the airfield prior to turning control of the drone over to the TV-2D recovery aircraft. The first Navy KDU-1 flight on 24 March 1955 was also the first flight operation by the newly established surface Regulus Guided Missile Unit FIFTY-THREE (GMU-53) stationed at Naval Air Station Chincoteague, Virginia. GMU-53 provided Regulus launch services and the pilots of Utility Squadron VU-4 RAM Detachment provided chase and recovery services. All GMU-53 flight operations launches were target drones for use in the East Coast Terrier evaluation program with USS Mississippi (AG 128), recently converted for evaluation of the new family of fleet air defense missiles. In September, 1955, GMU-53 was disestablished and its personnel combined with those from VU-4 RAM Detachment to form Guided Missile Group TWO (GMGRU-2) (see Chapter Eight).

The first Regulus Assault Missile Navy detachment pilots, attached to Composite Squadron VC-61, a photoreconnaisance/fighter squadron, began training with missile flights in June 1955. GMU-52 provided launch services and missile maintenance. For a brief period during this time, a small contingent of US Army officers and enlisted men joined the Navy training program at MCAAS. The Army team, with Captain Robert M. White, Jr., US Army as Officer-in-Charge, prepared and launched two missiles during the VC-61 training program.[33]

OPDEVFOR Phase C Flight Operations

OPDEVFOR operations continued in 1955 with four launches by the *Tunny* missile team in January. Two were successful Trounce operations and two were lost on launch due to booster misalignment. One of the successful flights had been launched under simulated tactical conditions, i.e., no drone flights or dress rehearsals. The launch took place in State Four seas without any problems.[34]

The first cruiser Trounce guidance flight was successfully launched on 13 January 1955 from Los Angeles. Trounce control signals from the cruiser were not received by the missile and control was assumed by the Trounce system at Pt. Mugu which successfully completed the mission. On 15 February 1955, another milestone in the Regulus program was successfully conducted. TM-1080 was launched from Los Angeles in the Hawaiian operations area as part of the first operational suitability test (OST) of the Regulus I program. The missile was controlled to an air burst detonation over Kaula Rock by pilots from VU-3 RAM Detachment (See Chapter Eight).[35] This was the culminating flight of the assault missile program evaluation with subsequent certification of tactical operations.

In March attention turned again to *Tunny* OPDEVFOR operations. Seven missiles were successfully launched and five were recovered, the other two were tactical missiles. Trounce IA operations in both the low-altitude (5,000 feet), short range (50 nautical miles) and high-altitude (35,000 feet), long range (100-180 nautical miles) operational modes were conducted with successful single boat operation out to 180 nautical miles. Two successful dual submarine Trounce IA guidance operations were conducted at 100 nautical mile range. Also in March, the first combined OPDEVFOR operation took place. GMU-52 launched a missile from *Hancock* and then *Carbonero* and *Cusk* each guided the missile during the 29 minute flight. This was the first and last time that the carrier-launched, submarine guidance combination was demonstrated.

The second milestone in the Regulus program during 1955 was the May OST with *Tunny* and the two Regulus I guidance submarines of Submarine Division FIFTY-ONE (see Chapter Ten). Cruiser OPDEVFOR operations resumed with the first missile launch from USS *Helena* (CA 75) on 17 August 1955. This was the first launch from the SR MK 3 launcher, the new fixed launcher installation for cruisers. Although the boosters ignited properly, the engine lost power immediately after booster separation and the missile nosed over into the sea. Two days later *Helena*'s missile team successfully launched and guided a missile using Trounce 1A.

1956

Chance Vought Phase A Operations

In mid-1955 the Marine Corps had expressed an interest in procuring Regulus missiles for use as a shore bombardment weapon launched from cruisers. The cruiser Trounce IA system utilized the AN/SPQ-2 search radar which operated at a different frequency than the submarine Trounce equipment. After several years of trying to develop dual frequency decoders for use on the missiles, BuAer finally recommended that the AN/SPQ-2 radar frequency be changed to that of submarine Trounce.[36] Unfortunately this meant that the Marine Corps MPQ-14A close air support radar guidance system, the one the Marines wanted to use to control Regulus, could not be used. Again after considerable debate, modifications to the MPQ-14A were made for test purposes. Three Phase A missile flights using the MPQ-14A radar for Trounce guidance were successfully completed in 1956. In early July, 1956, the Marine Corps abandoned the requirement for a Regulus I shore bombardment weapon system and the program was canceled (See Appendix I).[37]

On 31 August 1955, the Chance Vought began the design process for adaptation of the tactical missile nose section for the W-27 thermonuclear warhead on Regulus I.[38] The cylindrical shape of the W-27 warhead meant that the missile had to be modified by adding a slight bulge in the lower profile of the nose section. Concerned that this change in the symmetry of the missile would cause problems during the terminal dive maneuver, extensive wind tunnel tests were conducted. The results indicated that the new shape would have little, if any, effect during the terminal dive.[39] The first "bulged nose" tactical missile flight was conducted 28 June 1956 and was completely successful. Six additional warhead evaluation flights were conducted during 1956. Fifteen had been planned but after seven flawless flights the remaining eight were canceled. During one of these flights, a phototheodolite tracking camera recorded a most unusual and heretofore unseen event (outside of wind tunnels). Shock waves were recorded in one frame of the 35 mm motion picture film coverage of a terminal dive into the Atomic Energy Committee Salton Sea Test Range. The picture was published in a leading aviation magazine and major newspapers.[40]

A secondary design change, the missile wingfold hinge placement, was also made in 1956 to permit missile storage in a new class of Regulus launch submarines then under design. Previously designed with a 70-inch wing fold, the missile was now also to be produced with a 54-inch wing fold position.[41] Shock mitigation equipment on board the two smaller fleet-type submarines could not accommodate the 54-inch version because of the longer outer wing sections resulting from the new hinge position. Automatic wingfold mechanisms, already in use with the tactical missiles, were added to subsequent production fleet training missiles.[42]

A positive flight termination control system was also under study at this time. This feature was designed to allow greater physical separation between the control aircraft and the missile, permitting the missile to be aimed over a checkpoint on the target bearing line and then continue autonomously. This feature could also be used in conjunction with the Trounce system to extend missile range past that of the Trounce radar horizon.

The final mobile launcher design was completed and tested in 1956. The SR MK 4 launcher, designed and built by Long Beach Naval Shipyard, was a vast improvement on the earlier mobile SR MK 2 launcher. Light-weight construction and much more maneuverable, the SR MK 4 soon replaced the SR MK 2 at all land-based launch facilities.[43]

Navy Phase B and Fleet Training Operations

Commander James Osborn, Officer-in-Charge of GMU-50, reported that the reliability of the Regulus I tactical missile was 91%, finding it far better then expected and nearly that of the Mark XIV and XVIII submarine torpedoes. The overall probability of conducting a successful flight, compiled since the inception of the Regulus I flight program, was 82%. Review of the data indicated that expenditures could be predicted from probabilities derived over the entire program. The Lot X tactical missiles, the ones the Navy would be using for deployments, were found by

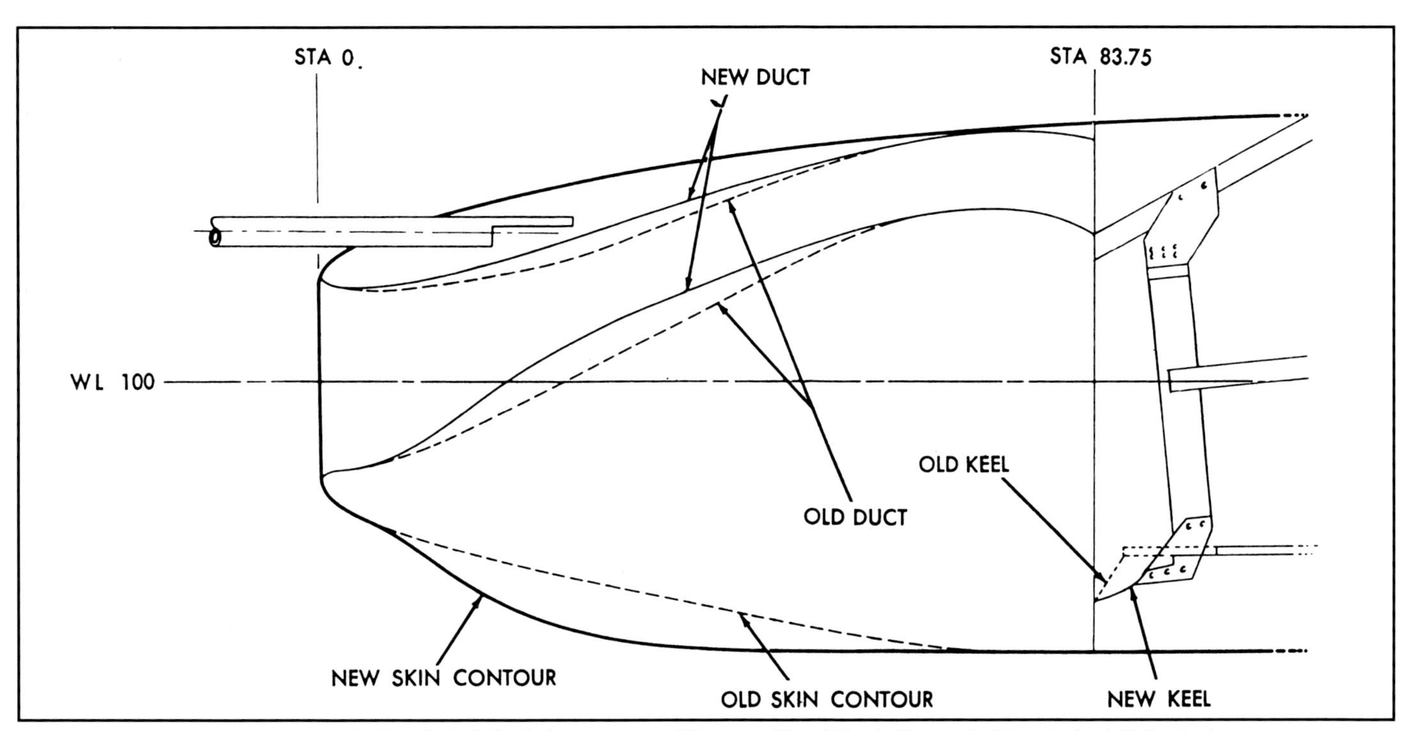

Regulus I tactical missile with the bulged-chin nose cone. (Courtesy of Loral Vought Systems Archives, Author's Collection)

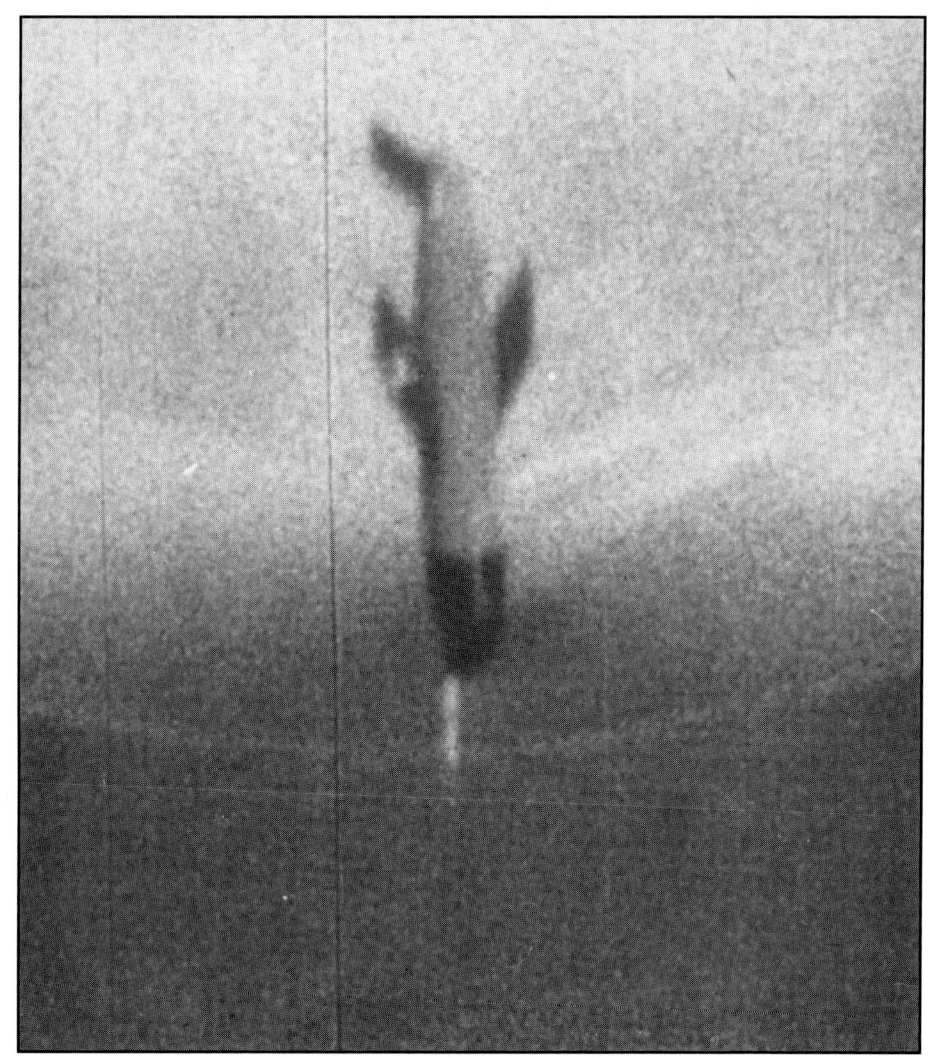

Regulus I in a supersonic terminal dive over the Atomic Energy Commission Salton Sea Test Range. Sandia Corporation was photographing the terminal dive sequence of this tactical missile from one and a half miles away with a 60-inch telescopic lense. Schlieren lines can be seen forming at the tip of the boom and the nose of the missile. Only one frame of the film, which was being shot at 95 frames per second, showed this effect. This was the first time that schlieren lines had been seen and photographed outside of a wind-tunnel environment. (Courtesy of Loral Vought Systems Archives, PeedCollection)

Osborn's analysis to be highly reliable units.[44]

Fleet operations in 1956 began with the first launch of a tactical missile from an aircraft carrier, *Hancock*, in WestPac (see Chapter Eight). GMGRU-1 began training Regulus assault mission pilot detachments for deployment on USS *Lexington* (CV 16) and USS *Bon Homme Richard* (CV 31). GMGRU-1 training operations continued at MCAAS, Mojave. GMGRU-2 flight operations on the East Coast were primarily target drone presentations for Sparrow III and Terrier missile testing. The third carrier to launch Regulus I was the Norfolk-based *Randolph*. GMGRU-2 provided a launch team and RAM aircraft detachment as the new mobile induced-pitch launcher was tested prior to the deployment of *Randolph* to the Mediterranean (see Chapter Eight). Cruiser flight operations continued with a successful launch and terminal dive to impact near Hawaii as *Helena* prepared for her deployment to Westpac. USS *Toledo* (CA 133) on the West Coast conducted her first launch on 16 May 1956. USS *Macon* (CA 132) on the East Coast conducted her first operation on 8 May 1956 (see Chapter Nine). Submarine Regulus operations had a new participant when the newly converted USS *Barbero* (SSG 317) conducted her first missile launch off of Pt. Mugu on 14 March 1956 (see Chapter Eleven).[45]

OPDEVFOR Phase C Flight Operations

OPDEVFOR cruiser flight operations were concluded in 1956 after nine launches from *Los Angeles* in a three week period. All flights were successful Trounce IA and Regulus assault mission operations. These were the first combined Trounce and Regulus assault mission flights, demonstrating the capability of cruisers to serve as launch platforms with pilots extending the guidance range over that of cruiser Trounce control alone. OPDEVFOR submarine flight operations were also concluded in 1956 after three launches from the *Tunny*, including an OST operation with deployment from Pt. Hueneme to French Frigate Shoals in the far western reaches of the Hawaiian Islands Chain. The launch on 22 July 1956 was highly successful as *Tunny* controlled the missile for the first 12.5 miles after which *Cusk* assumed control for the remaining 170 miles. The last OPDEVFOR flight was the next day. *Tunny* controlled the missile for the entire 85 mile flight which terminated with a highly accurate dive to impact.

The OPDEVFOR evaluation for Regulus I was mixed. Overall missile performance in temperate climates was satisfactory. Booster misalignment and slipper ejection failures needed to be remedied but considerable improvement in both systems had taken place over the course of the OPDEVFOR process. The support of Regulus operations was hampered by the lack of adequate support from the Navy logistics system. The personnel from the various support units were satisfactorily trained and able to maintain the systems.[46] Early in the OPDEVFOR evaluation, shipboard guidance equipment had been less reliable then the missile. The CP-98 Trounce Guidance Computer was difficult to calibrate and maintain. The Trounce Radar Course-Directing Centrals, the P1-X, AN/BPQ-1 and AN/SPQ-1, were deficient in regards to accuracy and reliability. In late 1956 the majority of these deficencies had been rectified although Trounce radar alignment remained a problem for several more years.[47]

The OPDEVFOR test conclusions for submarine Regulus operations were: both high and low-altitude single submarine Trounce guidance were recommended for service use; the two-submarine method of tactical delivery was not recommended for service use at this time; Trounce guidance equipment needed excessive maintainance and logistical support was barely sufficient; and the Mk 7 Mod 3 gyrocompass was inaccurate and seen as a source of many of the guidance system accuracy problems.

Surface ship delivery of Regulus I employing radio control Regulus assault mission tactics was recommended for service use. Though limited by the range of the escort aircraft, the RAM delivery system was accurate and reliable. Carrier flight operations were excessively disrupted by missile launches but the use of steam catapults lessened the problem to a great extent. Cruiser Trounce delivery was recommended for service use with a range of 125 nautical miles, with a range of 200 nautical miles considered feasible. Trounce guidance system problems were similar to those mentioned for the submarine system.

Endnotes

[1] *Chance Vought had urged BuAer to contract for a second source of autopilots since Sperry was just barely keeping up with production. After many discussions, Honeywell agreed to produce a prototype. It was flown in FTV-4, but the second source program was canceled soon thereafter due to budgetary constraints.*

[2] *Personal communication with Palmer Ransdell, field flight test engineer, February, 1993.*

[3] *Regulus I Flight Test Index, 1961, LVSCA A50-24, Box 9.*

[4] *Personal interview with George Sutherland.*

[5] *Personal interviews with Bill Albrecht and George Sutherland.*

[6] *Personnel communications with Captain Larry Kurtz, USN (Ret.), September 1990.*

[7] *Personal interview with Captain "Pappy" Sims, USN (Ret.).*

[8] *Personal interview with Palmer Ransdell.*

[9] *Personal interview with George Sutherland.*

[10] Evaluate Combinations of Surface Ship Launching of Regulus I with Final Control Exercised by Units other than the Launching Ship, Final Report on Project Op/S317/X11, 16 August 1957, page 1-1. Naval Historical Center, Operational Archives.
[11] "Beagle" was a guidance system similar to BPN and based on the use of a covertly placed buoy beacon system.
[12] Evaluate Combinations of Surface Ship Launching of Regulus I with Final Control Exercised by Units other than the Launching Ship, Final Report on Project Op/S317/X11, 16 August 1957, page 1-2. Naval Historical Center, Operational Archives.
[13] Personal interviews with Bill Micchelli.
[14] Personal interviews with Bill Micchelli and Captain Robert Freitag.
[15] RPS, page 11.16
[16] Amazingly enough, many historians consider this to be the date of "initial operational capacity" for the Regulus program. The Chief of Naval Operations, Admiral Arleigh Burke, testified in January 1959 that this was the IOC date. With only one launch submarine that could carry two missiles, this seems to be stretching the point.
[17] U.S. Naval Bombardment Missiles, 1940-1958: A Study of the Weapons Innovation Process by Berend D. Bruins, 1981, page 264. This is a doctoral dissertation and is available from University Microfilms International, Ann Arbor, Michigan. The code number is 8204465.
[18] Regulus I Flight Test Index, 1961, LVSCA, A50-24, Box 9.
[19] Evaluate Combinations of Surface Ship Launching of Regulus I with Final Control Exercised by Units other than the Launching Ship. Final report on Project OP/S317/X11, 16 August 1957. Operational Archives, Naval Historical Center.
[20] Regulus I Flight Test Index, 1961. LVSCA A50-24, Box 9.
[21] Ibid.
[22] Regulus BiPolar Guidance Feasibility Program, 31 January through 29 April 1955, undated, page 3, Pacific Missile Test Center Archives, NAMTC-MT-44-55.
[23] Ibid.
[24] Ibid.
[25] Ibid., page 2
[26] RPS page 4.10
[27] Regulus I Progress Report Number 11: 1 March to 30 June 1955. Page 1. LVSCA A50- 22, Box 5.
[28] Submission of Proposal on Advance Procurement for Regulus Target Drones. Letter from Chance Vought to BuAer dated 15 November 1951. National Archives, BuAer Conf. General Correspondence, Record Group 72, Box 121.
[29] Regulus I Flight Test Index, 1961, LVSCA A50-24, Box 9.
[30] Regulus I C-11 Catapult Launching Cart Study. Report Number 9670. 27 January 1955. LVSCA A50-24, Box 7.
[31] Wings for the Navy: A History of the Naval Aircraft Factory 1917-1956 by William F. Trimble, 1990, page 319. Naval Institute Press, Baltimore, Maryland.
[32] Personal interview with George Sutherland.
[33] Personal interview with Sam Lynne, Chance Vought MOE at the time. The Army Regulus I training program was an attempt to keep the Army up-to-date with guided missile concepts and operations.
[34] Regulus I Flight Test Index, 1961, LVSCA A50-22, Box 9.
[35] Personal interview with Sam Lynne.
[36] The AN/MPQ-14A Radar Course-Directing Central and Regulus I Compatibility Problem. Final Report on Project TED MTC EL-42716.2; 20 June 1957; page 1. Pacific Missile Test Center Archives.
[37] Bruins, page 267.
[38] RPS, page 25.
[39] Regulus I Progress Report Number 12, 1 July through 30 September 1955, page 2. LVSCA A50-22.
[40] Personal interview with Roy Pearson.
[41] The 70 or 50-inch dimension was the length of the inboard unfolded portion of the wing.
[42] RPS, page 27.
[43] The SR MK4 launcher is the only example of the many Regulus I launchers that has survived the scrapyard. In 1992 the author identified an SR MK 4 alongside the last remaining Regulus I tactical missile in the Suitland Silver Hill Storage and Restoration Facilities of the National Air and Space Museum.
[44] Regulus I Progress Report Number 17, 1 October to 31 December 1956, page 2. LVSCA, A50-22.
[45] By the beginning of the third quarter of 1956, ten aircraft carriers, four cruisers and seven submarines had been converted to Regulus I launch or guidance platforms. A large number of major afloat commands had Regulus I training or simulated tactical maneuvers taking place on a routine basis.
[46] Evaluate the Regulus Guided Missile for Service Use, 16 August 1957, Final Report on Project OP/S317/X11, Naval Historical Center, Operational Archives.
[47] Ibid.

Chapter Five: Regulus I Flight Testing and Development 1957-1966

With Operational Development Force (OPDEVFOR) acceptance of Regulus I for fleet use, Chance Vought Phase A and Navy Phase B operations continued to work on improving the performance of the missile and simplifying missile operations. The critical reviews of Trounce guidance by the OPDEVFOR report prompted careful analysis by both Chance Vought and the Navy for causes and solutions.

1957

Chance Vought Phase A Operations

In late 1956, Chance Vought had initiated a review of Regulus I survivability to target as part of the ongoing BuAer contract. The results were reported to BuAer in February 1957.[1] This exhaustive study compared the probability of Regulus I surviving against postulated Soviet defenses with that of the Navy's Douglas Aircraft A4D "Skyhawk" and A3D "Skywarrior." These aircraft were the contemporary carrier-based tactical nuclear bombers. On a high-altitude mission and no early warning by enemy radar, both Regulus I and these two aircraft had similar survival probabilities at ranges under 50 nautical miles. As range-to-target increased to 200 nautical miles, the A3D survival probability dropped quickly and at 500 nautical miles, the range limit for Regulus I, the missile still had a higher probability of survival than the A4D even though the missile would have required escort aircraft for guidance at this extreme range. The results were essentially the same if enemy early warning radar detected the missile or aircraft during the high altitude mission. Comparisons made at 5,000 feet altitude gave virtually the same results. Interestingly enough, and perhaps indicative of tactical operations later adopted, when the comparison was made with a 500 foot altitude run into the target, both aircraft and Regulus I had very high probabilities of success for targets up to 125 nautical miles from the coast. At greater distances the probability of success for the A3D dropped quickly while the A4D and Regulus I had nearly the same chances of survival. A typical Regulus assault mission profile for these calculations used FJ-3D escort aircraft with a combat radius shorter than the Regulus I range. Submarine or surface ship Trounce guidance range, assuming 35,000 foot altitude cruise to terminal dive, was slightly more than 228 nautical miles. The low-altitude profile at 8,000 feet had a guidance range of 109 nautical miles.[2] The report concluded that Regulus I would provide more than sufficient nuclear weapon delivery capability from 1957 to 1961.

Progress on the continuing improvement of the Regulus I weapon system emphasized extension of current missile capabilities. An estimated 50% increase in tactical range could be realized if the large fuel consumption rate during launch and climb-to-cruise altitude could be compensated for by additional fuel capacity. One solution was to increase the thrust of the JATO boosters. Chance Vought contracted with Philip Petroleum Company to design and build a 2.5 second, 45,000 pound thrust JATO bottle. While tests were conducted with dummy mass sleds, there is no evidence that any but the standard Aerojet General JATO boosters were ever used in the Regulus I program.[3] A second range extension method was the use of drop tanks. Three launches were conducted to flight test this concept. Two missiles were modified and two flights successfully conducted from Pt. Mugu, including one recovery with the external tanks still attached to the wing pylons. The third launch took place aboard USS *Franklin D. Roosevelt* (CVA 42) on 22 February 1957 (See Chapter Eight). All three flights successfully demonstrated the range extension capability but the external drop tank system was not adopted for operational use.

Chance Vought and BuAer were concerned with the sources of error that caused miss distances to increase with range when using Trounce guidance. A detailed study was completed by the contractor which indicated that during a typical tactical operation, 50% of the error in the impact area could be ascribed to errors in navigational accuracy when out of sight of land. Any type of missile guidance computer, such as the CP-98 used in the automatic Trounce guidance system, would only be as accurate as the quality of the information fed into it. Granted, the Trounce guidance system's inherent inaccuracies still needed considerable work, but the fact that the error in ship's

Fuel drop tanks were considered as one method to extend the range of Regulus I. Fitted inboard of the wing-fold, the tanks were used to replace the fuel consumed during launch. The system was successfully demonstrated but not adopted. (Courtesy of Loral Vought Systems Archives, Author's Collection)

Regulus I missiles were stored in canisters called transicontainers. The folded wing-span of the missiles, in the container, exceeded the width of many of the underpasses and roadways in the United States at the time. The solution was to place the canister on a flatbed truck trailer that permitted the canister to be rotated 60 degrees as shown here. (Courtesy of Loral Vought Systems Archives, Cannon Collection)

position contributed such a sizeable portion of the impact miss distance error underscored the need to improve ship's navigation. Inertial guidance was the obvious answer to these problems and was under intensive study not only as an aid to missile guidance but also for use in shipboard navigation.[4] As a result of these investigations, guidance system development studies were conducted by Chance Vought to enhance mid-course and terminal guidance assistance to the missile as it neared the target. These studies included Regulus Inertial Navigator (RIN) mid-course guidance and terminal guidance assistance systems such as Automatic Terrain Recognition and Navigation (ATRAN), Offset Beacon System (OBS), and the MARCH and DENNIS systems (See Appendix I).[5]

Chance Vought was given authority to proceed with the RIN system for Regulus II, the supersonic second-generation Regulus missile. RIN would be designed to also allow installation in Regulus I missiles. An additional low-cost inertial platform, Project SPADE (Stable Platform Augmented DENNIS Equipment) was proposed to BuAer but did not undergo further development.

Navy Phase B and Fleet Training Operations.

While most components and subsystems of Regulus I had been evaluated in low temperature environments, an complete tactical missile had not yet been subjected to this type of test. Since tactical operations would entail ramming the missile out of a heated hangar into sub-zero weather such as that found in the Northern Pacific and Atlantic Oceans, a full-scale test was considered essential. Guided Missile Unit FIFTY (GMU-50) and Pt. Mugu staff scientists placed TM-1229 in the climatic chamber of the US Naval Civil Engineering Research and Evaluation Laboratory at Port Hueneme, California, to test exposure to low temperatures as well as thermal shock conditions. Two large fans were used to blow air through the engine and engine compartment to accelerate cooling and heating of missile components. At 20 F degree intervals a pre-flight static check of major subsystems was performed and responses recorded. Upon completion of the -60 F degree checkout, the missile was removed from the cold chamber and, as frost accumulated on the missile, another checkout was performed. The missile remained at ambient temperature for 24 hours and 76% humidity, and then returned to the cold chamber again for exposure down to -50 F degrees. As ice formed throughout the missile the final pre-flight checkout was performed. The missile passed all tests and no major problems were encountered other than cracks developing in hydraulic system hoses after thermal cycling from -60 to 61 and back to -50 F. The hoses leaked small amounts of hydraulic fluid but did not fail. The test results indicated that the missile could be successfully readied for flight under these conditions.[6]

OPDEVFOR Phase C Operations

Phase VI of the OPDEVFOR evaluation of Regulus I was the use of the combination of a surface ship launch with final control by one or more submarines in sequence. From May 1956 to April 1957, three Regulus I missile flights and 22 manned drone aircraft flights were conducted for this evaluation. The conclusion was that this combination was acceptable for missile flights to a range of 220 nautical miles but with tactical limitations. The tactical limitation was not a matter of ability to transfer control between guidance centers. Transfer from one unit to another, provided they were within 75 to 100 nautical miles was rated as 90% probable. The limitation was that the circular error probable, the circle around the target within which 50% of the missiles would hit, had a diameter of 2,900 yards.[7] The report ended with a critical review of both cruiser and submarine Trounce Radar Course-Directing Central guidance equipment. While improvements had been made, much still needed to be done to make the system reliable and accurate. The CP-98 Navigational Computer, when calibrated, gave adequate automatic guidance commands but was found to be unreliable.[8,9]

1958

Chance Vought Flight Operations

With the production and delivery of Regulus I missiles scheduled to conclude at the end of 1958, Chance Vought turned its attention towards working with Navy scientists in solving the remaining issues concerning the operational use of the Trounce guidance system. A systematic review of the submarine Trounce guidance system was initiated on the East Coast by the Navy as the newest version of the Trounce radar course directing central, designated AN/BPQ-2, was readied for testing (See Appendix I). A similar test program was begun on the West Coast but focused its attention on the AN/SBQ-1 equipment in the current cruiser installation. Chance Vought personnel assisted the scientists at Pt. Mugu during this evaluation program with launches from *Toledo* and *Los Angeles*.

Navy Phase B and Fleet Training Operations

Navy Phase B operations were transferred to Guided Missile Unit FIFTY-FIVE (GMU-55) at Pt. Mugu and target drone operations continued for Sparrow III and Sidewinder missile evaluations. In May 1957, the West Coast Regulus submarines were transferred to Hawaii.[10] GMU-50 also moved and was renamed GMU-90.[11] All submarine Regulus I activities in the Pacific were conducted from Hawaii for the remainder of the program. Regulus I cruiser operations on the East Coast were concluded on 29 September 1958 with the last launch from *Macon*. With the termination of the cruiser Regulus program on the East Coast, Guided Missile Group TWO (GMGRU-2) operations came to a conclusion on 17 October 1958 with a final target drone flight for Sidewinder test firings from a F8U-1 Crusader. At this time GMGRU-2 had conducted 24 percent (198 flights) of the Regulus I program.[12]

In July 1958, East Coast submarine Regulus operations were transferred to Naval Air Station Roosevelt Roads, Puerto Rico to make use of the more suitable weather conditions. Regulus flight operations had always been hampered by the often unpredictable weather off the East Coast and the move to Puerto Rico would make training much more productive. In addition, the impending arrival of the new, supersonic Regulus II missile necessitated more air space for flight operations and the East Coast air corridors were far too congested.[13]

1959

Chance Vought Operations

On 3 February 1959 the first of a series of four flights were conducted from Pt. Mugu to evaluate operation of the Trounce IC equipment using the AN/BPQ-2 radar course directing central installation at Pt. Mugu. The AN/BPQ-2 was a modularized, simplified and more reliable version of the AN/BPQ-1 equipment. Trounce IC was the final version of the Trounce decoder, encoder and antenna equipment installed in the missile and permitted the selection of one of three pulse-pair combinations. At the end of the four flight series, operations moved on board *Toledo* where two successful flights took place off Santa *Barb*ara. Simultaneous with the West Coast cruiser Trounce and AN/BPQ-2 evaluation program, USS *Medregal* (SS 480) began the final series of East Coast submarine AN/BPQ-2 Trounce guidance system tests in the Caribbean. This flight test program culminated with a report that strongly endorsed the qualities of the AN/BPQ-2 equipment.[14] Trounce had at last matured into a fully operational and accurate guidance system for Regulus I.

Navy Fleet Training Operations

In mid-1959, the Office of the Chief of Naval OperationsO announced that the first deployments of the Polaris Fleet Ballistic Missile submarines would begin with the Atlantic Fleet in the Fall of 1959. The East Coast Regulus launch submarines, USS *Growler* (SSG 577) and USS *Barbero* (SSG 317), and one Regulus guidance submarine, *Medregal*, were reassigned to Pearl Harbor, Hawaii. The GMU-51 operations at Roosevelt Roads were discontinued with personnel and equipment transferred to GMU-90, the submarine Regulus support unit at Pearl Harbor, Hawaii. On 22 April 1959, Regulus I launch operations were concluded on board *Toledo*. On 23 October 1959, *Tunny* departed Pearl Harbor on the first Regulus missile deterrent strike patrol, beginning a nearly five year period where strategic targets in the Northern Pacific were continually covered by submarines carrying Regulus I missiles.

1960-1964

Navy Fleet Training Operations

Cruiser Regulus operations continued on board *Los Angeles* until 28 February 1961. Cruiser Regulus operations came to an end with the last launch by *Helena* on 3 March 1961 (see Chapter Nine). Regulus sumarine operations continued until July 1964 with the conclusion of the last Regulus deterrent patrol by USS *Halibut* (SSGN 587) (see Chapter Fourteen). The remaining missiles were expended on one way target drone missions. At 1437 hours, Monday, 6 June 1966, the last Regulus I launch operation was conducted by Chief Grover Wade, one of the original crewmen of GMGRU-1. Fourteen minutes later the missile was destroyed by a Terrier anti-aircraft missile fired from USS *King* (DDG 41) 50 nautical miles offshore. This was the 318th launch of Regulus I from the Bonham Auxiliary Landing Field, Kauai and the 1133rd launch in the program.[15]

Sixteen years after the first taxi-tests on the dry lakebed at Edwards Air Force Base, the Regulus I program ended. A total of 514 Regulus I missiles were built; 209 FTV/FTM's, 228 TM's and 77 TD/KDU-1's.[16] The recoverable missiles averaged slightly less than four flights each. More than one missile reached 10 flights and one, a target drone, was launched successfully 21 times (See Appendix VI).

Endnotes:

[1] *Regulus I Weapon System Analysis, Part 2: Survival Probability Analysis; pages 2-69. Report 10622-a. LVSCA ASO-22.*

[2] *Radar Navigation Manual, by Ernest B. Brown, 2nd Edition, 1975, Defense Mapping Agency, Hydrographic Center; page 1-14. Guidance range approximation can be calculated, in nautical miles, using the formula d= 1.22 x (square root of missile altitude in feet) assuming standard atmospheric conditions.*

[3] *Regulus I Weapon System Analysis, Part 2: Survival Probability Analysis; pages 2-69. Report 10622-a, page 1, LVSCA A5O-22.*

[4] *Regulus I Progress Report #14, 1 January to 31 March 1956, page 9. LVSCA A50-22, Box 5.*

[5] *MARCH was an infrared terminal guidance system while DENNIS was based on detection of high voltage power transmission lines. Neither system became operational.*

[6] *Low-Temperature Tests on Regulus I Missile TM-1229, 31 December 1957, PMTCA, NAMTC-LE-28-57.*

[7] *The W-5 warhead carried by Regulus I had a variable yield (see Appendix IV). By the time of the routine deployment of the Regulus submarines, the CEP of Regulus I was 1,100 yards((personal communcation with Captain Max Eckhart, USN (Ret.)). With a 50 kiloton warhead, equivalent to 50,000 tons of TNT, and a miss distance of 1,100 yards or 3,300 feet, the overpressure from an air burst at optimum burst height would have been 21 psi; a surface burst would have generated an overpressure of 13 psi. These kinds of overpressures would have been sufficient to severely damage multi-story reinforced concrete buildings within 5,000 to 6,000 feet of ground zero. Unsheltered aircraft would have been severely damaged at overpressures above 3 psi. Underground structures would have been damaged only if they were within 1,800 feet of ground zero. Crater depth for a surface burst would have been 74 to 100 feet; crater diameter would have been 300 to 520 feet depending on soil type.*

Regulus I had the capability of carrying a W-27 warhead with a yield of 2 megatons. With a miss distance of 3,300 feet, the overpressure generated by either an air burst at optimum height or a ground burst was over 200 psi, more then sufficient to eliminate probable targets.

These calculations were taken from "The Effects of Nuclear Weapons," by S. Glasstone and P.J. Dolan, 1977, US Government Printing Office.

[8] *Evaluate Combinations of Surface Ship Launching of Regulus I with Final Control Exercised by Units other then the Launching Ship; Final Report on Project OP/S317/X11. 16 August 1957. Naval Historical Center, Operational Archives.*

[9] *All of the submarine missile guidance officers have said that they routinely kept a manual plot of the missile flight path and frequently guided the missile manually.*

[10] *Detailed fleet unit operations can be found in the cruiser, carrier and submarine chapters.*

[11] *See Appendix II.*

[12] *Regulus I Flight Test Index, 1961, LVSCA A50-22.*

[13] *Personal communications with Captain Ralph "Snuffy" Jackson, USN (Ret.).*

[14] *Personal interviews with Captain Charles Wood, USN (Ret.), Captain of Medregal and Captain Max Eckhart, USN (Ret.), the officer in charge of wringing out the "bugs" in the BPQ-2 system.*

[15] *Last Regulus I Missile Fired from Kauai, 8 June 1966, Honolulu Star.*

[16] *The remaining fleet training missiles, target drones and tactical missiles were redesignated as RGM-6A, BQM-6C and RGM-6B respectively and used in target practice until 1966.*

Chapter Six: Regulus II Flight Test and Development 1952-1958

Aware of the need to build a supersonic version of Regulus I as soon as feasible, Chance Vought began a concurrent study of the supersonic variant in June 1951.[1] Given the company designator V-379, the project soon became known as Regulus II. Chance Vought's extensive experience with the Regulus I program indicated that a minimum of fundamental system changes would be needed for the Regulus II missile. Over the next year several designs were examined with considerable attention given to currently available and proposed engines since engine thrust and size would be the critical system change from the subsonic Regulus I. On 25 June 1952, BuAer released an invitation to bid on improvement of the Regulus I tactical missile to a range of 500 nautical miles with the last 100 nautical miles at supersonic speed. Funding would begin with the 1953 Fiscal Year. One month later the Bureau revised the missile specifications to include a 500 mile range at Mach 2.0 for the entire flight.[2]

By December 1952, after several presentations to BuAer, Chance Vought proposed a three phase program for the development of Regulus II. Phase I used the Wright Aeronautical J65 engine, a new fuselage and the Regulus I wing structure. Phase II replaced the Phase I wing with a newer, thinner wing with a resulting increase in speed to Mach 2.0. Phase III would be an increase in fuel load and range. Presented to BuAer officials 16 December 1952, the same day as the first launch of a Regulus I from the aircraft carrier USS *Princeton* (CV 37), this was the birth of Regulus II.[3]

Further refinements of the proposal took place over the next three months, resulting in a series of BuAer specifications that included launching the missile from the same launch equipment expected to be installed on Regulus I submarines. BuAer, however, refused to consider the use of the Regulus I wing and Chance Vought submitted for review on 10 June 1953, a revised phase program omitting this feature. Work to date on Regulus II had been supported by two amendments to Regulus I contract NOa(s) 53-285 (Letter of Intent) and on 3 July 1953, the Regulus II engineering design effort began.[4] On 4 September 1953 Chance Vought made a detailed cost proposal reflecting the work described in the earlier contract amendments. This proposal covered fabrication of four flight test missiles, spares, etc., as well as detailed design data for the tactical missile. The estimated cost for this work was $8,725,793.

The Air Force also had a supersonic cruise missile under development at this time, the North American Aviation Navaho (XSM-64). A test and development version, designated the X-10, had flown in October, 1953. Eleven test vehicles were built, completing 27 flights and setting a speed record of Mach 2.05 for turbojet-powered aircraft.[5]

On 9 April 1954 production of the Lot I four flight test missiles, GM-2001 to GM-2004, began under the Letter of Intent signed the previous September. On 2 August 1954 the Letter of Intent for Contract NOa(s) 53-285 was replaced by the actual contract specifically authorizing fabrication of four flight test missiles and design of the tactical missile prototype. This final contract came in at $8,155,252.[6]

A thorough review of the proposed Regulus II program was initiated by BuAer in October 1954 to further refine the requirements for the missile. This review pointed out a number of inconsistencies with the stated goals of the program and sought to resolve these issues. The initially specified Wright Aeronautical J65-W-6 turbojet engine had significant deficiencies that appeared to hamper its use for the operational engine. The General Electric J79-GE-3 turbojet engine, predicted to be available in 1957, was a good alternate choice but was still a gamble since sustained turbojet-powered flight at Mach 2.0 had yet to be demonstrated. The guidance system necessary for long range control remained a question mark since ship navigation accuracy was still believed to be on the order of 1-2 miles with an accurate land fix, hardly an auspicious beginning for an inertial navigation platform fix. Use of a homing beacon such as in the bipolar navigation system (See Appendix I) was unacceptable and airborne inertial guidance platforms had a long way to go. Though the initial specifications stipulated the use of hangars and launching equipment currently used by Regulus I, BuAer recognized that this was unrealistic and in October 1954 announced submarine specification changes to provide for a narrower but longer hangar. Wing and rudder fold mechanisms were added to the design at this time. Assembly of flight test missile GM-2001 began on 22 November 1954 and was completed in November 1955.[7,8]

The original Regulus II airframe design had several major changes made prior to its first flight. The rudder was extended to the full height of the vertical fin, nose trimmers were added to correct adverse pitching moments discovered in high-speed rocket flights of Regulus II models. A liquid ballast system was added to the flight test vehicles to permit shifting of the missile's center of gravity as dictated by missile speed. Since the tactical missiles would not be recovered, the ballast system was not incorporated in them. Additional structural changes prior to flight were those associated with the Navy's desire for use of a single booster rocket. The decision by Wright Aeronautical to redesign the engine afterburner shroud required the redesign of the tail cone, eliminating the original boat-tail shape.[9] The only structural change made after flight testing began was the addition of a long, narrow, ventral fin to the tactical missile after an in-flight breakup of during the first terminal dive maneuver.

Three primary guidance systems were proposed for Regulus II: Regulus Inertial Navigator (RIN), Trounce IB, and Automatic Terrain Recognition and Navigation (ATRAN).[10] Only RIN was flown on board the missile. RIN was a self-contained inertial guidance system built by AC Sparkplug. Development of RIN had begun in December 1955 as an improvement to the Regulus I guidance system and was quickly changed to be the primary guidance system for Regulus II with possible application to Regulus I (see Appendix I). Trounce IB was considered only as a back-up if RIN failed and ATRAN, an early radar map matching system, was not flown before program cancellation.

Three mid-course correction guidance systems were evaluated for Regulus II. The Offset Beacon System was similar to the Regulus I bipolar navigation system and never flown. Projects DENNIS and MARCH were early electromagnetic and infrared sensing systems, respectively, and did not demonstrate the required high altitude guidance capability before the program was canceled.

One frequently misquoted application of the Regulus II weapon system was the launch of a modified Regulus II from a B-52. This did not occur. On 20 November 1956, Chance Vought proposed the V-398 B-52 air-to-surface-missile, a variant of Regulus II, and in March 1957 issued a response to the Air Force decision to design a completely new missile with approximately one-half the weight of the V-398. No Regulus II missiles were launched from a B-52 and the missile that eventually fulfilled the Air Force requirement was the North American AGM-28B "Hounddog".[11]

Final Design

Regulus II was designed and built in three configurations (dimensions are given for the tactical version). The first seven missiles, production Lots I and II, were designated as XRSSM-N-9's with airframe numbers GM-2001 to GM-2007. These missiles were powered by the Wright Aeronautical J65-W-6 turbojet engine with 14,608 pounds of thrust at sea level. Manufacturer-imposed limitations on the J65 engine prevented flight above 35,000 feet and Mach 1.8 for prolonged periods. This was due to the use of aluminum compressor blades limiting the inlet air temperature to 200 F at 35,000 feet. The J65 engine was available at the start of the program thus was the engine of choice. These missiles were used in the taxi-take off phase of the flight test program and did not have provisions for booster rocket attachment.[12,13]

The fleet training missiles or XRSSM-N-9a's, airframe numbers GM-2008 to GM-2036, were powered by the General Electric J79-GE-3A engine with 15,600 pounds of thrust at sea level. This engine was capable of sustained Mach 2 flight at an altitude of 60,000 feet. The engine upgrade had been planned from the beginning of the program and design considerations allowed for the installation of the larger engine with a minimum of production line changes. These missiles were produced with booster attachment fittings.[14] As with Regulus I, Regulus II was launched from a short rail launcher using rocket assisted takeoff. Regulus II used one large booster designated as the 4KS with 115,000 pounds of thrust, built by Aerojet General. This propelled the missile to an airspeed of over 300 knots in four seconds. During the drone program a slightly more powerful booster, built by Rocketdyne, was used.

The tactical missiles or XSSM-N-9's, airframe numbers GM-3001 through GM-3038, were also powered by the J79 engine and were slightly more than 67 feet in length, including the nose boom. Wingspan was 20 feet, 9.2 feet

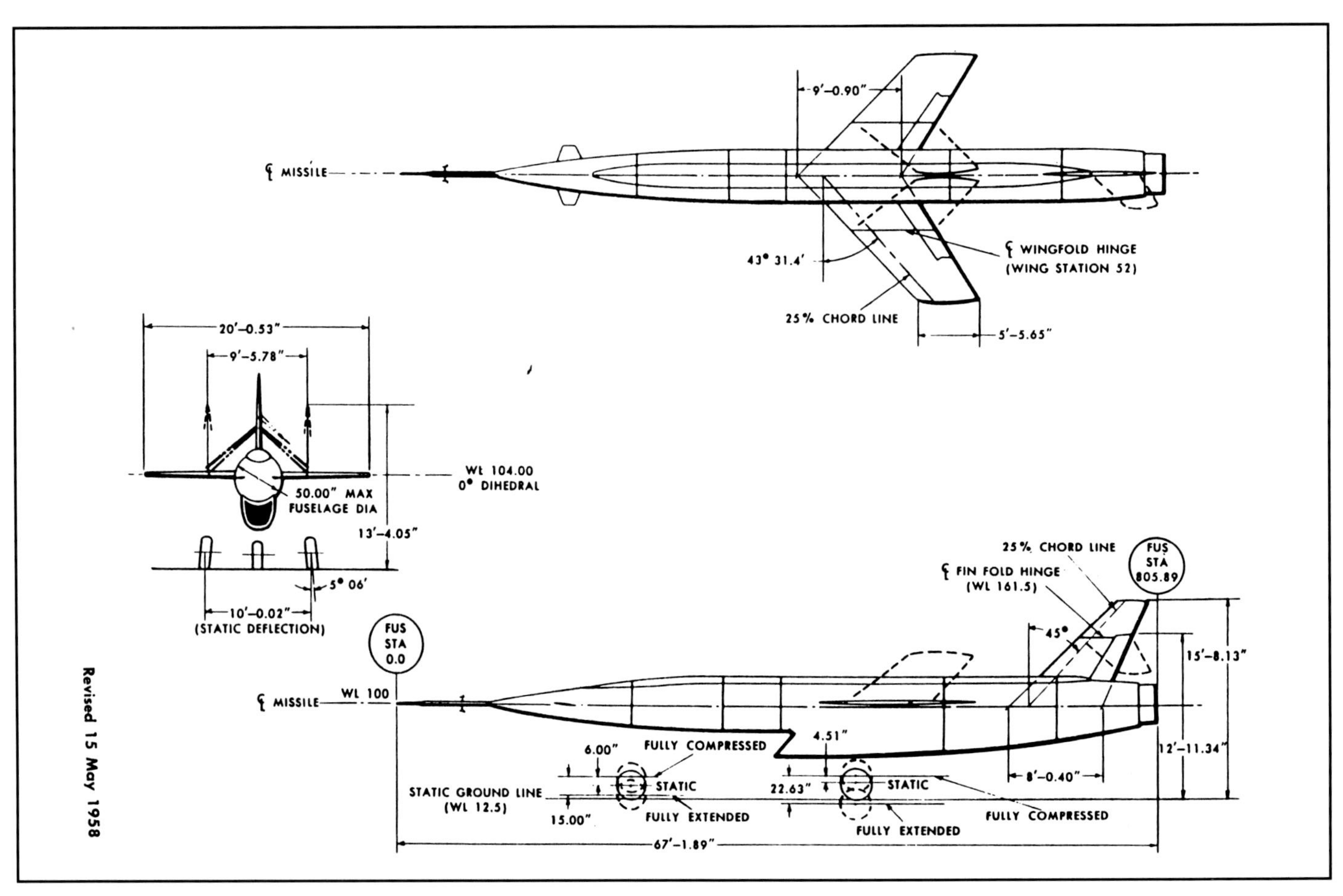

Revised 15 May 1958

Figure 6-1. Principal overall dimensions of Regulus II fleet training missile (X-SSM-N-9).

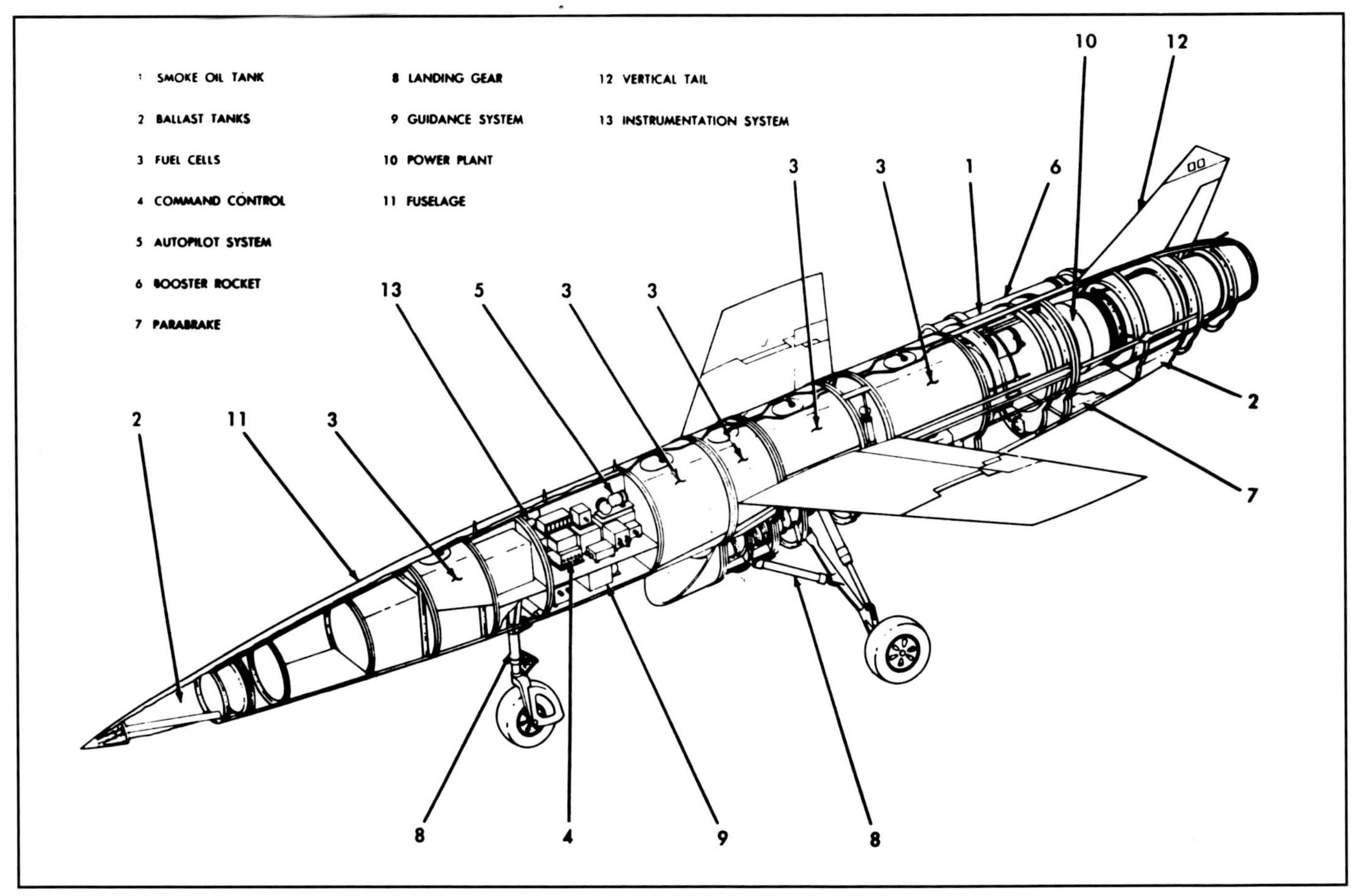

Figure 6-2. Regulus II fleet training missile (X-SSM-N-9) general interior arrangement.

FTM 2001 on the lakebed at Edwards Air Force Base. Engine access panels behind the wing are open. The duct on the fuselage just below the wing attachment point was used to bleed excess air from the air intake. FTM 2001 was one of seven missiles designated as XRSSM-N-9s. (Courtesy of Loral Vought Systems Archives, Albrecht Collection)

with wings folded, and a height of 17.5 feet with the vertical fin unfolded and the booster attached.[15] The tactical missile weighed 11,693 pounds empty and 21,331 pounds at launch, including booster and nuclear warhead. An additional feature in the tactical missiles was the anti-buzz plate. High-speed wind tunnel tests revealed pressure oscillation in the engine air intake duct during the terminal dive maneuver. The anti-buzz plate was lowered into the duct at the start of the dive to reduce these oscillations to an acceptable level. Tactical missiles had the landing gear removed and replaced by a warhead and extra fuel capacity Both the flight training and tactical missiles were later converted, in small numbers, into the KD2U target drone configuration.[16]

A variety of mission profiles could be flown with Regulus II. As originally designed, Regulus II had a Mach 0.94 cruise range, at 32,000 feet, of 1,000 nautical miles. At 5,000 feet, the range decreased to 600 nautical miles. With a Mach 2.0 cruise speed at 65,000 feet, the range was 635 nautical miles. With proposed additional external fuel capacity, in either expendable wing tanks or a saddle tank, the subsonic range could be extended to 1,800 nautical miles, while the supersonic cruise range could be extended to 1,200 nautical miles.[17] Neither of these range extension proposals were flown.

1956

The first Regulus II flight test vehicle, GM-2001, was transported to Edwards on 6 February 1956. This trip culminated almost four years of intensive design and wind tunnel testing of the new missile by Chance Vought and the Navy. Initial Regulus II testing at the Edwards Main Base facilities consisted of numerous ramp and reliability tests reminiscent of the Regulus I program. The experiences gained during the Regulus I program made this phase of the Regulus II test program proceed smoothly. Next came the low and high-speed taxi tests. These measured a multitude of control functions such as control deflection needed to raise the nose wheel at different speeds and landing gear tire integrity at high taxi speeds. With the take-off speed of Regulus II considerably higher than that of Regulus I, these tests, while routine, were still exploring new ground. Taxi tests commenced on 21 March 1956 using GM-2001. After five test runs, the Chance Vought flight test team, headed by George Sutherland, agreed that flight testing could begin as scheduled.

At 0711 on 29 May 1956, GM-2001 became airborne after a 2.3 mile take-off run. The landing gear retracted smoothly as the missile climbed to a cruise altitude of 10,890 feet at a speed of 356 knots. Afterburner ignition cycles were successfully completed and all control systems functioned perfectly. After 32.5 minutes of flight, GM-2001 proved to be reluctant to land as the missile briefly bounded back into the air before settling down and rolling to a stop one-half mile down the lakebed. The landing control aircraft, the same TV-2D's used with the Regulus I program at Pt. Mugu, could barely keep up with the high landing speed of the missile. Finding sufficiently fast chase and airborne control aircraft was a problem that was never really solved during the flight testing of Regulus II. The missile easily outran and outclimbed any of the available chase aircraft. Both the Douglas F4D "Skyray" and Chance Vought F8U-1 "Crusader" were utilized in the chase aircraft role with the F8U-1 finally selected for use in the rest of the program. Over the next six months, thirteen flight tests were conducted.[18,19]

The first crash of Regulus II occurred on 16 August 1956 during the seventh flight operation, with the loss of GM-2001 on landing approach. With the engine in afterburner and the landing gear extended, a normal landing appeared imminent. The throttle suddenly advanced to full override and four seconds later dropped to 77% rpm. This was insufficient thrust to maintain airspeed and the missile crashed onto the lakebed. All was not lost since the missile had successfully reached and held a speed of Mach 1.5 at 35,000 feet during the 29 minute flight. Flight characteristics at 35,000 feet were superior to those predicted by the high-speed wind tunnel tests. Review of the telemetry indicated that a failure had occurred in the throttle servo feedback wiring and corrections were made. The first supersonic flight with missile recovery took place on 7 September 1956.[20] The first high-altitude flight, reaching 54,000 feet and a speed of Mach 1.6, took place 23 October 1956 with GM-2003. The GM-2003 had the first full capacity fuel system that allowed sustained high speed flight. All objectives except automatic approach speed control were accomplished. Earlier wind tunnel tests had indicated possible problems in lateral stability at high altitude but these were not apparent on this flight.

The second test at high altitude with GM-2003 took place on 7 November 1956. This was the fourth flight of GM-2003 and the eleventh test flight of Regulus II. All out of sight controller command responses were positive as the missile climbed to 35,000 feet under military power and the Mach controller engaged as the missile accelerated to Mach 1.8. After reaching a cruise altitude of 43,000 feet, a right turn was

FTM 2001 on roll out after successful recovery on the lakebed 29 May 1956 after a 32.5 minute flight. FTM 2001 was lost in a crash during the seventh flight operation on 16 August 1956. (Courtesy of Loral Vought Systems Archives, Albrecht Collection)

A shock and acoustical energy test using a short burn version of the Aerojet 4KS 115,000 lb thrust JATO booster. GM 2007 is in a nose down attitude to prevent the booster rocket flame from contacting the ground directly below the missile. The J79 jet engine was at full throttle to fully simulate the expected launch conditions. The test was successful even though the hold-down lines were badly frayed at the end of the booster burn. (Courtesy of Loral Vought Systems Archives, Sutherland Collection)

commanded. GM-2003 responded with an uncoordinated turn coupled with a violent pitch up and then went into a steep, spinning dive. Telemetry signals were lost just after the beginning of the unplanned maneuvers. The missile crashed in a remote section of the Chocolate Mountains. Marines were called in to locate the wreckage and after several days of searching ran across an old prospector who told them he had found the wreckage. The missile had impacted on a steep, rocky slope so nothing of significance remained to help identify the cause of the crash.[21,22]

On 11 December 1956, GM-2005 experienced very similar maneuvers to those seen with GM-2003 during the high speed climbing turn a month earlier. The missile crashed and the wreckage was badly burned but inspection of all major structural elements indicated structural failure had not occurred prior to the onset of oscillations. Fortunately for the Navy and Chance Vought, complete telemetry data was available from GM-2005's flight. The Navy suspended all Regulus II flight operations on 15 December 1956 until the crash investigations determined the cause of these consecutive failures.[23,24]

The Air Force's Navaho supersonic cruise

Just as with Regulus I, each Regulus II missile had to have its center of gravity determined, fully fueled, to permit correct alignment of the booster rocket. Here a dummy test airframe used to qualify the launcher has had its center of gravity determined and is now ready for positioning on the mobile launcher located behind the test rig. (Courtesy of Loral Vought Systems Archives, Albrecht Collection)

missile made its first flight in November 1956. Twenty-six seconds after launch the missile crashed. Four months and ten launch attempts later, the second flight lasted for four and one-half minutes. The Air Force cancelled the program nine months later.

1957

By mid-January 1957, Chance Vought engineers had successfully isolated what appeared to be the cause of both accidents. Both flights had been part of the final phase of the high speed test program, designed to investigate flight at the Mach 1.8 limit of the J65 engine using the automatic flight profile controllers to be used in the tactical missiles. In both flights the missiles experienced excessive pitch oscillations while making a high speed right-hand turn at a high rate of climb. Common to both flights was use of the constant Mach number controller during this turn maneuver. It was found that the Mach number controller was sensing an erroneous speed due to the spacing and location of the static pitot ports in the nose boom coupled with airflow variations due to jet-stream winds. Correlation of telemetry reports of wind speed and Mach number variations supported this conclusion. Computer simulation reproduced many of the telemetry signals from GM-2005.[25] The confined airspaces of the Edwards flight test area had predicated the use of high-speed climbing turns, a maneuver the original design team had not contemplated. The recommended fixes were to move the static port orifice locations on the nose boom, reduce the Mach number control system gains to eliminate response to short term signal variation, and modify the lateral directional stabilization system to prevent gyroscope cross-coupling. A revised flight program was submitted to BuAer 5 March 1957 along with proposed fixes and Regulus II was released for further testing at the end of April 1957.

Flight testing resumed 8 May 1957, when GM-2002 reached 50,000 feet and Mach 1.8 while performing a similar flight profile to that of GM-2003 and GM-2005. Although the problems appeared to be resolved, a second flight test was planned. On 4 June 1957, the first flight of GM-2006 began with a normal taxi take-off and initial flight path with the missile reaching 35,000 feet in 4.6 minutes. When the missile attained Mach 1.1, a new problem developed: roll oscillations that lasted for 10 seconds, stopped, and then started again. Disengagement of the constant altitude controller had no effect. The lateral autopilot modification was turned off, resulting in increased amplitude rolling and pitching with loss of control 16 seconds later. Plunging from 35,000 feet, the missile impacted in the mountains near Trona, California. Telemetry data indicated that gyro coupling and gain settings were still the source of the problem. To confuse the issue even more, the next flight, the third for GM-2002 and the 17th in the program, went perfectly, executing the climbing turn at Mach 1.8 using the Mach number controller and maintaining stable cruise at 57,000 feet. The same gain settings were used as on last flight of GM-2006 when it crashed. Clearly the problem was a subtle one.

The next two flights of GM-2002 were partially successful, though the high speed oscillations again occurred. Both flights were recovered and the telemetry analyzed. The next solution decided upon was to further modify the yaw gyro. GM-2007 flew its first mission 5 September 1957 and was a complete success. This turned out to be the solution and the program did not lose another missile due to high speed lateral oscillations.

Each of the first 20 flight operations were taxi take-offs at Edwards. The first of the Lot III missiles, GM-2008, designed to be powered by the J79 engine, was also slated to be the first used for a JATO boosted launch. Five years of experience with Regulus I dual booster launching had convinced the Chance Vought engineers that the ejection of the launch rail slippers and boosters were problems that they did not want to see repeated. Bill Albrecht, a veteran of the Regulus I program and now a field engineer on Regulus II, remembers clearly the solution to the problem. He designed a booster cradle that housed both the launch slippers and the booster as one piece of equipment. In this design, the booster assembly, once free of the launcher, was held in place on the missile by the thrust of booster. Separation of the booster assembly took place when the vertical component of booster thrust became less then the weight of the booster assembly. This disengaged the aft attachment point, allowing gravity and aerodynamic drag to rotate the booster assembly away from the body of the missile. As drag pulled the booster assembly aft and away from the missile, the forward support hook would disengage and the assembly would fall clear. Albrecht remembers encountering considerable skepticism with this design, but once demonstrated in wind tunnel tests, the idea proved to be fool-proof. There were no booster-missile separation failures during the entire Regulus II test program and none recorded for any of the nearly 100 drone program launches.

Previous experience with Regulus I booster alignment permitted an elegantly simple system to be used on Regulus II. Just as with Regulus I, an A-frame hoist was used to balance the missile to determine the center of gravity (c.g.). This point varied slightly from missile to missile depending on guidance equipment and whether the missile was a tactical or flight test vehicle. Once the c.g. was determined, the location was recorded in a weight and balance data handbook that accompanied each missile to the field. With Regulus II, when booster installation was to take place, index plates with adjustable pointers were installed on each side of the missile over the c.g. and the pointers adjusted to be one-half inch below the c.g. A line was attached on each side and brought back to the booster nozzle fitting. A yoke that fit across the booster exit nozzle throat provided the second alignment point and by adjusting the angle of the adjustable exit nozzle, the lines could be made to fall across the appropriate point on the index plate.

The 4KS 115,000 lb thrust JATO booster built by Aerojet General was ready for testing by mid-1957. As with Regulus I, a shock and acoustical energy test program was run using GM-2007 mounted on a sled at an angle such that the booster thrust would be horizontal and clear of ground interference. With the J79 in afterburner, the JATO bottle was ignited. It was a short burning version of booster because the ignition shock was the only parameter being tested. No major surprises were encountered.[26]

The first boosted launch of Regulus II took place at Edwards on 13 November 1957 with the flight of GM-2008. The launch was perfect, with the missile rising quickly and smoothly, showing no attitude changes due to booster alignment errors or any problems due to booster ignition shock forces. The 49 minute flight, reaching 35,000 feet and Mach 1.1 was uneventful and the recovery was successful even though the parabrake was destroyed by the hot engine exhaust.

1958

On 30 January 1958, the first RIN flight operation was conducted. After a normal boosted launch, GM-2010 became unstable after 18 seconds under inertial guidance so control was switched to the out of sight controller station for the remainder of the flight. The rest of the test objectives were completed successfully. Review of telemetry indicated that one of the RIN system gyroscopes had tumbled and caused the con-

The first boosted launch of Regulus II took place on 13 November 1957. The 49 minute flight, reaching an altitude of 35,000 feet and a top speed of Mach 1.1 was routine. (Courtesy of Albrecht Collection)

trol problem. Six weeks later the second RIN flight was launched, again using GM-2010. The inertial guidance system operated for 102 seconds before one of the gyros was driven to its stops and locked up. Control was automatically switched to manual and the missile was recovered intact.[27]

One of the major concerns in flight testing Regulus II was the location of a suitably instrumented test range for such a high-speed, high-altitude and long range missile. One of the guidance systems on the developmental horizon was terrain following radar, the automatic terrain recognition and navigation system (ATRAN). ATRAN would provide updates to the inertial

GM 2010 is readied for launch early in the morning of 30 January 1958 displaying an olive drab paint scheme. GM 2011 was successfully launched but became unstable after 18 seconds of control during this first test of the Regulus Inertial Navigator. (Courtesy of Loral Vought Systems Archives, Kilpatrick Collection)

GM 3001 is being readied for launch on 16 July 1958. This was the first flight of the tactical version of Regulus II and the first to undergo the terminal dive maneuver. The warhead section was actually well aft of the nose, beginning at the white line just above the "star and bars" insignia. This flight ended with GM 3001 breaking up in flight shortly after beginning the terminal dive. (Courtesy of Loral Vought Systems Archives, Sutherland Collection)

guidance system by providing radar map matching input at specified flight path way points. A review of the Air Force's Cape Canaveral range being used for the first ballistic missile testing, as well as the Air Force Snark and Matador programs, indicated that the costs and restraints dictated by the Air Force, as well as the unvarying terrain of the flat ocean surface, occasionally interrupted by islands, were not suitable for radar map matching tests. The Sea Test Range at Pt. Mugu was likewise inappropriate. Virgil Ketner, the Regulus II program civilian manager at Pt. Mugu, turned his attention inland from Pt. Mugu. He toured a large area of both inland California and *Nevada*, limiting the outer edge of his search to the 500 nautical mile, Mach 2 flight range of the missile. After exploring most of *Nevada* he focused his attention on Antelope Dry Lake near Tonopah. Other than the fact that the lakebed was littered with bomb casing debris from World War II training missions, it was perfect. Ketner contacted Chance Vought to obtain its reaction prior to reporting to BuAer. Harry Hislop, a Chance Vought Regulus II engineer, was sent to inspect the site and make his recommendations in light of the rather stringent operational evaluation requirements. Hislop agreed with Ketner. The geography between a launch location at either Pt. Mugu or Edwards, and the landing/recovery site at Tonopah seemed an ideal area for testing the radar map matching system. The most desirable flight envelope envisioned was to launch at Pt. Mugu towards the sea, make a 180 degree turn to become "feet dry" near Vandenburg Air Force Base (then known as Camp Cook), fly over Edwards and land the missile at Antelope Dry Lake. A tactical missile flight could be terminated at Dugway, Utah, diving into the marshes and bury itself out of sight, an obvious security advantage.[28] The first use of the Inland Test Range by Regulus II took place 5 May 1958 with the launch of GM-2016 at Pt. Mugu. This was also the first all-Navy flight operation with the Guided Missile Unit FIFTY-FIVE (GMU-55) launch team. The flight was successful and recovery at Tonopah uneventful. The first flight of Regulus II at sustained Mach 2 speeds took place on 3 July 1958. Successful in all respects, the flight demonstrated a sustained high-speed cruise 46,000 to 52,000 feet, using the Mach number controller. The highest speed yet obtained in the Regulus II program, Mach 2.1, was achieved during the 37 minute flight. Stability was better than expected from computer simulations.

Less than two weeks later came the first tactical missile terminal dive flight using GM-3001, the first XSSM-N-9 airframe to be flown. Operational objectives centered around demonstration of a vertical dive from above 45,000 feet and cruising speed of Mach 2.0; obtaining aerodynamic and engine data for use in the design refinement of the terminal dive controller; and demonstration of the structural integrity of the XSSM-N-9 design during a terminal dive. GM-3001 was launched 16 July 1958 at Edwards. The flight profile was complex but all maneuvers prior to the terminal dive were successfully executed. At 36.6 minutes the "dump" command was issued with the missile cruising at Mach 1.87. Earl Holcomb, a safety-chase pilot for Chance Vought, remembers this first tactical operation quite vividly. He joined up on the missile at 46,000 feet in the latter part of the flight plan to observe the missile's performance as it entered the terminal dive maneuver. Jim Hayes, a Chance Vought test pilot flying secondary safety-chase, was at a lower altitude observe the terminal dive portion of the flight. As Holcomb saw the missile approaching, the afterburner in his F8U-1 flamed out. He was unable to get a re-light until he had descended 10,000 feet. The missile was a considerable distance ahead of him by this time so did not observe the beginning of the terminal dive and did not see Hayes' aircraft. The missile pitched over into the terminal dive. He recalled seeing two objects impact the ground, one some distance ahead of the other. Since he had not heard or seen anything from Hayes, he was concerned that the missile might have collided with Hayes' aircraft and both had plummeted to the ground. Holcomb made several calls on the radio before he finally got a re-

First Regulus II launch at Point Mugu 15 August 1958. This was the first flight to evaluate the Inland Test range and GM-2016 was successfully recovered at Tonapah, Nevada. Notice Regulus I on launcher. (Courtesy of Leue Collection)

GM 3003 lifts off for the first successful tactical operation flight test on 1 November 1958. The lower ventral fin was added only to the tactical missiles to increase directional stability during the terminal dive maneuver. (Courtesy of Loral Vought Systems Archives, Albrecht Collection)

King County *pierside at Port Hueneme on 5 December 1958. Main deck details clearly show the hangar installation forward of GM 2008 installed on the launcher. Directly behind the missile is a conning tower mockup fully instrumented to evaluate the effects of engine and booster rocket exhaust and temperature. (Courtesy of Pacific Missile Test Range Collection)*

ply. Hayes said that when he spotted the missile it was about to run him over. He made a violent evasive maneuver to avoid the missile, causing his plane to go into a spin. He was too busy regaining control to answer answer Holcomb's calls.[29]

Review of the photo-theodolite film after the flight indicated that GM-3001 started the terminal dive maneuver precisely as planned. Twenty-two seconds into the dive, a slight left slide-slip occurred, followed by control correction. A small piece of the missile was lost and then the side slip was repeated with corresponding control correction but by this time the amplitude of the right-hand roll rate was increasing and at 30 seconds into the dive, exceeded the telemetry instrumentation recording capability. One second later the roll rate reversed and went off the instrument scale. At 32 seconds into the dive the sequence reversed again and another small piece of the missile was lost. Then another sequence reversal took place and the missile rolled 180 degrees. Thirty-three seconds after the dump command the right wing failed at

A prototype Halibut *missile compartment being transported from the site of fabrication to the* King County *for installation during refit for the LEHI Project. The hangar is in an inverted position with the larger opening at the bottom being the missile hatch. (Courtesy of Mare Island Naval Shipyard Collection)*

the wing fold, damaging the left wing on the outboard trailing edge. Large angles of attack followed and a severe bending of the nose boom was noted. Complete structural failure took place one-half second later.[30] Despite the high-speed impact, quite a bit of the missile was recovered intact in the impact area. Analysis of the phototheodolite film indicated the anti-buzz panel had separated at the beginning of the roll oscillations. The oscillations were due to yaw-roll transients. The roll transient was probably caused by the temporary blockage of the boundary layer bleed duct by the anti-buzz panel as it failed.[31] Two fixes were recommended before the next tactical operation. Wind tunnel tests indicated that additional directional stability would be gained during the terminal dive maneuver by adding a small lower ventral strake or fin. Wind tunnel tests indicated that a relatively modest fin would be adequate. This meant that the current launch equipment could still be used in both the flight test program and aboard the three submarines under construction. The second fix was to fabricate the anti-buzz panel out of steel instead of magnesium. Torsional loads during the early stages of the dump maneuver proved to be beyond the strength of the magnesium panel.[32]

On 16 September 1958, the first and only launch of a Regulus II from a submarine took place. GM-2016 was launched from USS *Grayback* (SSG 574), commanded by Lieutenant Commander Hugh Nott, off of Pt. Mugu (see Chapter Eleven). Holcomb was flying ABLE and Bill Coleman, another Chance Vought test pilot, was flying BAKER in the TV-2D. Twenty-five seconds after launch Holcomb turned control of the missile over to the out of sight controller station at Pt. Mugu. As the GM-2016 flew its planned course, Holcomb and Coleman proceeded across the mountains to rejoin the missile over Harper's Dry Lake in preparation for landing. The recovery intercept took place as planned and Holcomb rolled the missile out on a turn for Runway 26 at Edwards. He advanced the missile's throttle to military power and commanded the missile landing gear down. Much to his surprise, the gear failed to extend, the first such failure in the program. He cycled the landing gear selector with no response from the missile. He transferred control to the standby recovery TV-2D with no response. With control of the missile returned to him, Holcomb tried one more time to drop the landing gear without success. Holcomb turned the missile towards Rosemond Dry Lake for a wheels-up landing. The missile touched down gently, slid along the lakebed, and then nosed down, coming to rest intact. A small fire started but the missile did not burn.

After nearly four months of investigation and decisions on modifications to prevent the tactical missiles from breaking up during the terminal dive maneuver, the second tactical missile flight test took place on 1 November 1958 with the launch of GM-3003 from Edwards. The Edwards out of sight controller transferred command to Tonopah at 14.5 minutes into the flight. The terminal dive was a complete success with the missile impacting at Mach 1.6 and disappearing into the soft soil. When a team went to look for the wreckage, none was found. A slight crater lip was the only evidence of impact.[33,34] The third tactical mission followed eighteen days later with the flight of GM-3002. The missile was under inertial guidance all the way to target and made a successful terminal dive. The Regulus II tactical missile appeared to have its problems solved and was now ready for the final Navy test and evaluation phase.

The fifth flight test of the inertial guidance system was on 3 December 1958. The operation was highly successful with the missile assuming a great circle navigational heading at launch and acquiring the required track after only ten minutes of flight. Successfully recovered at Tonopah, the recent success of the tactical missiles and the inertial guidance system seemingly demonstrated that the program had solved its teething problems and was ready for fleet operational evaluation and acceptance.

Part of the Regulus II fleet acceptance process was the establishment of a Navy Regulus II training facility and program. GMU-55 had as-

GM 2008 is brought up on deck and onto the launcher. The entire system was automatic, including wing unfolding, except for folding and locking the nose boom into place. (Coutesy of Cannon Collection)

The missile control room aboard King County *during a countdown run through on 9 December 1958. L to R: Lt. P. Fullinwider, Missile Officer; Lt. Cmdr. E. G. Jones, Safety Observer; Lt. G. E. Mueller, Assistant Missile Officer; Lt. C. R. Knutson, Test Conductor and a civilian observer. The launch was aborted when an indicator lamp failed and the parabrake parachute popped out before launch. (Courtesy of Fullinwider Collection)*

On December 1958 the second and final shipboard launch of Regulus II took place. King County *is anchored off of Pt. Hueneme, just prior to launch. (Courtesy of Cannon Collection)*

GM 1008 leaps off the launch with normal launch altitude. Note the television camera platform directly beneath the missile. (Courtesy of Cannon Collection)

Smoke begins to engulf King County *as GM 1008 clears the ship and heads towards out over the ocean before turning inland. (Courtesy of Cannon Collection)*

King County *nearly engulfed in smoke from the JATO booster exhaust. The flight was completed successfully with a perfect recovery at EAFB. (Courtesy of Cannon Collection)*

Artist's conception of the Permit Class submarine design to house four Regulus II missiles. This concept was not implemented due to the cancellation of the program in December 1958. (Author's Collection)

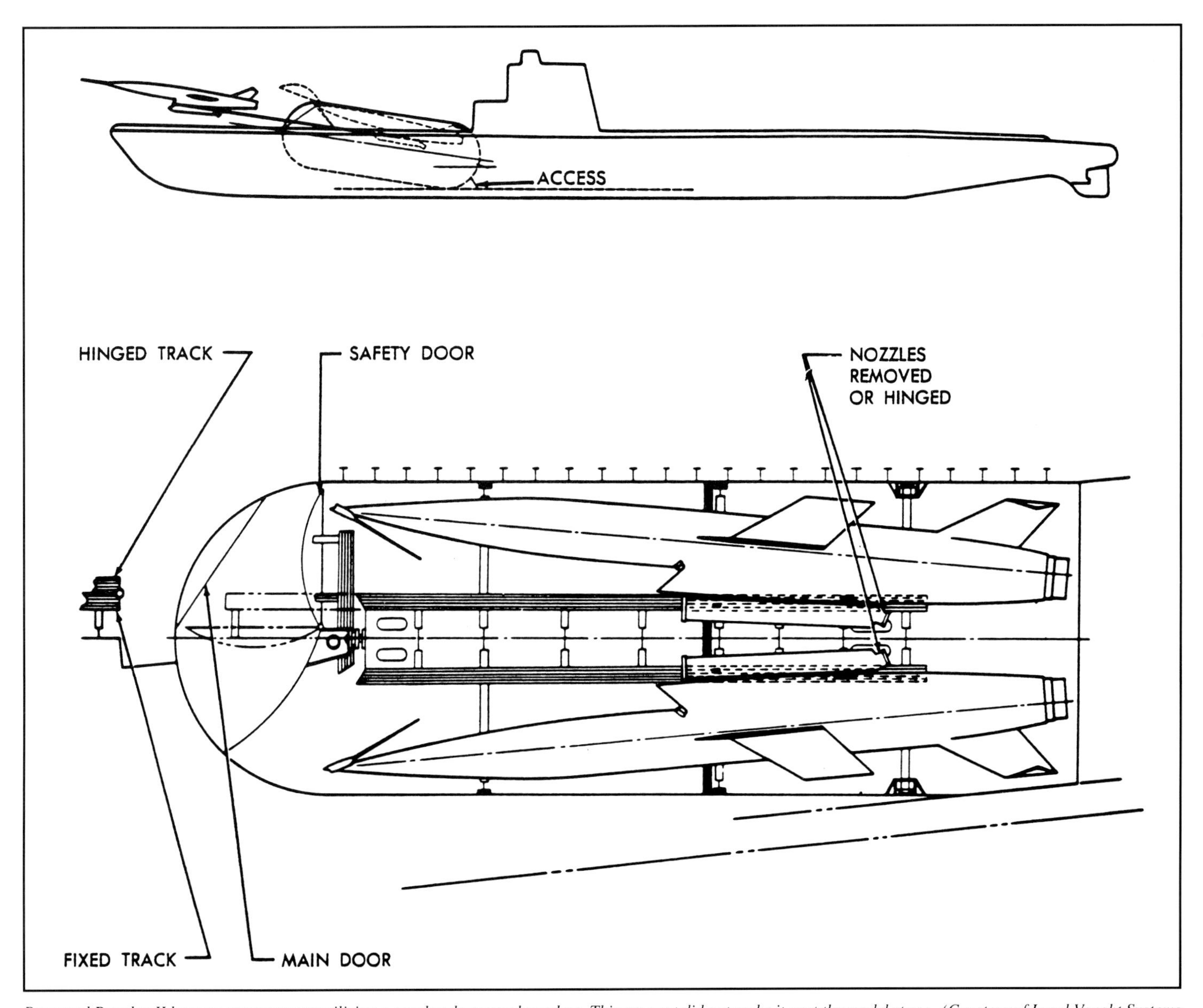

Proposed Regulus II hangar arrangement utilizing a quadruple rotary launcher. This concept did not make it past the model stage. (Courtesy of Loral Vought Systems Archives, Author's Collection)

sisted in launching several missiles, including the launch from *Grayback*, but the first complete GMU-55 operation took place on the USS *King County* (AG 157). Given the code name Project LEHI, the program involved the conversion of a World War II Landing Ship Tank's cargo hold into a replica of the missile hangar of a *Halibut* class guided missile submarine. In 1956, striving to ensure that everything would go as smoothly as possible in outfitting the first of four proposed nuclear powered submarines with the Regulus II weapon system, project officials at Pt. Mugu recommended that a complete steel mock up of the missile hanger and launch equipment be built for design evaluation. This would insure that the blueprint specifications and clearances were correct for the submarine installation when the time came. The resulting system, placed in a suitable surface ship, could then be used for training Regulus II launch crews.

The Commanding Officer of *King County* during conversion was Lieutenant Sumner Gurney. In June 1958, he was relieved by Lieutenant Joseph Metcalf III. Metcalf, fresh from teaching at the Naval Academy, found that none of his five officers and few of the enlisted men had been to sea before. The Executive Officer had been on a midshipmen cruise but that was all. The first trip out of San Francisco Bay was to the Farallon Islands. They cruised back and forth near the islands, standing watches, getting seasick and basically learning how to operate the ship. They repeated these trips as often as the ship's conversion process would permit.[35]

The first missile related operation was the firing of dummy "mass" sleds into the Carquinez Straits mudflats near the Mare Island Naval Shipyard. They fired both Regulus I and II dummy sleds to check launcher capability. Hangar door clearance for the nose of the Regulus II was found to be too tight so a slight dip was placed in the tracks used to move the missile from the hangar to the launcher and the problem was solved. Clearly the concept of Project LEHI was beginning to pay off.

King County sailed at 0800 hours on 9 December 1958 for her first Regulus II launch. Aboard for this event were Prospective Commanding Officer of USS *Halibut* (SSGN 587), Lieutenant Commander Walt Dedrick and his Assistant Missile Officer, Lieutenant B.M. Eppler. In addition, a full complement of area newsmen and photographers were present to record this milestone in the program. With fifteen minutes remaining to launch, the parabrake parachute ejected prematurely, causing the launch to be scrubbed. The malfunction was traced to an oversize light bulb in the missile launch control console which had caused a short circuit. The operation was rescheduled for the following day.

At 1188 hours on 10 December, the second and last shipboard launch of a Regulus II took place, as Lieutenant Ray Arison of GMU-55 pushed the launch button and GM-2008 roared off the *King County* launcher, enveloping the ship in white smoke. Sensors in the submarine sail mock-up aft of the missile monitored the blast and heat effects from the JATO booster rocket exhaust and prolonged exposure to the exhaust of the J79 engine during engine run up. GMU-55 passed control of the missile to a TV-2D after the first several seconds of flight. The TV-2D passed control to the out of sight controller controller at Pt. Mugu. In the interim, an

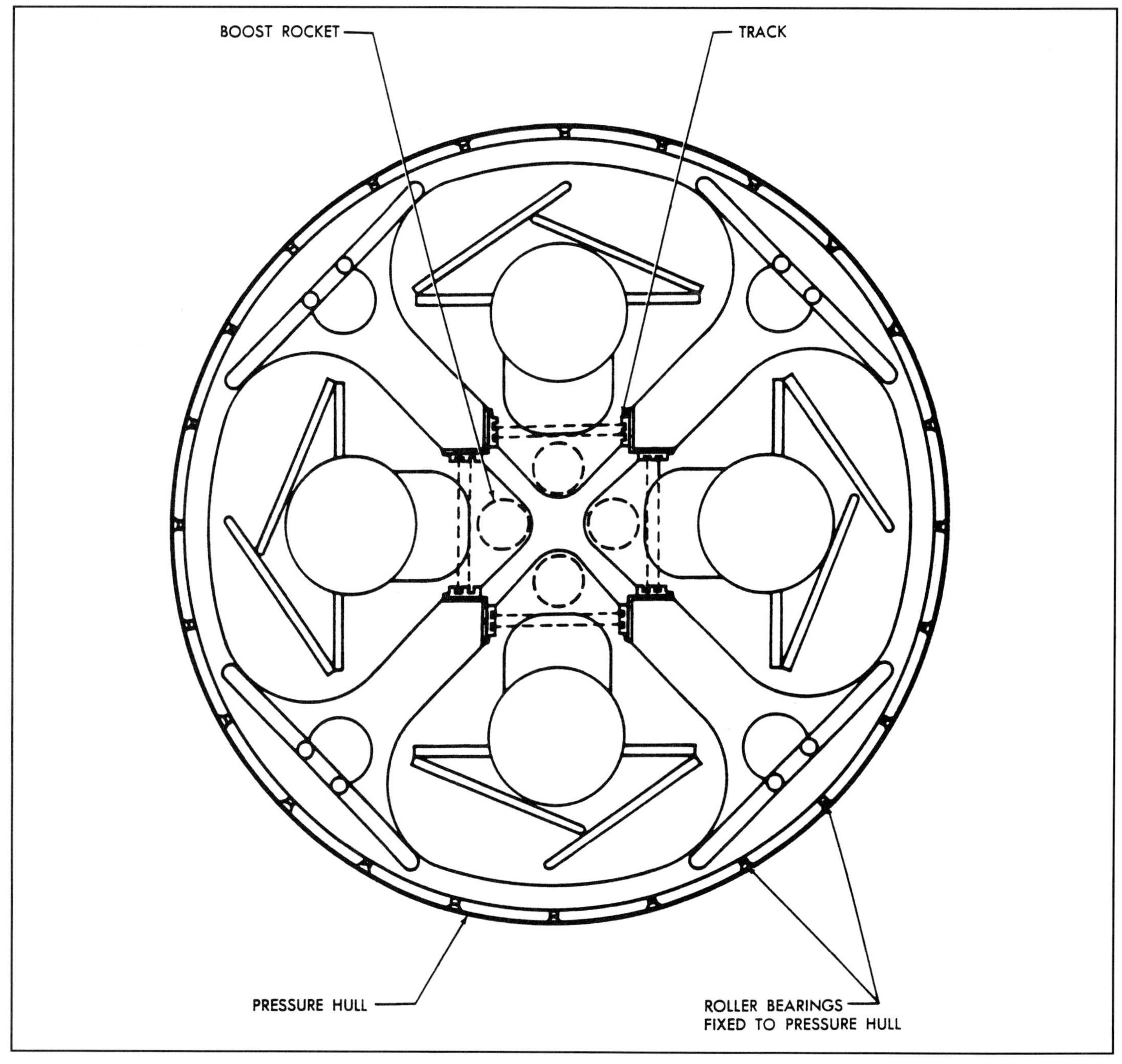

Front view of proposed Regulus II hangar arrangement utilizing a quadruple rotary launcher. This concept did not make it past the model stage. (Courtesy of Loral Vought Systems Archives, Author's Collection)

F8U-1 chase plane swooped in to try to keep the missile in visual contact. At T plus 14 minutes control of the missile was passed to the out of sight controller station at Edwards. At T plus 17.5 minutes GM- 2008 was cruising at 41,000 feet at a speed of well over Mach 1.0. The F8U-1 chase plane was having difficulty keeping the missile in sight, and radioed to Edwards to throttle GM-2008 back a bit. At T plus 22 minutes, control was again transferred, this time to the out of sight controller at Tonopah, a mobile ground controller stationed next to the recovery runway. This controller brought the missile into the final approach pattern and then gave control to a TV-2D recovery aircraft. An excellent touchdown and roll-out was achieved, a perfect ending to a very significant mission.[36]

The fourth tactical missile, GM-3004 was successfully launched on 11 December 1958. GM-3004 flew a 627 mile flight from Edwards to the Dugway, Utah, impact site and successfully completed the terminal dive maneuver without any sign of flight instability. With eight consecutive successful flights, including three tactical missions, Regulus II seemed to be on the way to the fleet.

Cancellation

On 12 December 1958, Secretary of the Navy Thomas S. Gates announced the cancellation of the Regulus II missile program.[37] When asked why the program was being terminated on the eve of reaching operational status, Secretary Gates explained that rapid advances being made in the Polaris fleet ballistic missile program warranted the full attention of available funding focused on that program. In reality, the Secretary of Defense, Charles E. Wilson, had authorized the funding for Polaris but on the condition that the funding come from within the Navy's current budget.[38] The Navy had estimated over $100 million could be saved with termination of Regulus II at this point in the program, most of this savings coming from cancellation of ship modifications.[39] For Regulus II's projected operational deployment, four nuclear attack submarines, one nuclear powered cruiser, two new construction diesel submarines, and six existing conventionally powered cruisers were to have been either constructed or converted to launch Regulus II. Of this number, two diesel powered attack submarines and one nuclear submarine had been constructed or were in construction, the *Grayback*, USS *Growler* (SSG 577) and *Halibut*. Previously authorized, but now canceled were three Permit class nuclear

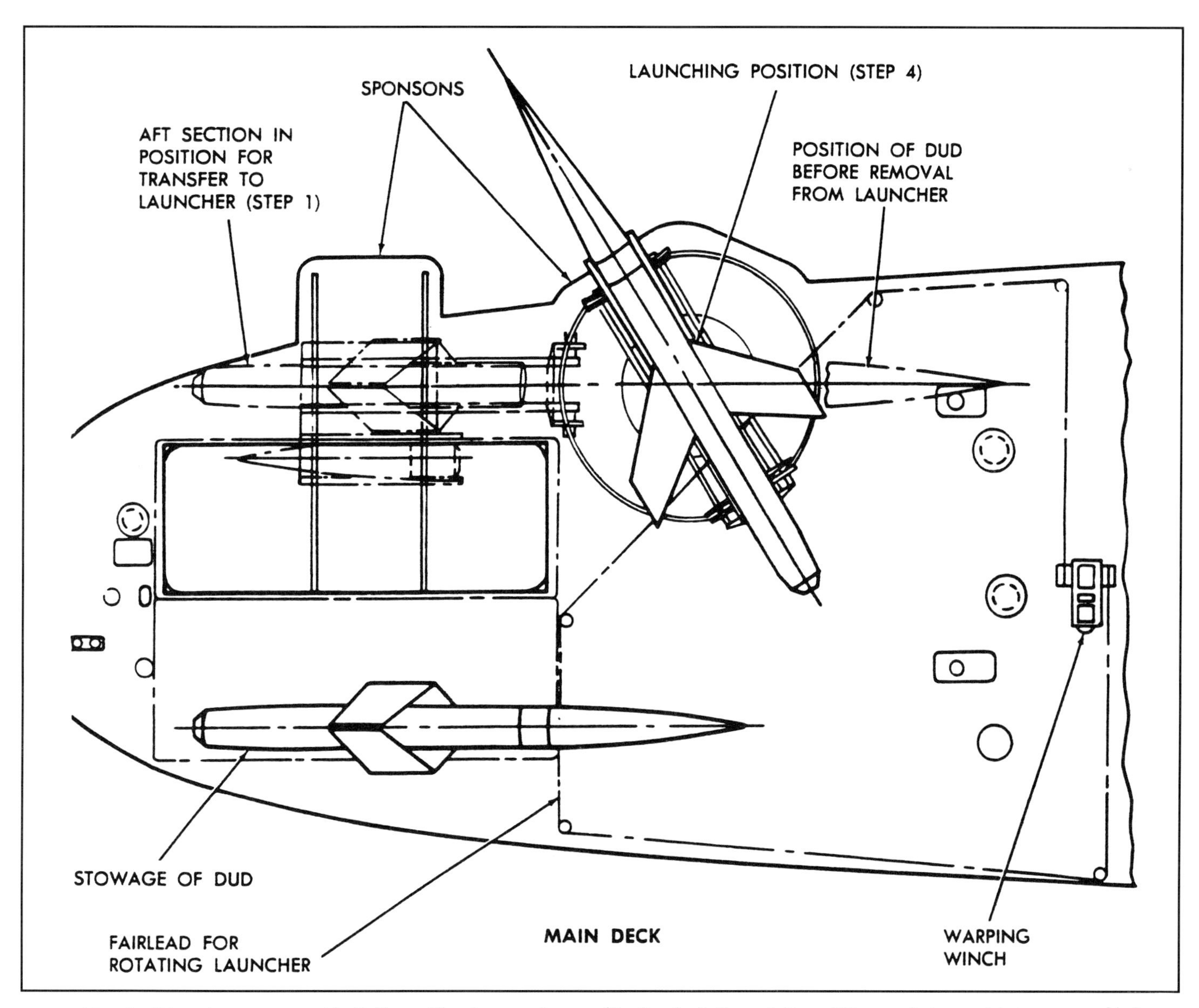

Proposed Regulus II launcher arrangement for Baltimore Class heavy cruisers used for Regulus I. Four missile could be stored. A second design, presumably for use beginning with USS Long Beach *(CGN 9), placed five missiles in a main deck hangar. (Courtesy of Loral Vought Archives, Author's collection)*

submarines, USS *Permit* (SSGN 594), USS *Pollack* (SSGN 595), and USS *Plunger* (SSGN 596). The USS *Long Beach* (CGN 9), under construction at the time, was a nuclear powered guided missile cruiser that had been designed to carry five Regulus II missiles in an amidship hangar. *Long Beach* would now be completed without Regulus II equipment. Six guided missile cruisers: USS *Albany* (CG 10), USS *Fall River* (CG 11), USS *Chicago* (CG 12), USS *Columbus* (CG 13), USS *Oregon City* (CG 14) and USS *Bremerton* (CG 15) had been due to be converted to carry four Regulus II missiles each, these conversions were likewise cancelled.[40]

The cancellation of Regulus II was not a capricious decision. In mid-1955, the newly appointed Chief of Naval Operations, Admiral Arleigh A. Burke, had reviewed the current cruise missile projects; Regulus I, Regulus II and Triton, a ramjet powered missile still in an early developmental stage. While funding was available, all three systems were continued while Burke agreed to work with the Army in developing an intermediate range ballistic missile. If the fleet ballistic missile program ran into insurmountable difficulties, these systems needed to be kept as alternatives. Within a year the Navy was working on its own Polaris submarine launched ballistic missile. In August 1957, technical difficulties in the ramjet powerplant of Triton caused cancellation in favor of additional spending on Polaris. Simultaneous with this cancellation, Admiral Burke approved the construction of three nuclear-powered Regulus II submarines. In April 1958, Admiral Burke approved installation of Regulus II in 6 cruiser conversions over the next three years.[41]

On 19 September 1958, the Naval Warfare Analysis Group released a report on the use of Regulus I and II as an interim deterrent weapon system during the development of the fleet ballistic missile system. At this point the build up in the Polaris program to a sufficient number of submarines was projected to take well into 1965. An alternative would be to use Regulus I and II deployed on converted Victory hulls to bridge the deterrent gap. Careful analysis had indicated that the Victory ships would be easily "lost" in the maritime traffic and in the event of a nuclear war, enough would survive the initial hours of the war to launch their missiles. Targets would be either coastal cities or targets up to 1000 miles inland.[42]

Within the ensuing three months the three nuclear-powered submarine program was cut to one and then placed on hold. By mid-November, it was evident that with the stunning advances made in the Polaris program, the considerable costs of Regulus II submarine and cruiser construction, let alone a dispersed surface ship program, required a choice to be made; Polaris or Regulus II but not both. With the aid of historical hindsight, one can easily argue that the correct decision was made. At the time, however, to many individuals in the Defense Department and Congress, it seemed that the Navy was placing too great an emphasis on the promise of Polaris rather then the reality of Regulus II.[43] One month after the program cancellation, Admiral Burke testified before the Senate Armed Services Committee at the request of Senator Lyndon B. Johnson. Admiral Burke praised the overall Regulus II program, pointing out that 90% of the inertial guidance development work had been completed and 50% of the fleet test and evaluation program was finished. Deployment to the fleet, scheduled for fiscal year 1960 had been on track. The bottom line was that Polaris had a "much longer anticipated service effective life, due to the relative vulnerability of Regulus II to anti-aircraft defensive measures."[44]

In 48 missile flight test operations, Regulus II recorded a 83% launch reliability rate when a missile reliability of 50% was considered excellent. Of the 48 missile flights, 30 were successful in that all primary objectives were achieved; 14 were partially successful and four were failures. The last eight flights were successful. These launches included both Chance Vought and Navy operations, indicating that the program had truly matured. Forty-seven Regulus II airframes in various stages of construction remained at cancellation, with seven fleet training missiles ready for flight and seven in storage. In addition there was one tactical missile ready for launch and five in storage.[45]

Endnotes

[1] *Regulus I had been conceived as a subsonic missile for relatively quick fleet introduction while supersonic ramjet technology for use in the Triton and Rigel missiles was being evaluated. Chance Vought realized that to stay competitive, Regulus needed to grow into a supersonic cruise missile.*
[2] *Regulus II Program History, page 1, LVSCA A50-24, Box 7.*
[3] *Ibid., page 2, 1.1.-1.2.*
[4] *Ibid. page 1.3.*
[5] *"The Evolution of the Cruise Missile," Kenneth P. Werrell, Air University Press, 1985, page 93.*
[6] *Regulus II Program History, page 12.1.1, LVSCA ASO-24, Box 7.*
[7] *Ibid., page 9*
[8] *Ibid., page 10.1*
[9] *Ibid., pages 6.4-6.10*
[10] *Regulus I Progress Report #14, 1 January to 31 March 1956, page 17, author's personal collection. Regulus II program work was often part of the concurrent Regulus I contract funding.*
[11] *V-398 B-52 Air-to-Surface Missile Charts for WSEG Presentation, Report Number 10596, 20 March 1957. LVSCA ASO-22, Box 5.*
[12] *Regulus II Progress Report Number 18, 1 January through 31 March 1957, page 15, LVSCA A50-22, Box 5.*
[13] *Regulus II Program History, page 10.1, LVSCA, A50-24, Box 7.*
[14] *Regulus II Progress Report Number 18, 1 January through 31 March 1957, page 15, LVSCA A50-22, Box 5.*
[15] *The dimensions are given for the tactical missile only. The major difference between missile airframes was that the J65 powered flight test missile was 55.83 feet long, not including the noseboom.; the J79 powered missiles, both the flight test and tactical variants, were 57.5 feet long, not including the noseboom. The extra length was necessary to accomodate the J79 engine.*
[16] *Ibid.*
[17] *Regulus II Program Progress, 30 January 1957, pages 2-4. Chance Vought document provided by George Sutherland.*
[18] *Personal interview with George Sutherland.*
[19] *Regulus II Program History, undated, Exhibit 11C, LVSCA A50-24, Box 7.*
[20] *Personal interview with George Sutherland.*
[21] *Ibid.*
[22] *Regulus II Program History, undated, page 11.7, LVSCA A50-24, Box 7.*
[23] *Ibid., page 11.8*
[24] *Regulus II flight test operations conducted 13 launches in 1956 with seven successful recoveries, one missile lost during recovery and three missiles lost during flight. Four missiles were used with GM-2001 launched six times and GM-2003 launched four times.*
[25] *Report of Accidents, XRSSM-N-9, GM-2003 (FTM-3), XRSSM-N-9, GM-2005 (FTM- 5), 19 January 1957, Chance Vought Report 10777, LVSCA ASO-22, Box 7.*
[26] *Personal interview with George Sutherland.*
[27] *Ibid.*
[28] *Personal communication with Virgil Ketner.*
[29] *Personal communications with Earl Holcomb.*
[30] *Regulus II Analysis of Terminal Maneuver of GM-3001, 22 September 1958, Chance Vought Report E8R-11734, page 10-13, LVSCA ASO-24, Box 5.*
[31] *Ibid.*
[32] *Ibid., page 47.*
[33] *Personal interview with Ace Yeager, Chance Vought primary safety-chase pilot on this flight.*
[34] *Archival film of this terminal dive shows the missile appearing to enter the ground in one piece with a small fireball of unused fuel igniting and then quickly extinguishing.*
[35] *Personal interview with RADM Joseph Metcalf III, USN (Ret.)*
[36] *Ibid.*
[37] *REGULUS II Guided Missile Weapon System; directive concerning (U). From: Secretary of the Navy; to: Chief BuShips; Chief, BuAer; Chief, BuOrd. OP-514B/11h ser 444P51, 12 December 1958. Naval Historical Center, Operational Archives.*
[38] *"Politics and Force Levels: The Strategic Missile Program of the Kennedy Administration," Desmond Ball, University of California Press, 1980, page 61.*
[39] *Regulus Cancellation Ends 10-Year Program, 29 December 1959 Missiles and Rockets, page 24.*
[40] *Regulus II Weapon System Program Review Conference, 12-14 August 1958. Briefing "slides" with this information are at the back of the booklet, unpaginated. LVSCA A50-24, Box 7.*
[41] *Memorandum for ACNO (General Planning) (OP-90), subject: Regulus II Chronology and Funding History (U), Op-514D/mb, Ser 0456P51, 18 December 1958, pages 1-7. National Archives, Bureau of Aeronautics, Guided Missile Divison, Record Group 72, Box 96.*
[42] *Memorandum for the Director, Naval Warfare Analysis Group; subject: A Regulus II Interim deterrent weapons system, 19 Sept. 1958, (NWG)121-58, pages 1-28. National Archives, Bureau of Aeronautics, Guided Missile Division, Record Group 72.*
[43] *"The Development of Navy Strategic Offensive and Defensive Systems" by Captain D.A. Paolucci, USN (Ret.); May 1970; U.S. Naval Institute Proceedings, Volume 96:204-223.*
[44] *The Regulus Program, testimony to the Senate Armed Services Committee, January, 1959; Admiral A. Burke, CNO. page 6.*
[45] *The final 1959 budget proposal for Regulus II had been $139 million to purchase 68 missiles. Earlier proposals had envisioned production of as many as 113 missiles for $219 million. Memorandum for ACNO (General Planning) (OP-90), subject: Regulus II Chronology and Funding History (U), Op-514D/mb, Ser 0456P51, 18 December 1958, pages 1-7. National Archives, Bureau of Aeronautics, Guided Missile Divison, Record Group 72, Box 96.*

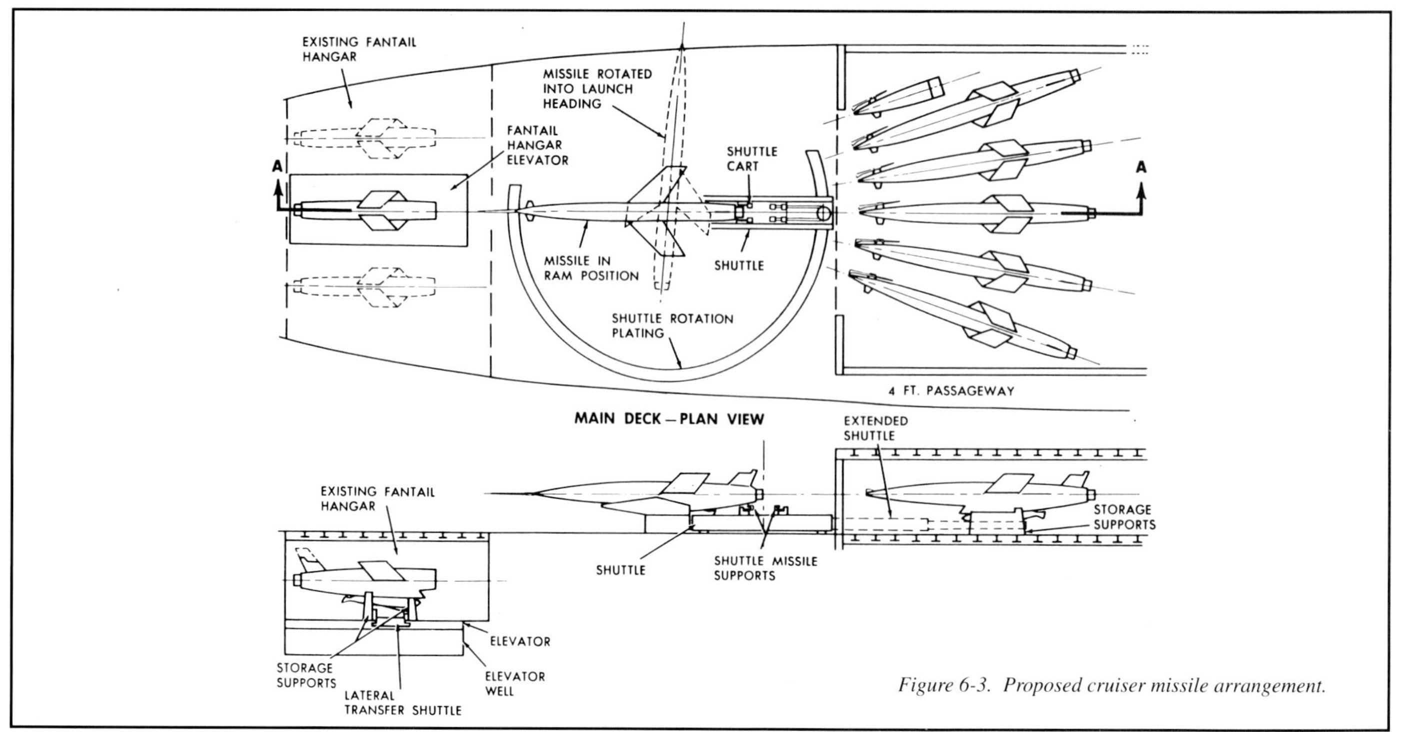

Figure 6-3. Proposed cruiser missile arrangement.

Chapter Seven: Regulus II After Cancellation

Immediately after cancellation of the Regulus II program, the Navy had to decide how to dispose of the 24 fleet training missiles and 23 tactical missiles that were either partially or nearly complete.[1] Simultaneously, Chance Vought began to look for options that would keep the missile assembly lines in operation as well as find other uses for the stockpiled missiles that would include employment for the Regulus II field operations personnel.

Navy Target Drone Program

Use of Regulus II as a high-speed target drone to challenge the new air-to-air guided missiles such as the Navy's Sidewinder IC, Talos, Sparrow II and the USAF ASG-18/GAR-9 was an easy choice to make since all the Regulus II operations facilities at Pt. Mugu, including chase and recovery aircraft, were still available. Conversion to the KD2U-1 target drone configuration included addition of radar tracking beacons and near-miss indicators so that drones did not have to be expended by direct hits. The first target drone launch was in January 1959.

Lieutenant Don Barrett of Guided Missile Group ONE, the chase and recovery support unit for Regulus II, was the primary chase pilot for these missions from 1962 to 1963. Barrett recalls that for the Air Force ASG-18/GAR-9 air-to-air missile, produced as the Hughes AIM-47A Falcon, test flights were particularly interesting because the "shooter" aircraft was the Convair B-58 "Hustler," the most advance Air Force bomber. The KD2U-1 was modified to carry a passive Luneberg Lens in a small wing-tip pod, increasing the radar "signature" or reflectivity of the drone. The B-58, based at EAFB, had been extensively modified with a prototype F-108 aircraft cockpit complete with an ASG-18 radar. On a typical mission the B-58 would depart Edwards and proceed to Yuma, Arizona. From Yuma, on a direct course to Oceanside, California, the B-58 climbed and accelerated, attaining the desired speed and altitude upon reaching Oceanside. Meanwhile, Barrett, flying an F8U-1 airborne control aircraft, brought the KD2U-1 to 35,000 feet at Mach 0.92 while crossing the Santa Barbara airway opposite Pt. Arguello. He then turned the drone south for the target presentation. At this time, control of the drone was transferred to the out of sight control operator at Pt. Mugu. As the target drone was brought to full afterburner and commenced a climb to 50,000 feet, Barrett cleared the range. Upon reaching 50,000 feet, the drone was traveling at Mach 2.0. The result was a head-on presentation for the B-58 starting just north of Oceanside with a closing rate of Mach 4.0 and a 100 mile separation. The B-58 was not cleared to fire at the drone until it was within the Pacific Missile Range's live firing area. At this time, the B-58 radar operator would track, identify and lock onto the KD2U-1 in preparation for missile launching. Once cleared to do so by the Range Safety Officer, the B-58 could then fire the GAR-9, one of the few times the drones were actually live targets for shoot down. All this had to take place in slightly more than two minutes due to the high target closure speed.[2]

On 13 December 1965, GM-2048 was flown as an MGM-15A drone for the 10th and final time. The drone was fired at five times during the flight as it traveled at Mach 2, at 58,000 feet and finally guided to a landing on San Nicolas Island. This was the end of the

Out-Of-Sight-Control (OOSC) station at Venice Airport, Florida. Gerry Parsons, at the console and Bill Sheperd, prepare for flight operations. While escort planes were still necessary for safety reasons, OOSC operations with the KD2U-1 drones were extremely successful during the Florida operations. (Courtesy of Loral Vought Systems Archives, Parsons Collection)

KD2U-1 (GM 2008) being readied for the first flight from the Venice Airport facilities on 3 September 1959. The BOMARC intercept was recorded as a near miss, and the missile was recovered sucessfully at Eglin Air Force Base. (Courtesy of Loral Vought Systems Archives, Parsons Collection)

Regulus II drone operations at Pt. Mugu. GM-2048, first launched in May 1964, was the remaining survivor of the Regulus II missiles modified to serve as drone targets. The Navy launched 17 drones on 64 flights, several drones accumulated as many as 10 flights, others only one. GM-2048 is currently in storage at the New England Air Museum, Windsor Locks, CT.[3,4]

Regulus II and the Air Force BOMARC Interceptor Program

In March 1959 Chance Vought received an inquiry from the Air Force concerning the possibility of using Regulus II as a high-speed target for training BOMARC (IM-99A) ground-to-air interceptor missile crews at the Air Proving Ground, Eglin Air Force Base (Eglin), Florida.[5,6] BOMARC-A was a Mach 2.8 anti-aircraft interceptor with a range of 250 miles and a ceiling of 60,000 ft. Boosted by a hypergolic propellant liquid-fueled rocket to supersonic speed, two ramjets then propelled the missile to intercept incoming fleets of bombers. BOMARC flight testing had begun in February 1955 with unguided flight test vehicles. At the time of the proposal, 140 missiles in five squadrons were planned and the Air Force needed to train the operational crews with realistic targets.[7]

Chance Vought received the go-ahead from BuAer and prepared a proposal for KD2U-1 reconfiguration and operations in Florida. With the Air Force desiring launch operations to begin 1 August 1959, time was of the essence in getting the program underway. The Air Force gave Chance Vought four months and six drone flights to prove that KD2U-1 could be a reliable target drone for BOMARC training purposes. Chance Vought's Venice, Florida drone operation was divided into three parts under the overall direction of Robert Grafton, a former missile operations engineer for both the Regulus I and II programs. Dick Witte was manager of the Venice unit where the missiles were maintained and launched. Gerry Parsons was manager of out of sight control operations. He controlled the drone from launch to approximately 150 miles north and out over the Gulf of Mexico. The second out of sight control operator, George Aitchison, took over control from his station at Fort Walton Beach and controlled the missile to the vicinity of Eglin AFB where he turned it over to the recovery pilots. E.H. "Ace" Yeager, a veteran test pilot from the Regulus I and II programs, was manager of the recovery team along with pilots Joe Engle, Jim Hayes and Glen Paulk, all veterans of Regulus operations at Edwards. The F8U-1 chase aircraft and TV-2D recovery aircraft were stationed at Eglin as was the F8U-1 high-speed airborne control aircraft.[8]

The target drone launch site was a strip of land between the Venice, Florida, airport and the beach. A clearing surrounded by trees was used for engine run-up tests with a nearby building able to house three drones for maintenance and repair. The drones were launched on a southwest heading from the airport launch pad, turned to a northwesterly course for the specific mission profile and then were recovered at Eglin AFB. The BOMARC missiles were controlled by the new computerized Semi-Automatic Ground Environment complex at Gunter Air Force Base, Montgomery, Alabama, about 150 miles away. This was one of the first uses, by the Air Force, of a centralized computer system to coordinate the interception of enemy bombers by guided missiles. After recovery, the KD2U-1's were placed in their shipment containers and trucked the 450 miles from Eglin to Venice for refurbishment and subsequent reuse.[9]

The first missile arrived 2 July 1959, having been flown to McDill Air Force Base in an Air Force C-124 "Globemaster" and trucked to the Venice facilities. It was GM-2008, the missile launched from USS *Grayback* (SSG 574) 10 months earlier. Since BOMARC-A was designed to target large bomber forces and not individual cruise missiles, modifications were made to the missiles to convert them to suitable target drones. These changes involved installation of target scoring aids and radar signature enhancement. Target scoring aids included wide-vision photographic equipment as well as miss distance indicators. Luneberg lens devices were added to enhance radar signature.[10,11]

C.J. Bitter recalls quite clearly the shipment of Regulus II launch equipment from Pt. Mugu via Chance Vought facilities at Dallas, Texas. Organization and construction of the Venice facilities were nearly finished with the single remaining task of installing the launcher on the newly constructed pad at Venice. Bitter recalls that early one morning in late June the phone rang at his Eglin AFB office. The caller identified himself as an employee of the Louisville and Nashville Railroad and wanted to talk with someone from Chance Vought. Bitter identified himself and the caller informed him of a train derailment at Montgomery, Alabama, that had left two railroad cars identified as carrying Chance Vought equipment, upside down in a ditch. Bitter replied that he was leaving immediately.

When he arrived at the wreck, he saw that the railroad employee had not exaggerated. The train cars were indeed upside down in a ditch; and, worse yet, they had been twisted to such an extent that the doors would not open. Luckily the two cars were wooden-sided. After borrowing a fire axe from the railroad team at the site, Bitter chopped access holes in both cars and crawled in with a flashlight to assess the damage, expecting the worst. Much to his surprise, all the heavy equipment was suspended on chains, hanging down from what had been the floor of the car. The Chance Vought shipping department had apparently reloaded the cars at Dallas. They had cut holes in the floors and chained all the heavy equipment to the main frame of the cars. All of the smaller equipment had been crated and equally well secured. Except for the lubricating oil that had run out of the mobile gas turbine compressor truck and some bent safety rails and walkways on the launcher, no significant damage had occurred.

As he emerged from the last of the train cars, he realized that he still had a mountain of red tape to clear and only two days to do so and get the equipment to Eglin and Venice. Bitter called the railroad management and told them the situation. He was willing to sign off on any liability for the damage if the railroad would get the shipment to its destination within the next two days. Within hours a crane had arrived, new railroad trucks were placed on the rails and the cars righted and placed back on the track. Towed to the nearby yard, they were delivered on time to Venice.[12]

On 3 September 1959, GM-2008 climbed away in a picture perfect launch. The BOMARC intercept was recorded as a near miss and GM-2008 recovered successfully at Eglin. One week later came the second launch with the second KD2U-1, again a good flight with successful recovery. With one and a half months of the four month trial period gone and two of six successful launches under their belts, the Chance Vought team felt they had the beginnings of a successful program. On 17 September the picture changed dramatically. GM-2008, on its eighth flight and second target drone mission, suffered a direct hit by a BOMARC A and was completely destroyed.[13] On 8 October 1959 the third drone flight was successfully launched. The first BOMARC A fired had to be destroyed by the Eglin Range Safety Officer but the second one, fired 20 seconds later, recorded a near miss and the drone was recovered successfully. On 15 October 1959, a KD2U-1 was lost due to fuel exhaustion O.5 miles from the Eglin runways. Only one target drone remained with one launch in the six-flight series yet to be conducted. Since the program had been much more successful than the Air Force had anticipated, the original six month contract was extended to January 1960. Three weeks later this extension was increased to 15 months as the BOMARC A launch pads at Santa Rosa Island had to be redesigned, delaying the sixth launch by several months. On 22 March 1960 the flight program resumed with target drone inventory increased to four. The last flight of the KD2U-1 BOMARC program took place on 31 September 1961 with the missile lost on recovery. Only three missiles of the thirteen brought to Venice remained. After a series of three and six month extensions of the original three month contract, totaling 27 months of operations, including 46 KD2U-1 launches, the Air Force did not require KD2U-1 target drone services as QF104 and QB-47 aircraft were becoming available for use as drones.[14]

A Regulus II KD2U-1 drone gets its center-of-gravity determined at Venice facilities prior to attachment and alignment of booster rocket. The slight hump seen just behind the front fins was a dorsal compartment that ran the length of the missile, providing increased space for instrumentation such as miss indicators and wide angle cameras for recording missile intercepts. (Courtesy of Loral Vought Systems Archives, Parsons Collection)

Naval Air Station, Roosevelt Roads

Regulus II KD2U-1 drone operations then moved to Naval Air Station Roosevelt Roads, Puerto Rico. Yeager continued in the program as a pilot and was also Field Manager for the entire operation. Flight operations began in February 1962. The KD2U-1 was a target drone for Tartar, Terrier and Talos anti-aircraft missiles fired from destroyers and cruisers. Twenty KD2U-1 drones were launched during the 12 month duration of the program, but detailed record of drone operations are not available.

The flight program for the Regulus II drone operations, both in Florida and Puerto Rico, had an impressive safety record. There were no aircraft or personnel losses during the two and one half years of operations, a tribute to the skill of both the Chance Vought maintenance and operations staff as well as those of the Air Force and Navy.[15]

NATO Sea and Land-Based Regulus II

Within a week of the program cancellation, Chance Vought approached BuAer with a proposal for the production and deployment of Regulus II in Europe under the control of NATO. At the time of cancellation, 47 missiles were either complete, in storage or 98% fabricated. All tooling and production facilities were intact and significant quantities of test and support equipment had been placed in storage. Requesting a sixty day delay in completing the program termination process, Chance Vought wanted to transfer all of the production and support equipment to a NATO country or countries. The entire production of missiles and operation of the European based system would be NATO controlled. BuAer agreed and the proposal was presented to NATO in March 1959.[16]

Chance Vought proposed both a land and sea-based European Regulus II deterrent force. The land based force would be developed in two phases. Phase I would utilize the existing airframes and support systems to train and equip two squadrons of eight missiles each in either mobile or fixed-base installations. Phase II would provide for the production of Regulus II in a NATO country or countries and permit expansion of the deterrent program to the limits necessary for NATO's defense posture. With missiles based in France, West Germany or Turkey, virtually all of Eastern Europe and the Soviet Union, well past Moscow, could be covered by this land-based "counter-force." The sea-based program was focused on the aircraft carriers of the French Navy. One or more would be turned into Regulus II launching ships with provisions for a dozen or more missiles and several launch stations.[17] The NATO presentation was well received but due to political differences and the question of control of the nuclear warheads by the United States, the program was not taken any further. It is somewhat ironic that 27 years later, Air Force ground-launched cruise missiles were deployed in similar numbers with a nearly identical basing concept.

Endnotes

[1] *Regulus II Program for NATO Application, 1959, page 17. National Archives, Bureau of Aeronautics, Guided Missile Division, Record Group 72, Box 89.*
[2] *Personal communications with Captain Thomas D. Barrett, USN (Ret.).*
[3] *Endurance Record Set by NMC Target Drone, article in "The Missile", the Pacific Missile Range newspaper, 17 December 1965, Vol 15., #50.*
[4] *Personal communications with Bob North, Curator, New England Air Museum.*
[5] *BOMARC was an acronym for BOeing Michigan Aerospace Research Center, the lead designers of the missile.*
[6] *Regulus II Drone Program for U.S. Air Force, letter to Chief, Bureau of Aeronautics from Chance Vought Aircraft, Incorporated, 24 April 1959. National Archives, Bureau of Aeronautics, Guided Missile Division, Record Group 72, Box 89.*
[7] *The History of the US Nuclear Arsenal, James Norris Gibson, 1989, pages 165-167, Brompton Books Corporation, Greenwich, CT.*
[8] *The two F8U-1 "Crusader" high-speed chase aircraft were the number 3 and 9 airframes in their program. The number 3 aircraft had been used to qualify the Crusader design (from the Chance Vought Vangard, 19 September 1960, page 3).*
[9] *Personal communication with Gerry Parsons.*
[10] *1st Regulus Now Here for 'Flying Tests', 2 July 1959, Venice Gondolier, page 1.*
[11] *Technical Proposal for Six Flight KD2U-1 Target Drone Program, BOMARC/SAGE Weapon System Evaluation, April 1959, pages 1-25. National Archives, Bureau of Aeronautics, Guided Missile Division, Record Group 72, Box 89.*
[12] *Personal interview with C.J. Bitter.*
[13] *The missile was intercepted 154 miles offshore at 54,000 feet. The complete interception and firing operation had been controlled from Gunter Air Force Base, Alabama. The entire exercise took seven minutes.*
[14] *Chance Vought is Leaving, Last Shot was Saturday, 5 October 1961, Venice Gondolier.*
[15] *Personal interview with Ace Yeager.*
[16] *Regulus II Program for NATO Application, 1959, page 17. National Archives, Bureau of Aeronautics, Guided Missile Division, Record Group 72, Box 89.*
[17] *Ibid.*

Chance Vought KD2-U preparation and launch team poses in front of two KD2-U drones at the Venice, Florida, Launch site. The missiles were recovered at Eglin AFB and trsnsported to Venice to be refurbished for the next flight. (Coutesy of B.N. Smith Collection)

The KD2U was transported to the launch site using the mobile launcher. The work platforms incorporated on teh launch provided access to all parts of the drone for last minute service needs. (Loral Vought Systems Archives, Parsons Collection)

Toledo *launches her last Regulus I missile, FTM-1330, off the California coast 17 April 1959. This was the 901st launch of a Regulus I missile. (Author's Collection)*

Part III

Regulus I Deployment Aboard Aircraft Carriers and Cruisers

Chapter Eight: Regulus I and Carrier Aviation

Naval aviation played an important role in the development and deployment of Regulus I. Carrier pilots supported submarine and cruiser training programs as well as deployment of Regulus I missiles and guidance aircraft in the Regulus Assault Missile program. Six months after the inception of the idea in the Office of the Chief of Naval Operations, the Navy began training escort pilots.[1]

Regulus Assault Missile Program

1952

The first Navy-controlled evaluation flight took place on 17 October 1952. Lieutenant Billy May and Lieutenant George Monthan flew the TV-2D as primary control, May flying as the ABLE pilot and Monthan as BAKER. Takeoff was uneventful as they guided the missile to the range at 6,000 feet. With the turn into the assault run, May brought the missile down to 150 feet above the lakebed. Five miles from the target, Monthan broke away from the missile on May's command. As the missile passed over the target, May selected "camera on" and then "camera off" to record the simulated detonation point. Due to an instrumentation error, the "camera off" signal shut off the fuel supply and the missile crashed.[2,3]

The next flight was also the first shipboard launch of a Regulus I. The converted seaplane tender, USS *Norton Sound* (AVM-1), was a testbed for a wide variety of the missile programs at NAMTC. The mobile "desert" launcher built by Freuhauf Trailer Company, was installed on the ship's fantail.[4] FTM-1015 was launched successfully on 3 November 1952 with Chance Vought pilots controlling the missile. Target accuracy was not good but the rest of the flight confirmed that shipboard Regulus launch operations could easily be conducted.

The first full test of the assault missile concept with launch from an aircraft carrier took place in December 1952, during the refit of USS *Princeton* (CVA 37) after her tour in Korean waters. The hydraulic catapults on *Princeton* could not generate the 180 knot end-speed needed to launch Regulus I so the mobile launcher from the desert operations was brought on board. This was not particularly popular with the captain of *Princeton* since none of the launch and support equipment had been designed with this in mind. On 16 December 1952, with Lieutenants May and Monthan serving as primary and secondary controllers respectively in F2H-2P's, Chance Vought personnel assisted a team from Guided Missile Training Unit FIVE with the first Regulus I operation from an aircraft carrier, nine months after decision to evaluate the assault missile concept. While target accuracy was marginal, the concept of launching Regulus from a carrier had been tested and found to be feasible, albeit somewhat cumbersome. The mobile launcher had not been designed for use aboard ship, being far too large for routine use in the relatively confined below-deck hangar spaces. From an operational viewpoint, the launcher and missile would displace too many aircraft. The position of the launcher on the flight deck was dictated by the availablity of power and meant that one of the two catapults would be out of service during missile operations. Needless to say, carrier flight operations officers were less then enthusiastic about this new wrinkle to aircraft flight operations.[5]

1953

The first all-Navy Phase B flight, took place on 6 February 1953 with Lieutenant May flying ABLE and Lieutenant(jg) David Leue flying BAKER. Both booster bottles failed to eject and

FTM-1035 is maneuvered on board Hancock *off the San Diego coast. F9F-6 control plane piloted by Lt. H. J. Hays of VU-3 is being readied for launch on the port catapult. The missile launcher is the SR MK 2, a twelve caster leviathan that was extremely difficult to position. (Courtesy of Loral Vought systems Archives, Wilkes Collection)*

FTM 1035 lifts off from the Chance Vought desert launcher on Hancock's *flight deck, 19 October 1954. Much easier to maneuver both below and the flight deck, this launcher was still too cumbersome and took up too much below deck space. For the RAM program to succeed aboard carriers, a new launcher design was vital. (Courtesy of Loral Vought Systems Archives, Wilkes Collection)*

the resulting drag caused the missile to stall on landing approach and crash. In the course of the next several months the assault missile evaluation program successfully transitioned to Grumman F9F-6P "Cougars" as the guidance aircraft.

In May 1953, a key figure joined the evaluation program. Lieutenant Commander Larry Kurtz reported to Pt. Mugu to assist in the development of the training program. Kurtz was intimately involved in the Korean War Assault Drone Program and he brought needed operational experience to the team. Fresh from temporary duty as Naval Air Development Center, Johnsville, *Pennsylvania*, where he had been involved in the qualification flights for the new F9F-2P drones especially configured for the program, Kurtz immediately qualified in TV-1 and TV-2D aircraft.

By early Fall 1953, the Navy realized that it was time for this informal training and evaluation program to become formalized into an organizational unit. Utility Squadron THREE (VU-3) was stationed at Naval Air Station North Island, San Diego, and was already responsible for the F6F target drone operations for fleet training on the West Coast. On 12 September 1953, Kurtz was assigned as Officer-in-Charge of VU-3 Regulus Assault Missile Detachment (RAM Det). VU-3 RAM Det drone training activities interfered with the rest of the North Island flight operations so soon thereafter it was transferred first to Ream Field and finally settled in at the Naval Auxiliary Air Station, Brown Field, to the southwest of San Diego.[6]

1954

Soon VU-3 RAM Det pilots were providing support for the Trounce and BiPolar Navigation guidance systems development. After participating in seven missile flights, all successful, in the first quarter of 1954, VU-3 RAM Det was awaiting the availability of production tactical missiles to continue training. By mid-June 1954, numerous objections had been voiced within BuAer concerning the assault missile concept and its assignment on board aircraft carriers. Specific objections included: that the space consumed resulted in the loss of too many aircraft; restrictions placed on control aircraft due to formation flying with the missile; pilot work load was too heavy, i.e., guiding the missile and navigating was too difficult; inability to recover the tactical missile if the mission was aborted or unable to reach target; and disruption of carrier flight operations. Program advocates answered each objection in turn: space requirements would be reduced substantially by installation of a permanent missile checkout station and/or catapult launch of the missile. Moreover, any new weapons system would take up space; control aircraft maneuverability would be a primary area for improvement; none of the pilots had complained of work overload; all unmanned atomic weapons system at this time were unrecoverable; the magnitude of flight operation disruption would decrease as launch teams and methods were refined.[7] Additional carrier conversions would focus on installation of cabling, etc., to permit rapid hookup of the missile support vans. With Rear Admiral John H. Sides, Director, Guided Missile Division, OP-51, supporting the program, the Operations Development Force (OPDEVFOR) evaluation went forward as planned and the conversion of USS *Lexington* (CV 16) was begun.[8]

By August 1954, 15 assault missile operations with missiles had been flown by Navy pilots. The first OPDEVFOR launch was on 20 August 1954 from the *Norton Sound*. Lieutenant Commander Kurtz and Lieutenant Plog, both flying F9F-6Ps, controlled the missile during a successful pass over San Nicolas Island. The missile was damaged on recovery. During the next seven months of the OPDEVFOR program, 16 launches were conducted: eight from USS *Tunny* (SSG 282), four from the heavy cruiser USS *Los Angeles* (CA 135) and four from the aircraft carrier USS *Hancock* (CV 19). While the submarine and cruiser operations provided valuable training in simulated target passes and demonstrated the utility of the concept, flight operations from *Hancock* were critical to acceptance of the concept by OPDEVFOR staff.

On 18 October 1954, personnel from Guided Missile Unit FIFTY-TWO (GMU-52), the surface Regulus I operations support unit, successfully launched the first Regulus missile,

FTM-1034, from *Hancock*. This first *Hancock* launch used the Short Rail Mark 2 (SR MK 2) launcher. This had 12 metal casters and was a heavy and cumbersome monstrosity built at the Naval Aircraft Factory, Philadelphia, with apparently little attention to operational utility. Nonetheless, GMU-52's Executive Officer, Lieutenant Commander John Callahan, had to work with the equipment that was available. Callahan recalls with great clarity the events preceding the first *Hancock* launch. A week before, Admiral Felix Stump, Commander in Chief, Pacific Fleet, and his staff toured *Hancock* while she was in port. A walk-through of the upcoming missile operation was conducted for Admiral Stump. Several days earlier the details of maneuvering the SR MK 2 with two flight deck tractor "mules" had been worked out using an unfueled missile. All procedures appeared to work reasonably well. For the walk-through, a fully fueled missile was installed on the launcher without regarded to the added weight. The fuel weight caused the SR MK 2 to sag in the middle. As a result, the missile-launcher combination hung-up on the hangar door sill between Hangar Bays One and Two, much to the embarrassment of the GMU-52 personnel and consternation of the Admiral's staff. Not a good start to the OPDEVFOR process to say nothing of having been less than impressive to Admiral Stump.

One further incident sealed the fate of the use of the SR MK 2 on board carriers. During routine manuevers, *Hancock* began a high-speed turn just as the launcher, with a fully fueled missile, was towed onto the port elevator. The tractor's brakes could not hold the loaded launcher and the entire assembly began to roll towards the elevator edge with the tractor driver frantically pushing the brake pedal through to the deck. When the caster wheels hit the elevator coaming, the launcher stopped, and the ship completed her turn without losing the missile or personnel overboard.[9] On 19 October and again on 20 October 1954, GMU-52 successfully launched fleet training missiles from *Hancock* using the more maneuverable desert launcher designed by Chance Vought. VU-3 RAM Det pilots completed the simulated missions and both missiles were recovered at San Nicolas Island.

1955

On 15 February 1955, VU-3 RAM Det pilots participated in the first launch of a tactical Regulus I from USS *Los Angeles* (CA 135) as she cruised off the Hawaiian Islands (see Chapter 9). This was the first Operations Suitability Test (OST) for the Regulus program and combined both a test of the tactical missile and shipboard launch operation with a simulated mission profile. TM-1080 had all the components of the W-5 nuclear warhead in place except for the nuclear material which was replaced with concrete. Lieutenant Commander Kurtz and Lieutenant Plog deployed to Naval Air Station Barbers Point, Oahu, Hawaii, flying F9F-6P's. They rendezvoused with the fleet 50 nautical miles offshore. Using radio command control, Kurtz picked up the missile immediately after launch for a flight to Kaula Rock, southwest of Niihau . Cruise to the target was at 24,000 feet.

The top speed of the F9F-6P was well below that Regulus I, so after closing formation with the missile, Kurtz accelerated to his top speed and then throttled the missile back. When they neared the target, Kurtz sent the arming command and guided the missile down for a low-altitude, high-speed run into the target. Detonation of the warhead occurred 300 yards short due to air turbulence causing Kurtz's thumb, resting on the control switch, to slide it closed (the switch was later modified).[10] The OST was a clear success. *Los Angeles* sailed to WestPac carrying three Regulus I missiles armed with W-5 nuclear warheads. This deployment was the first nuclear missile deployment of the US Navy. Though aircraft and pilots of a RAM Det were not yet assigned to WestPac, several highly successful *Los Angeles* Trounce guidance operations had taken place prior to deployment. One could argue that this deployment was more a test of operational feasibility than one of tactical availability but in any event, the first deployment of a US Navy heavy cruiser armed with missiles carrying nuclear warheads was certainly something to be reckoned with.

The last of the boosted launches from *Hancock* took place on 15 March 1955. This launch was additionally significant as it turned out to be the only aircraft carrier-launched, submarine-guided flight operation of the entire Regulus program. This test utilized two Regulus guidance submarines, USS *Cusk* (SS 348) and USS *Carbonero* (SS 337), in sequence, to control the missile after launch. The test was a success and expanded the possibilities for Regulus operations from carriers. The range of the F9F-6P was only 250 nautical miles while the missile's range was 500 nautical miles. By using carriers as the launch platform, standing well off the enemy coast, submarines could serve as

Head-on view of the STAR cart launcher system for Regulus I. Sam Lynne, on far left near the wing tip, the Chance Vought missile operations engineer for VU-3 RAM DET, played a key role in ensuring tha the STAR cart program succeeded. The STAR cart served as an adaptor for launching the tactical missiles from a catapult since they did not have landing gear. Once clear of the flight deck, the cart fell into the water and was not recovered. (Courtesy of Loral Vought Systems Archives, Lynne Collection)

STAR cart launch sequence. On 19 July 1955 the first launch of a Regulus I missile, FTM-1058, using the STAR cart took place. Easily maneuvered and taking up much less deck space then early launchers, the STAR cart made the carrier RAM program feasible. (Courtesy of Loral Vought Systems Archives, Pearson Collection)

guidance relays to attain maximum range. At a cruise altitude of 35,000 feet, the control range for the Trounce guidance submarines was a circle 456 nautical miles in diameter. Positioned between a carrier 430 nautical miles offshore, two submerged submarines could guide the missile to a coastal target with little chance of being detected.

Upon completion of the *Hancock* JATO boosted Regulus launch evaluations, aircraft operations officers united in their opposition to placing Regulus I on carriers. Even though launch operations were now interrupted for only 30 minutes instead of the hours required at first, flight deck handling of the available launchers was still a nightmares. Chance Vought had anticipated this problem and developed a new launch method, the steam assisted Regulus (STAR) cart for use with the new powerful Type C, Mark XI, steam catapult on *Hancock*.[11] This catapult had already demonstrated reliability and more than sufficient power to launch a fully fueled and armed Regulus I missile. The STAR cart provided a launcher that was also the checkout and handling dolly for the missile. Equally as important was elimination of the large JATO bottles that the carrier captains felt were disasters waiting to happen. The cart and missile still occupied one aircraft equivalent of space but this was an improvement over the launchers of four months earlier.

After two inert-mass sled STAR cart launches successfully demonstrated the concept, the first STAR cart Regulus launch took place on 19 July 1955. The launch was completely successful as the missile flew straight off the deck and began climbing immediately as the expendable cart plunged into the ocean. The next day the performance was repeated and the system was considered operational.[12]

Hancock departed San Diego 5 August 1955 carrying four Regulus I tactical missiles armed with W-5 atomic warheads. The four missiles and STAR carts, plus missile checkout equipment occupied space equivalent to five aircraft.[13] The RAM detachment did serve a dual purpose however. Originally a photoreconnaisance aircraft, the F6F-6P camera equipment could be quickly removed and replaced by the radio command control guidance equipment. The other drawback was that the W-5 warhead had to be converted from the Mk 5 gravity bomb while on board *Hancock*, The accompanying security problems further aggravated the carrier flight operations staff. As a result Regulus was not a welcome addition to *Hancock*'s arsenal.[14]

The first carrier Regulus assault missile OST took place on 8 August 1955, 250 nautical miles to the west of the Hawaiian Islands. Lieutenant Monthan flew primary airborne control with Lieutenant Peck and Lieutenant(jg) Stone flying as backup. TM-1081 was equipped with a war reserve W-5 nuclear warhead minus the nuclear components but still containing the equivalent of 3,000 pounds of high explosive. After a smooth STAR cart launch, Monthan turned the missile 180 degrees and climbed to a cruise altitude of 35,000 feet while proceeding towards the target, Kaula Rock.

Fifty nautical miles out, Monthan let down from 35,000 to 100 feet for the approach to target. Visibility was variable with scudding clouds. Without airborne radar to locate the island and operating under radio silence to simulate war time conditions, dead reckoning was the sole navigation system. At about ten nautical miles out, Monthan spotted the target, lined up the missile in azimuth, corrected for wind drift, and armed the warhead. At five nautical miles to go, he performed the 45-degree breakaway maneuver, selected the pop-up altitude of 1,500 feet, and "pickled" the warhead as the missile crossed over Kaula Rock. All three pilots later saw a picture at Commander-in-Chief, Pacific Fleet,

TM-1128 is readied for launch early on the morning of 20 February 1955. This was the third and last STAR cart launch from Hancock during this WestPac deployment. (Courtesy of Hogan Collection)

headquarters showing the OST explosion. The warhead had detonated directly over the target.[15]

The first Regulus I operation in WestPac, and the second of *Hancock*'s deployment, was on 14 December 1955 with Monthan flying as primary airborne controller. The STAR cart launch was normal and after a pass on a small target rock east of Okinawa, four VF-121's F9F-8 "Cougars" were vectored to intercept and shoot down TM-1129 since it could not be recovered. Monthan had to slow down the missile to insure a successful interception by the fighters.[16] Two months later, a third OST was conducted. Lieutenant Peck was flying the primary guidance aircraft and recalls having to virtually set the missile up for the interception as again the fighters vectored to intercept the missile found that it was not an easy task.

Hancock's presence in WestPac with Regulus I nuclear capability probably did not go unnoticed as the tensions the Communist and Nationalist Chinese concerning the Quemoy and Matsu Islands lessened. This deployment in *Hancock* marked the end of her involvement in the assault missile program since she did not deploy again with either missiles or a RAM Det aboard. No other West Coast aircraft carriers were deployed to WestPac with Regulus I on board although one, the *Lexington* did launch a Regulus I during training off the West Coast.

Guided Missile Group Formation

One month after *Hancock* deployed to WestPac, VU-3 RAM Det was disestablished along with GMU-52. The two units were then combined on 16 September 1955 into Guided Missile Group ONE (GMGRU-1) commanded by Commander R.C. Millard at Naval Air Station North Island, San Diego. Guided Missile Group ONE was tasked with providing launch and recovery services for cruisers and carriers

in addition to missile recovery services during submarine operations. Similarly, Guided Missile Group TWO (GMGRU-2) was formed on the East Coast. For sake of continuity, GMGRU-2's history will discussed after GMGRU-1.

1956

On 30 January 1956, Lieutenant George Gregory, Lieutenant Commander John Callahan and Lieutenant(jg) Bill Allen were involved in a Regulus crash that highlighted the need to find a more suitable training area. After completing the RAM training mission, Gregory, the primary guidance pilot, commanded a turn to bring the missile to a heading for recovery at Marine Corps Auxiliary Air Station Mojave, California. The missile failed to respond and started a slow climb on its own into the overcast. Since they were heading towards Los Angeles at 400 knots, Gregory quickly passed control to Callahan but still the missile failed to respond. Callahan and Allen, who was along as an observer and had no guidance controls, remained above the clouds with the missile while Gregory went down to look for a uninhabited place to dump the missile.

Gregory radioed to Callahan that now was as good a time as ever and he triggered the destruct signal, causing fuel cutoff and a dive sequence to begin. Following the missile through the cloud cover, Callahan saw that it was headed for an orange grove near the town of Porterville. Unfortunately, a ranch house looked to be too near the probable point of impact but at this point there was nothing anyone could do. Luckily the missile struck the ground 600 feet away from the building, digging a trench 25 feet long and 8 feet deep.[17]

Lieutenant Robert Blount, a relatively new arrival to GMGRU-1, was on the ground listening to the whole show on a radio in the hangar. Shortly after they received confirmation from Gregory that apparently no one on the ground had been hurt, Blount walked over to the Group offices to see what else was going to happen. The squadron duty officer for the day was Ensign Ken West. On top of the situation from the beginning, he had called the Porterville Sheriff's office. As Blount walked into the office he heard West asking if they had any reports of an aircraft crashing nearby. West then said, "Don't worry about the pilot, it was a guided missile, not one of our aircraft." West hung up the phone, turned to Blount and said, "The sheriff just yelled `A WHAT??' and hung up." The public was relatively uninformed about guided missiles, in particular Regulus I, and now the cat was out of the bag. As can easily be imagined, the phone lines lit up as the press descended on the story. The result was a banner headline in the Mirror News, "Pilots Guide Runaway Missile from Homes."[18]

On 7 May 1956, tragedy struck GMGRU-1 during a Regulus I recovery at San Nicolas Island. Lieutenant Arnie Wagner and Lieutenant Al Rice were the primary recovery team in the TV-2D with Rice flying as ABLE pilot and Wagner as BAKER. After a successful STAR cart launch from *Lexington*, an autopilot malfunction developed that caused the missile to flight with a slight left bank attitude. Rice and Wagner decided to attempt to recover the missile anyway. Having skillfully synchronized the turn with the approach to the San Nicolas Island runway, Rice thought he had the recovery under control. As usual his TV-2D was tight on the wing of the missile as it touched down. Unfortunately the missile landed short, throwing up a distracting cloud of dirt. Wagner began to bank away and caught the TV-2D's starboard wing tank on the runway. The plane cartwheeled and broke into several sections, killing both pilots. Lieutenant(jg)s Al Thayer and Bill Allen were flying backup chase in F9F-6s and remember to this day the helpless feeling as they watched the accident unfold.[19]

On 20 July 1956, the majority of GMGRU-1 personnel, aircraft, equipment, embarked on USS *Philippine Sea* (CVS 47) for transportation to the Hawaiian Islands. This move was to facilitate the use of the RAM Det teams as carriers departed for WestPac from Hawaii while at the same time the group would have a less populated training area. The remainder of GMGRU-1, the Continental US Detachment (ConUS Det), with Lieutenant Commander Len Plog as Officer-in- Charge, was frequently used to augment Pt. Mugu's research and development support programs. In addition to conducting KDU-1 target drone operations for *Norton Sound* Terrier and VX-4 Sparrow III evaluations, they flew safety chase and recovered missiles launched by other fleet units.

Naval Air Station Barbers Point was GMGRU-1's main base while Bonham Auxiliary Landing Field (Bonham ALF), Barking Sands, Kauai, a World War II Army Air Corps field, provided the launch and recovery site for missile operations.

1958-1960

The 100th Regulus I launch by GMGRU-1 took place 25 February 1958 and was the 713th flight in the Regulus I program. This level of launch activity illustrates the point that the level of GMGRU-1 training and launch activity was consistent with a ready reserve assault missile capability. Increasingly important, however, was the submarine launched Regulus recovery support. GMGRU-1 Bonham Detachment was transferred to Guided Missile Unit NINETY (GMU-90), the submarine Regulus I support unit. GMU-90 Bonham Detachment continued to service missiles recovered at Bonham as well as launch missiles for GMGRU-1 pilot and submarine guidance training.

In 1958 the GMGRU-1 CONUS Det participated in the only two shipboard launches of Regulus II, one from the USS *Grayback* (SSG-574), 18 September 1958, and one from USS *King County* (AG-157) 10 December 1958 (see Chapter Seven). Over the next 22 months, GMGRU-1 continued to fly missions, but only as part of operational readiness exercises for carriers on their way to WestPac. Virtually all of the simulated missile attacks on the carriers were accomplished undetected by carrier radar.

GMGRU-1, including the ConUS Det, was disestablished 1 September 1960, two weeks short of five years from its date of activation. GMGRU-1's assets and functions were transferred to Utility Squadron ONE (VU-1) at Barbers Point, Oahu.[20] Flight operations continued, both in support of submarine Regulus training flights since five submarines were now conducting round-the-clock Regulus deterrent missions and also as anti-aircraft target practice for the Pacific Fleet. With the end of the submarine Regulus program in mid-1964, the remaining fleet training and tactical missiles were expended as targets. Recovery of the fleet training missiles continued during this phase of the program, but tactical missiles were not modified with landing gear and so were expended on one-way missions.[21] The last Regulus I flight took place on 6 June 1966 from ALF Bonham, 16 years from the start of flight operations at Edwards in 1950.[22] The ConUS Det functions were transferred to

Table 8.1 Carrier RAM Deployments of Guided Missile Group ONE

Ship	Date	Missile Operations[a]
USS Lexington (CV 16)	May 1956	none
USS Shangri-La (CV 38)	November 1957	one, cruiser
USS Lexington (CV 16)	April 1956	one, cruiser
USS Ticonderoga (CV 14)	October 1958	none
USS Shangri-La	March 1958	two, cruiser and submarine
USS Lexington	July 1958	none

[a] These were RAM flights after launch from either a cruiser or submarine.

Guided Missile UNIT FIFTY-FIVE, the submarine Regulus II support unit.

GMGRU-1 RAM Deployments

Between May, 1956 and July, 1958, six RAM detachment deployments were made aboard Pacific Fleet aircraft carriers (See Table 8.1) The first missile operation during deployment took place on 31 March 1957. Lieutenant(jg)s Bill Allen and Duane Myrin, GMGRU-1 RAM Det "Alpha" from the USS *Shangri-La* (CV 38) flew as safety-chase for a tactical missile launched and guided from the heavy cruiser USS *Toledo* (CA 133). The target was the island of Okino Daito Shima, part of the Ryuku Islands south of Japan (see Table 8.1).

By getting behind the missile and aligning it with the target, Allen could see that it was drifting left and that the *Toledo's* Trounce guidance team had not been able to compensate for the cross winds at the 35,000 foot altitude. Myrin agreed with Allen that they should assume control of the missile and guide it to the target. Without time to confirm this decision with the cruiser, the pilots completed the terminal dive manuever on target and returned to the carrier. Allen was soon explaining his decision to Vice Admiral W.M. Beakley. With an array of observation ships below and in the apparent path of the missile, Allen's decision met with the Admiral's approval.[23]

The second RAM Det missile operation in the Western Pacific took place less then three months later. The heavy cruiser USS *Helena* (CA 75) launched a tactical missile and controlled it to the Trounce guidance line-of-sight limit. GMGRU-1 RAM DET "Golf", with Lieutenant Commander V.F. Forsberg, Lieutenant Robert Blount and Lieutenant (jg) Al Thayer had deployed on board *Lexington*. With only one guidance aircraft available, Lieutenant Robert Blount assumed radio command control for the remaining 120 nautical miles to Okino Daito Shima and detonated the missile on target.

In early October, 1957, prior to deployment, the RAM detachment aboard USS *Ticonderoga* (CVA 14) participated in an undetected missile attack on the carrier that ended in a pass at 200 feet altitude, 1,000 yards astern of the ship. This was the first cruise missile attack against a carrier and demonstrated the pronounced vulnerability of aircraft carriers to cruise missile tactics. When *Ticonderoga* arrived in her operational area, the Commander, Seventh Fleet, decided that RAM Dets should be shore-based while in WestPac. Upon arrival Det "Mike" was flown off to Naval Air Station Naha, Okinawa and Naval Air Station Cubi Pt., Philippines for the remainder of the cruise.[24]

The final RAM Det missile operations conducted by GMGRU-1 took place during the March 1958 deployment of the *Shangri-La*. GMGRU-1 RAM Det "Charlie" participated in two missile launches during this deployment, both during the Seventh Fleet Review for the annual SEATO exercise off Okinawa. This was a major demonstration of the Regulus I weapon system for senior officers from all the SEATO countries aboard three aircraft carriers, two heavy cruisers and six escorting destroyers.

On 22 May 1958, RAM Det "Charlie" pilots guided the first Regulus I launched in the Western Pacific from a submarine, the *Tunny* (see Chapter Ten). The next day was overcast with a low ceiling, threatening to cancel the scheduled missile launch from *Toledo*. The forecast was for more of the same for the next several days. After radio consultation with the two RAM pilots and Regulus I launch personnel, the decision was made to proceed as a demonstration of the all-weather capability of the RAM program. The final countdown began without visual contact between the cruiser and aircraft. At T-minus 30 seconds, the aircraft were still not in radio contact nor visible below the cloud cover. At T-10 seconds, the two aircraft suddenly dropped through the overcast and formed up on the missile as it was launched.[25]

Lieutenant Fred J. Orrik was the primary safety pilot and was flying close formation in the cloud cover as the cruiser Trounce guidance team controlled the missile to the target island, Okino Daito Shima. As he neared the target area, Orrik remembers noticing that he was beginning to pull slight negative "g's". He ignored it until he realized that the missile had begun its terminal dive maneuver under Trounce guidance, unannounced to him by the *Toledo* missile guidance team. He quickly pulled away and returned to the carrier. Ships in the target area described the detonation as directly on target.[26]

The sixth and final GMGRU-1 RAM Det deployed to WestPac in *Lexington*, departing Pearl Harbor 17 July 1958. No missile operations took place. After the *Lexington* deployment, the RAM program primarily flew flying safety chase for cruiser and submarine missile operations. GMGRU-1 pilots were stationed with RAM Det "Sierra" at Naval Air Station Naha, Okinawa, for these training operations until the phase out of the program.

Guided Missile Group TWO

1955

In July 1954, VU-3 RAM Det on the West Coast had been divided to provide pilots and experienced missile maintenance personnel for the newly formed RAM Det to be attached to Utility Squadron FOUR (VU-4), Naval Air Station Chincoteague, Virginia. Lieutenants Dave Leue and William Kelly were the first Regulus guidance pilots with VU-4. Lieutenant Al Monger and Lieutenant(jg) Austin O'Brien joined soon thereafter as the core of the RAM Detachment pilot staff.

By March 1955 GMU-53 and VU-4 RAM Det were ready for the long awaited first launch. Naval Air Station Chincoteague was located in the middle of civilization compared to either of the West Coast launch facilities and the first East Coast Regulus I launch turned into quite a media event with television and news radio crews everywhere. Postponement of the launch for 24 hours due to high winds and a faulty hydraulic line heightened the anticipation. On 24 March 1955, State Highway 175, which passed directly underneath the flight path, was closed to morning traffic for five minutes prior to launch. All traffic and boating in the vicinity of the booster impact area in a nearby marsh was stopped 30 minutes prior to launch.[27] The two hour countdown went smoothly. On schedule, the missile, the first Navy launch of the KDU-1 target drone version of Regulus I, roared off the launcher in its characteristic cloud of billowing white smoke. The thirty-minute flight was uneventful and Lieutenant William Kelly brought the missile to a safe landing dead center on the runway, ending a perfect flight and starting the East Coast program in fine fashion.[28] The remainder of 1955 consisted of 15 target drone flights, the majority of which were presentations for the very active Terrier and Sparrow test and evaluation programs underway at NAS Chincoteague. On 23 September 1955, VU-4 RAM Det and GMU-53, the East Coast surface fleet Regulus maintenance and launch unit were disestablished. Their personnel and equipment were recombined as Guided Missile Group TWO (GMGRU-2) with Commander William Coley as Officer-in-Charge.

1956

In May 1956, GMGRU-2 pilots received a welcomed change of pace from the target drone launch program. In quick succession, three fleet Regulus launch platforms became available on the East Coast. The only Regulus cruiser on the East Coast, USS *Macon* (CA-132) had recently been converted and began her training program on 8 May 1956 with the successful launch of a fleet training missile. *Macon* departed for the

On 15 June 1956, the first shipboard launch using the induced pitch launcher (IPL) was conducted on board Randolph. *The IPL was an alternative to the STAR cart and was carried on board* Randolph *during her Mediterranean deployment though no further launches took place. (Courtesy of Elislrac Enterprises Collection)*

Eastern Mediterranean soon thereafter with three Regulus I tactical missiles, armed with W-5 nuclear warheads (See Chapter Nine).[29]

On the heels of *Macon*'s departure, flight operations commenced on aboard USS *Randolph* (CV 15), newly converted to conduct Regulus flight operations. *Randolph* did not have steam catapaults at this time but a new missile launcher had been developed since the earlier *Hancock* missile tests. This induced pitch launcher (IPL) was smaller and considerably more maneuverable. A missile could be stored on the IPL and so *Randolph* could carry five missiles during deployment. The control van and missiles were located in Hangar Bay 1 with hard-wired umbilical connections to the starboard side. Altogether the Regulus I equipment effectively occupied the same space as an entire Douglas AD-6 "Skyraider" squadron, much to the dissatisfaction of Commander Dewitt Freeman, the Aircraft Handling Officer. Freeman remembers having mixed emotions on the subject because he had been the first Navy pilot to control a Regulus missile in flight.[30]

Randolph's launch team from GMGRU-2 launched only two missiles from the carrier. The first, FTM-1263, was launched on 15 June 1956. Though damaged on recovery, this first flight from an East Coast carrier was a complete success. The second and what ended up being the final launch from *Randolph*, was TM-1204. A successful RAM flight profile and low-level air burst detonation was completed on 17 June 1956.

Randolph deployed to the Eastern Mediterranean on 14 July 1956, carrying four Regulus I tactical missiles and GMGRU-2 RAM Det 36 with Lieutenant A.J. Monger, Officer in Charge, and Lieutenant (j.g.)s Austin O'Brien, and S. Olmstead and Ensign R. Pekkanen. Three FJ-3s were used as control aircraft. No Regulus missile flight operations were conducted during the deployment but Monger recalls that training flights in the FJ-3 guidance aircraft were part of the normal routine.

Randolph operated off the coast of Egypt during the Suez Canal Crisis. Monger recalls watching the contrails of the British and French aircraft as they passed overhead on the way to bombing runs on Suez Canal targets. On 18 February 1957, *Randolph* returned to Chincoteague ending the first and last Atlantic Fleet carrier-based Regulus deployment.[31]

The third Regulus launch platform was USS *Barbero* (SSG 317). While submarine missile launch and support duties were supplied by GMU-51, GMGRU-2 provided Regulus I recovery services. With *Randolph* and *Macon* deployed in the Mediterranean, GMGRU-2 resumed Regulus target drone services to the Atlantic Fleet. RAM Det training was included in many of the recovery phases of these target drone missions.

One of the more amazing photographs of a Regulus I missile in flight was taken on 18 September 1956, at Naval Air Station, Chincoteague. Lieutenant David Leue was control pilot and after completing a RAM pass at 500 feet, requested permission to make a low-level run along the runway during a Regulus I demonstration for newly designated admirals. Permission was granted and Leue brought the missile down to 30 feet for a truly low-level RAM pass at 250 knots along the main runway.[32]

1957

In late February 1957, GMGRU-2 provided a fleet training missile, pilots and launch team for a unique demonstration program named Operation SNOWBALL. Lieutenant Larry Dion, Lieutenant Austin O'Brien and Lieutenant William Laurentis reported aboard USS ***Franklin D. Roosevelt*** (CVA 42) with a fleet training missile reconfigured to carry external fuel tanks for extended range. Operation SNOWBALL was a multi-service demonstration of equipment capability in the extreme winter weather of the North Atlantic. The extended range idea had been successfully tested on the West Coast nine days earlier but the combination of a new system and

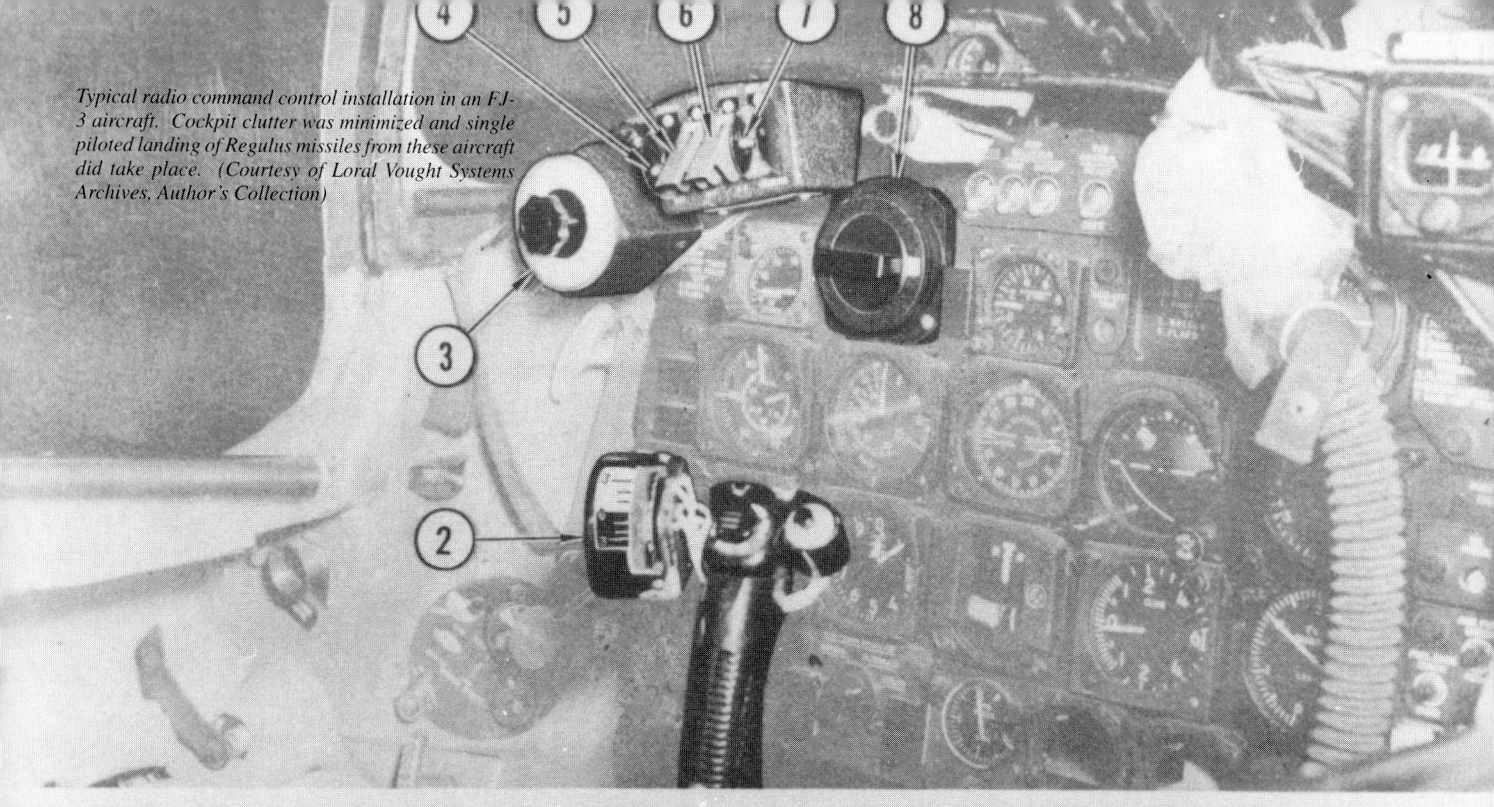

Typical radio command control installation in an FJ-3 aircraft. Cockpit clutter was minimized and single piloted landing of Regulus missiles from these aircraft did take place. (Courtesy of Loral Vought Systems Archives, Author's Collection)

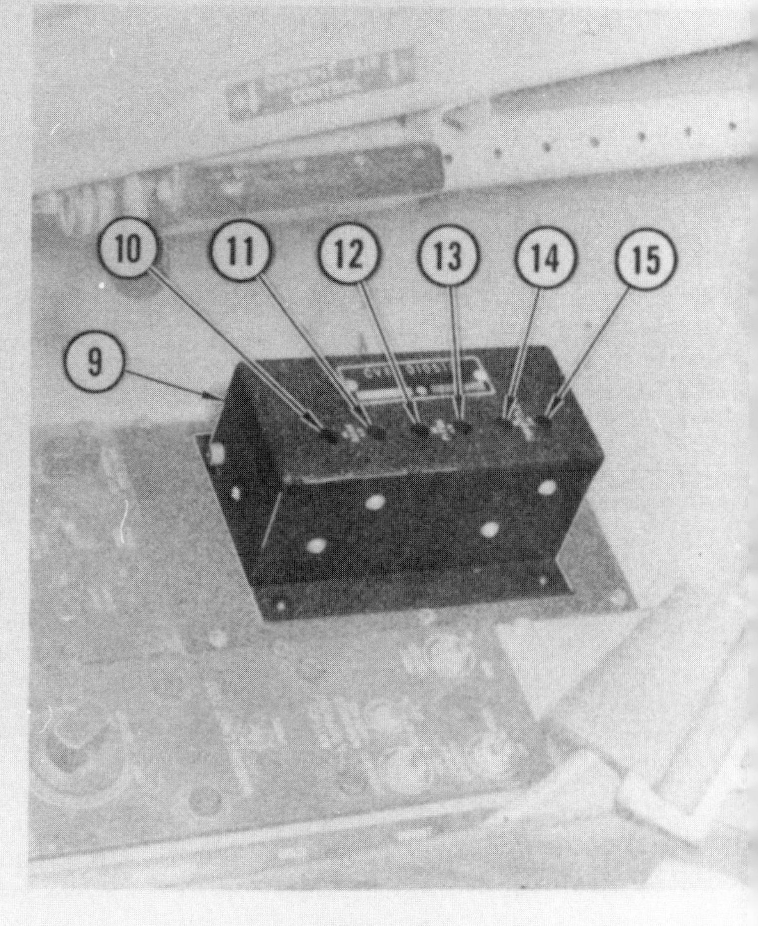

1. Throttle Control
2. Pitch Control
3. Bank-Turn Control
4. Carrier Switch
5. Power Switch
6. Flash-CW Switch
7. Execute Switch
8. Channel Select Switch
9. Control Adjustment Potentiometer Assembl
10. Bank Adjustment Screw (High)
11. Bank Adjustment Screw (Low)
12. Pitch Adjustment Screw (High)
13. Pitch Adjustment Screw (Low)
14. Throttle Adjustment Screw (High)
15. Throttle Adjustment Screw (Low)

The induced pitch launcher used curved guide rails to change the initial horizontal motion of the missile into launch altitude. This permitted much shorter rails and a more compact launcher, as well as directing the majority of engine and booster exhaust behind the missile rather than onto the flight deck. The large number of onlookers visible on Randolph's island soon regretted their decision as they were engulfed by the acrid fumes of the rocket booster exhaust. (Courtesy of Enterprises Collection)

One of the more amazing photographs of a Regulus I missile in flight was taken on 18 September 1956, at Naval Air Station, Chincoteague. Lt. David Leue was control pilot and after completing a RAM pass at 500 feet, requested permission to make a low-level run along the runway during a Regulus I demonstration for newly designated admirals. Permission was granted, and Leue brought the missile down to 30 feet for a truly low-level RAM pass at 250 knots along the main runway. (Courtesy of Purkrabek Collection)

the cold weather did not make any of the pilots overly optimistic that the tanks would stay on, let alone perform properly in flight. Nonetheless, the three chase aircraft were launched on 22 February 1957, followed by the missile 15 minutes later. The weather was clear all the way to Chincoteague. After several exchanges of guidance control between the pilots, the missile finally responded to Lieutenant Dion's commands as he guided it to a cruise altitude of 35,000 feet. With no instrumentation to indicate the quantity of fuel remaining in the external tanks, Dion made an educated guess and jettisoned the tanks. Excess fuel trailed from the tanks as they tumbled away and all three pilots realized that the long range objective of this flight was now in jeopardy. The rest of the flight was uneventful and as the formation neared NAS Chincoteague they were met 10 miles north of the airfield by a TV-2D landing control aircraft piloted by Lieutenant Leue. Rather than make the approach from the north end of the runway, a fateful decision was made to fly the extra twenty miles to make a standard southern approach since the external tanks should have provided the necessary extra fuel. Unfortunately the calculations were off by about three miles. On final approach with gear down and all the cameramen at the airfield anxiously awaiting an historic landing, FTM 1274, out of fuel, settled into the trees and was destroyed. This flight had lasted 75 minutes and was the longest flight in the Regulus I program.[33]

1958

GMGRU-2 was redesignated as Guided Missile Service Squadron Two (GMSRON-2) on 1 July 1958. Since the carrier assault missile program had been canceled, missile launching services were relegated to GMU-51. GMSRON-2 continued to provide recovery services as well as many long hours of drone services during evaluation of the improved submarine Trounce guidance system.[34] GMSRON-2's last Regulus I launch took place on 30 October 1958 in support of the Sidewinder missile evaluation program. Unfortunately after a normal launch and climb to 22,000 feet, radio carrier signal was lost and the missile self-destructed. With the decision to deploy Polaris Fleet Ballistic Missile Submarines first in the Atlantic, *Barbero* and the newly commissioned USS *Growler* (SSG 577) were moved to Hawaii. With cruiser Regulus activity already completed, GMSRON-2 redirected its efforts to other missile programs.[35]

Endnotes

[1] For additional details of naval aviation's role the RAM and drone programs, see: "Take Control: Guided Missile Groups ONE and TWO and the Regulus Missile," by LCDR Ernie Mares, USN(Ret.) and David K. Stumpf, Ph.D., and "The End of Regulus" by Walter Fink; both articles in The Hook, 20:1; Spring, 1992, pages 14-27 and 36-40 respectively.
[2] Regulus I Flight Test Program Weekly Activities Memo #114, Regulus Flight Test Program, 20 October 1952, LVSCA A50-18, Box 1.
[3] The Lot I RCC guidance system suffered from frequency drift caused by vacuum tubes overheating. This was resolved in the Lot II equipment (See Appendix I, RCC).
[4] The "desert" launcher was designed by Chance Vought and used at EAFB and Pt. Mugu for boosted launches. It was the first short rail launcher, using 12 foot launc^h rails. Though not officially designated as the Short Rail Mark 1, the remaining launchers used in the program are numbered.
[5] Personal interview with Captain William "Pappy" Sims, USN (Ret.).
[6] Personal interview, Captain Larry Kurtz, USN (Ret.).
[7] First Endorsement on OP-34 ser 00527P43 of 27 May 1954 dated 18 June 1954. Subject: REGULUS Program (Project RAM); shipboard installation during FY1955. Naval Historical Center, Operational Archives.
[8] Ibid.
[9] Personal interview, Commander John Callahan, USN (Ret.).
[10] Personal communication with Captain Larry Kurtz, USN (Ret.),
[11] Wings for the Navy, A History of the Naval Aircraft Factory, 1917-1956; William F. Trimble, 1990, Naval Institute Press, page 319.
[12] Personal interview with George Sutherland.
[13] VC-61 RAM Det "George" reported aboard with Lieutenants George Monthan and Paul Peck; Lieutenant(jg)s Don Stone, Roy Mock, and Bernie Welch and three F9F-6P control aircraft.
[14] Personal interview with Captain George Monthan, USN (Ret.)
[15] Ibid.
[16] Personal interview with Rear Admiral Paul Peck, USN (Ret.).
[17] Personal communication with George B. Gregory.
[18] Personal communication with LCDR Robert Blount.
[19] Personal communication, Commander Al Thayer, USN (Ret.)
[20] Air Planning Concerning Disestablishment of Guided Missile Group ONE (GMGRU-1). CNO ltr OPNAV 5440 OP-507C1 ser 7120P50, 29 June 1960. Naval Historical Center, Operational Archives.
[21] Personal communication with Lieutenant Commander L. "Pat" Kilpatrick, USN (Ret.).
[22] Honolulu Star-Bulletin, 8 June 1966.
[23] Personal interview with Commander William D. Allen, USN (Ret.).
[24] Personal interview with Captain Roy Kraft, USN (Ret.).
[25] Personal communication with Rear Admiral Harry Reiter, USN (Ret.), commanding officer of Toledo.
[26] Personal communication with Captain Fred Orrik, USN (Ret.).
[27] The Virginian Pilot, 24 March 1955, page 1.
[28] Personal interview with Lieutenant Commander William Kelly, USN (Ret.).
[29] Personal communication with Vice Admiral Vernon L. Lowrance, USN (Ret.).
[30] Personal communication with Rear Admiral Dewitt Freeman, USN (Ret.).
[31] Personal interview with Rear Admiral A.J. Monger, USN (Ret.).
[32] Leue had to wait 36 years to receive a copy of the photograph located during research for this book.
[33] Personal communication with Captain William D. Laurentis, USN (Ret.).
[34] Photocopy of GMGRU-TWO Aviation History Summary obtained from Robert Lawson.
[35] VU-4 RAM Det/GMU-53, GMGRU-2 and GMSRON-2 combined, launched a total of 185 Regulus missiles with five lost at launch, eleven during flight and five on recovery.

TM 1214 on a STAR cart launcher during checkout aboard Saratoga *May 1957 in New York Harbor. Two launches were conducted from* Saratoga, *using fleet training missiles, on 30 May and 4 June 1957, including a Sidewinder live fire demonstration for President Dwight D. Eisenhower.*

Chapter Nine: Regulus I and Heavy Cruisers

Regulus I was designed from the start as a submarine launched cruise missile. The April 1952 decision by the Office of the Chief of Naval Operations to evaluate the Regulus assault missile concept opened two new launch platforms in aircraft carriers and heavy cruisers. While the immediate response from carrier aviators was that the system would take up too much deck space and hamper launch operations, the cruiser navy quickly jumped onto the bandwagon. Placing Regulus I, with a nuclear warhead, aboard a cruiser gave new meaning to the concept of shore bombardment, serving to augment the newly acquired and more mundane cruiser role of task force air defense.[1] The initial concept was to use cruisers simply as launch platforms. Guidance installations would be kept to the minimum necessary for launch and short range guidance since aircraft from task force carriers would be used to guide the missile to the target. Conversion costs would be kept to a minimum by making all of the installations portable and converting the former seaplane hangar into a Regulus I maintenance and storage hangar.[2] Eventually four heavy cruisers of the Baltimore Class received Regulus I missile installations. The USS *Los Angeles* (CA 135), USS *Helena* (CA 75) and USS *Toledo* (CA 133) deployed with the 7th Fleet in Western Pacific waters while USS *Macon* (CA 132) deployed with the 6th Fleet in Mediterranean waters and NATO naval forces in the North Atlantic.

USS Los Angeles (CA 135)

Los Angeles was the first cruiser selected for Regulus I conversion, entering Mare Island Naval Shipyard in May 1954 for installation of missile equipment. Captain William W. Outerbridge was enthusiastic about the addition of guided missile capability to *Los Angeles*.[3] All missile checkout and maintenance equipment was carried in temporarily installed vans positioned in the converted seaplane hangar. Three missiles were carried in the hangar; two in cradles on the starboard bulkhead of the hangar and one on the missile handling dolly. Originally designed for use aboard carriers and on land, the Short Rail Mark 2 (SR MK 2) launcher was temporarily installed on the fantail.[4]

The first Regulus I launch by a *Los Angeles* missile team took place on 28 October 1954 as part of Phase III of the Operational Development Force (OPDEVFOR) evaluation. Pilots and aircraft from the VU-3 RAM Detachment based at Naval Air Station, Brown Field, San Diego, took control and successfully recovered the missile after a RAM pass over San Nicolas Island. Two more radio command control guidance flights were launched by *Los Angeles* in the closing months of 1954 as part of OPDEVFOR operations. One was lost after booster ejection, the other was successfully guided and recovered.[5]

1955

On 13 January 1955, the first cruiser-Trounce guidance flight was conducted from *Los Angeles*. The many hours spent practicing with manned drone aircraft paid off as the missile was successfully guided through a medium-altitude flight pattern and recovered. Two weeks later the *Los Angeles* Trounce guidance team controlled two missiles launched from Pt. Mugu. The first was a tactical missile with a dummy warhead. After a 43 minute flight off the coast of Southern California, the missile was guided over San Nicolas Island and the signal for terminal dive issued. The accuracy of the terminal dive to impact was not ascertained due to the lack of phototheodolite coverage. Several hours later a second missile was launched from Pt. Mugu, guided successfully by *Los Angeles* and recovered at San Nicolas Island.[6]

Los Angeles deployed to the Western Pacific (WestPac) in February 1955 carrying three nuclear armed Regulus I missiles. On 15 February 1955, *Los Angeles* participated in the first fleet Regulus I missile launch during an OST in the Hawaiian Islands area (see Chapter Eight). The warhead was a war reserve weapon with a depleted uranium core.[7] Lieutenant Commander Larry Kurtz and Lieutenant Len Plog, flying out of Naval Air Station Barbers Point, Oahu, picked up the missile shortly after launch and successfully guided it to a terminal dive at Kaula Rock, off of Kauai Island.[8] Shortly after completion of OST operations, *Los Angeles* departed for WestPac, becoming the first naval vessel to deploy with Regulus I missiles aboard.

Six months prior to the WestPac deployment of *Los Angeles*, the first of several conflicts had flared up between Nationalist Chinese and Communist Chinese forces over two groups of islands in the Formosa Strait. In early January 1955, 100 Communist Chinese aircraft bombed the Tachen Islands, north of Quemoy and Matsu, the site of an earlier aggression. On 18 January 1955, an island nearby the Tachens was overrun by Chinese Communists. Chinese Communist airfield construction continued across the Formosa Strait from Quemoy and Matsu. By late February the tension had risen considerably. The 7th Fleet was patroling the Formosa Straits with the *Los Angeles* presenting not only her conventional shore bombardment capability but also the new and much more powerful potential of Regulus I.

Flight Test Missile 1032 during engine run-up prior to launch. Several crew members can be seen beneath the missile working on two umbilical cables. This was the first launch from Los Angeles, *with FTM-1032 reaching 17,000 feet altitude and a maximum speed of 560 knots. Flight time was 27 minutes. The missile was successfully recovered at San Nicolas Island after a successful RAM pass over the airfield. The launcher is a modified SR MK 2 that could be trained as well as elevated and was installed only on* Los Angeles. *(Courtesy of Zelibor Collection)*

Figure 9-1. Typical cruiser overhead missile storage installation with a lower missile berth visible below. Three missiles could be stored in the missile hangar, with associated warheads and boosters.

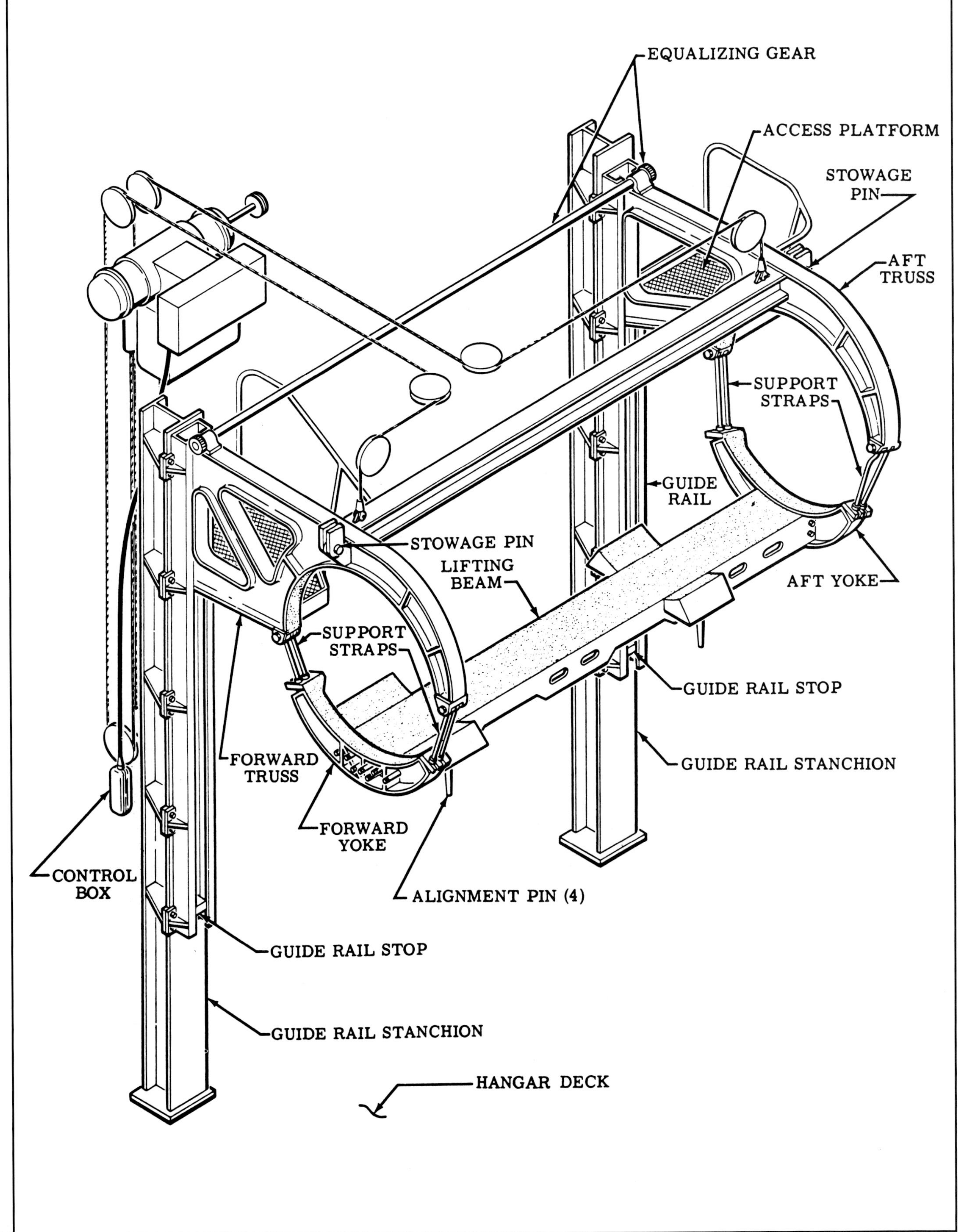

CRUISER	RADAR	COMPUTER	COMMAND CONSOLE	PLOTTER	GYRO	DRAI
Los Angeles	SPQ-2	CP-98 (XN-2)	WPA-1	Manual	MK 19	MK 6 Mod 1
Helena	SPQ-2	CP-98 (XN-2)	WPA-1	Manual	MK 19	MK 6 Mod 1
Toledo	SPQ-2	CP-98 (XN-2)	WPA-1	Manual	MK 19	MK 6 Mod 1
Macon	SPQ-2	CP-98 (XN-2)	WPA-1	Manual	MK 19	MK 6 Mod 1

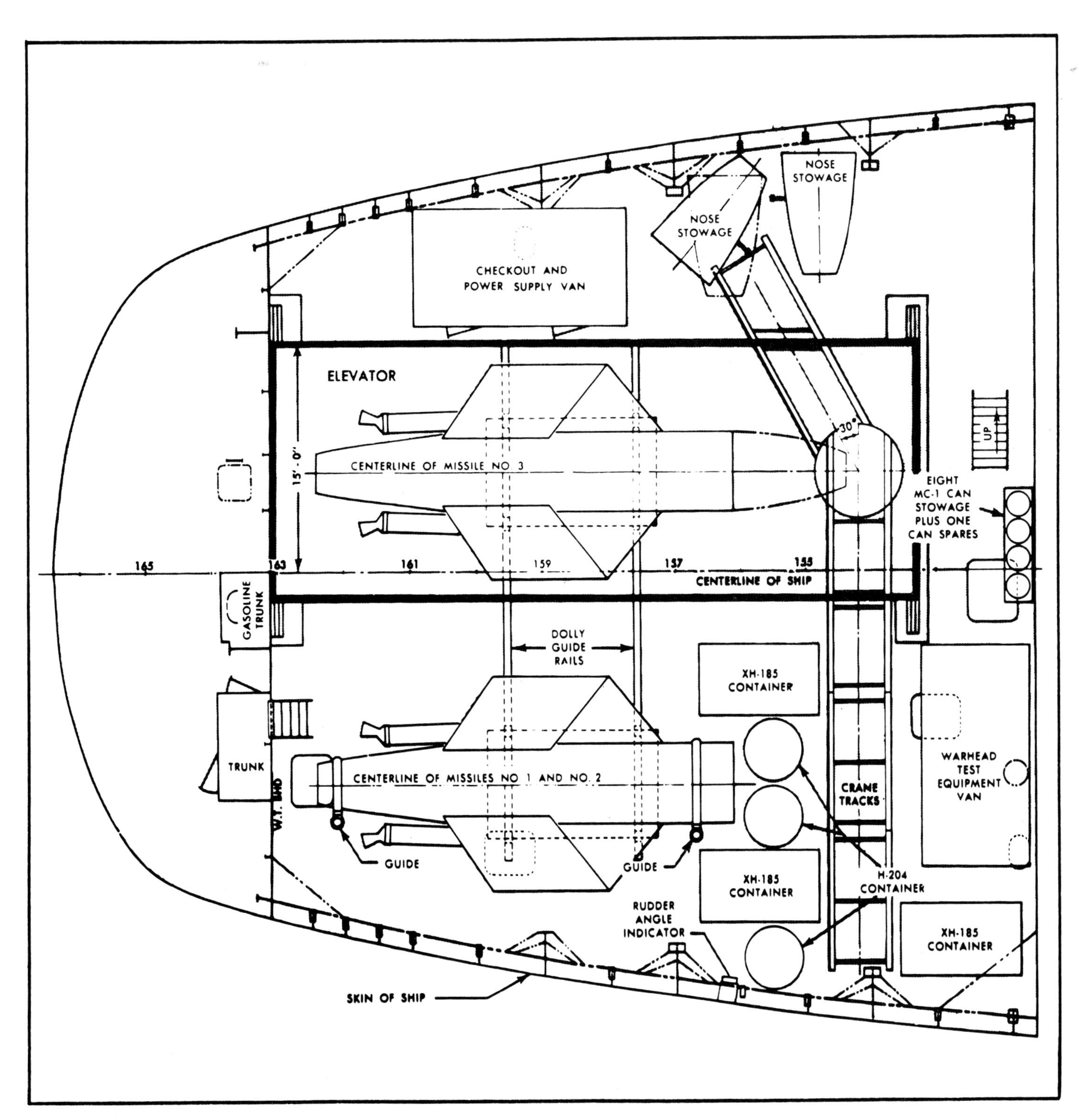

Figure 9-2. Portable hangar installation layout for Los Angeles. *the first deployment of* Los Angeles, *the portable installation was remo\ and replaced with the permanent installation that was installed on* Helena, Toledo *and* Macon.

On 15 February 1955, TM-1080 was launched from Los Angeles during her Operational Suitability Test in Hawaiian waters prior to her first Regulus deployment. This was also the first operational launch of a Regulus I tactical missile. (Courtesy of Kurtz Collection)

The *Los Angeles* returned from WestPac in August 1955. After a overhaul and refit period, during which she received a completely new SR MK 3 launcher and permanent installation of missile checkout and maintenance equipment, *Los Angeles* was ready to again participate in the OPDEVFOR cruiser Trounce guidance evaluation program. On 18 and 19 October 1955, *Los Angeles* successfully launched four missiles. One week later, her Trounce guidance team set a record for the longest Regulus flight to date, 56 minutes with a new Trounce guidance range record of 210 nautical miles. The missile had been launched from Pt. Mugu.[9]

1956-1961

On 21 March 1956, the *Los Angeles* missile team launched the first of what was to be a string of 11 successful flights in the OPDEVFOR program. At the end of these tests, OPDEVFOR qualified cruiser Trounce operations as acceptable for service use to a range of 125 nautical miles and feasible to 200 nautical miles. Cruiser Regulus assault missile operations had been jointly conducted on most of these flights and were also qualified for service use.[10]

After a Trounce guided OST launch near Hawaii on 5 June 1956, *Los Angeles* again deployed to WestPac armed with three Regulus tactical missiles. One was expended in an OST flight on 18 August 1956 utilizing Trounce guidance and no safety-chase aircraft. One week later, *Los Angeles* was resupplied with a tactical missile flown out to her resupply base by an Air Force C- 124 "Globemaster." This was the first

The Los Angeles launch team manuevers one of two JATO bottles into position for attachment to TM-1133 on 19 July 1956 during her second WestPac deployment. TM-1133 was successfully launched and guided for 29 minutes using Trounce 1A before being intentionally expended. This was the twentieth launch for Los Angeles. (Courtesy of Zelibor Collection)

such resupply effort overseas and was considered to be an excellent demonstration of forward resupply capability of the Regulus program.[11]

Los Angeles deployed an additional six times to WestPac over the next six years.[12] In late August 1958, *Los Angeles* was recalled to port during training exercises off the coast of Southern California. The Nationalist Chinese occupied island of Quemoy was again under heavy bombardment by the Communist Chinese. Originally scheduled to deploy four hours after having returned to port to take on critical supplies, the departure was delayed 24 hours so that the three fleet training missiles on board could be exchanged for three tactical missiles armed with the W-27 thermonuclear warhead. *Los Angeles* steamed to Hawaii at 27 knots and after a short layover proceeded to the Formosa Straits to join Task Group 77.2. The crisis had cooled somewhat by the time she arrived but one month later *Los Angeles* was again in the thick of things as she took up station in the Formosa Straits 14 miles from Quemoy. She served as the aircraft control ship during the largest resupply operation of the crisis. One day later *Los Angeles* turned south to rejoin the remainder of Task Group 77.2.[13]

The last Regulus missile launch from *Los Angeles* took place on 28 February 1961. In slightly over six years, missile teams on *Los Angeles* had conducted 43 flight operations with only two missiles lost at launch.[14] The *Los Angeles* was decommissioned 15 November 1963 at Long Beach Naval Shipyard, California.

USS Helena (CA 75)

In June 1955, *Helena* became the second cruiser to receive a Regulus I installation. From the experiences gained on the *Los Angeles*, the missile checkout and launcher installation for *Helena* were to be permanent, but for the first test launches the missile checkout and warhead vans were still used. A vast improvement on the original *Los Angeles* installation was the SR MK 3 launcher. This launcher was a modification of the one used on USS *Tunny* (SSG 282) and was permanently mounted on the port side facing starboard as a fixed launcher. Provision was made for storing a second missile on the starboard deck in the case of a missile malfunction on the launcher. Three missiles were carried below deck in the converted seaplane hangar.[15]

The first Regulus I launch by *Helena* took place 17 August 1955 and was part of the OPDEVFOR Phase III Cruiser Trounce evaluation program. The missile engine shutdown just after booster separation and the missile was lost. Two days later, the second launch was successful and the missile was recovered after a 27 minute flight under Trounce guidance control from *Helena*. Two days prior to deployment to WestPac, *Helena* successfully launched and guided a tactical missile on an OST operation off the coast of California. On 13 January 1956 *Helena* deployed to WestPac carrying three tactical missiles armed with W-5 nuclear warheads.

1957-1960

On 10 April 1957 *Helena* departed for WestPac carrying four Regulus tactical missiles, the three in the missile hangar and one on the launcher covered with a tarpaulin. This missile was to be launched near Hawaii as part of the

USS Los Angeles, *in the Saigon River at Saigon, South Vietnam, in October 1959, on the first anniversary of the Republic of South Vietnam, is reviewed by President Ngo Dinh Diem. (Courtesy of Lucas Collection)*

Regulus I tactical missile rigged for display on Los Angeles *in November, 1959 during her last cruise to WestPac carrying Regulus I missiles. Notice this is one of the "bulged chin" tactical missiles capable of carrying either a W-5 or W-27 warhead. (Courtesy of Whitt Collection)*

ship's OST. On 24 April 1957 the OST missile was launched but the missile failed to turn after the boosters ejected and continued in a steep climb until it stalled and splashed. The conclusion was that prolonged storage on deck was not an acceptable procedure. On 17 June 1957 *Helena* departed Okinawa for her first WestPac OST launch. On board were a host of Far East newspaper reporters as well as several American reporters. The preparation and launch went exactly as planned. The first 160 nautical miles were controlled by Trounce guidance from *Helena* and then aviators from the USS *Lexington* (CV 16) switched to radio command control and guided the missile the remaining 120 nautical miles to the target island, Okino Diato Shima. The dummy warhead detonated in an airburst over the target as reported by the destroyers stationed offshore. *Helena* returned to Long Beach, California, on 19 October 1957.

On 19 November 1957 *Helena* launched a fleet training missile and controlled it on the first of three guidance legs. After 14 minutes of flight and at a range of 112 nautical miles, *Helena* turned over Trounce control to USS *Cusk* (SS 348) one of two West Coast Regulus guidance submarines. After 70 miles of flight *Cusk* turned Trounce control over to USS *Carbonero* (SS 337) the second guidane submarine for the final 90 miles to a simulated terminal dive point off of San Nicolas Island. The missile was successfully recovered. This flight demonstrated the versatility of Regulus guidance options but was not used again. In December *Helena* began a major overhaul that included modernization of her Regulus equipment to the standard of the new *Los Angeles* installation the year before.[16] The overhaul was completed in March 1958.

On 3 August 1958, after refresher training, including successfully launching five missiles, *Helena* departed for Westpac. Renewal of Chinese Communist shelling of the Nationalist Chinese fortifications on Quemoy Island brought *Helena* into the area. On 7 September 1958, she steamed to within 10 miles of the Chinese mainland providing cover for the resupply ships from Taiwan. *Helena* returned to Long Beach on 17 February 1959.[17] After a two month installation of new missile navigation computers and five months of missile launches, *Helena* returned to WestPac in January 1960 for her last Regulus deployment, a six month cruise ending in June 1960. *Helena* launched her last Regulus missile on 1 March 1961, a tactical missile with a war reserve warhead minus the nuclear components. It was successfully guided to target. During her career as a Regulus I cruiser, *Helena* launched 32 missiles, losing five at launch and two during flight.[18]

USS Toledo (CA 133)

Toledo was the last of the West Coast cruisers to receive the Regulus I installation. On 16 May 1956 her missile team successfully launched their first missile, the first of eight successful launches in a row.[19] The flights averaged 230 nautical miles, demonstrating the long range capability of Regulus I combined with cruiser-Trounce guidance.[20] After two successful tactical missile flights, *Toledo* departed for WestPac on the first of four WestPac cruises. *Toledo*'s missile team launched one tactical missile in WestPac during these four cruises. This launch was part of the annual SEATO review held off of Okinawa in May 1958 (see Chapter Eight).

Toledo conducted Regulus flight operations through April 1959, participating in a variety of the contractor evaluation programs including the initial evaluation of the positive flight termination system, and the only cruiser evaluation of the AN/BPQ-2 radar installation. The last Regulus I launch from *Toledo* was on 17 April 1959

Loading sequence for Regulus I missiles onboard USS Helena *(CA 75). Missile is transferred to the cruiser fantail using the dollies seen in the center background. The cruiser's crane then transferred the missile to the missile elevator. To the right can be seen the rail tracks that the missile dolly sat on during transfer to either the launcher or the "dud" storage position to starboard. (Courtesy of Brisco Collection)*

Missile stored in the upper berth of the hangar. With the missile storage being directly over the ship's propellers, these storage facilities had to absorb considerable acoustical energy and vibration. (Courtesy of Brisco Collection)

The missile to the left is being readied for storage in the upper position on the starboard side of the hangar. The missile on the elevator will be stored underneath it, and the third missile will be stored on the elevator. (Courtesy of Brisco Collection)

Helena *underway with a Regulus I fleet training missile rigged out on the fantail. (Courtesy of Brisco Collection)*

with successful recovery at Pt. Mugu. This was the 901st launch of the Regulus I program. During the course of her Regulus I operations, *Toledo* conducted 21 flight operations with only one loss at launch.

USS Macon (CA 132)

Macon received her Regulus I missile launcher and checkout equipment during overhaul from January to March 1956. The missile team, a detachment from Guided Missile Group TWO (GMGRU-2), the surface fleet Regulus support unit, had launched thirteen missiles prior to joining *Macon*.[21,22] *Macon*'s first missile launch took place on 8 May 1956 off of the Virginia Capes. *Macon* had spent the two weeks prior to this launch practicing guidance operations with manned droned F9F aircraft. The drone operations were unsuccessful for the most part as weather and equipment problems in the aircraft prevented all but a few flights from providing useful guidance operation feedback.

May 1956 was an extremely active month for the *Macon* missile team with six launches in eight days. All of this activity was in response to the growing concern over the Soviet intervention in the simmering Suez Canal situation. *Macon* sailed for the Eastern Mediterranean in early July 1956, becoming one of the first American naval vessels to appear in that area during the Suez Crisis. *Macon* was involved in the evacuation of American personnel from the Suez Canal area before returning to Norfolk in November 1956.[23]

1957-1958

The next Regulus I flight from *Macon* was conducted during Operation Springboard, the annual Caribbean naval exercises of the Atlantic Fleet. These flight operations were also carried out as part of Phase VI of the OPDEVFOR evaluation of the Regulus I missile. Phase VI involved the launch of the missile by one ship with mid-course and terminal guidance by a second guidance ship, in this case a submarine. If routinely successful, this guidance relay concept would permit utilization of the full range of the missile. Flight operations took place off the southern coast of Cuba with units of Submarine Division SIXTY-THREE.

On 14 March 1957 two fleet training missiles were launched by *Macon* and controlled in series by USS *Barbero* (SSG 317) and USS *Torsk* (SS 423) 220 nautical miles to a target off Leeward Point near the US Naval Base at Guantanamo Bay. The two flights had an average target miss distance of 1,000 yards. *Macon* also conducted two cruiser Trounce operations, both successful, including one tactical missile operation.[24]

In early September 1957, *Macon* deployed to the North Atlantic as part of Operation STRIKEBACK. For nine days, beginning on 19 September 1957, nearly 300 ships from six NATO countries operated in waters from Norway down the coast of Europe and into the Mediterranean as they responded to simulated Soviet aggression throughout these theaters. *Macon* simulated several launches against targets in Eastern European countries.[25]

Nine months later, in June 1958, *Macon* departed Boston to participate in the LANTFLEX fleet exercise and then toured several Mediterranean and European ports. On 9 July 1958, *Macon* began a one week visit to Oslo, Norway. Three Soviet naval officers from the Soviet Embassy attended an official reception given aboard *Macon* on 12 July 1958. One was quite frank in his interest in the various radar antennas, the helicopter and the Regulus I missile on the launcher. The Assistant Naval Attache from the American Embassy suggested that a ship's tour was in order and the Soviet officers were escorted through the ship. They expressed passing interest in the Regulus I, asking how many were carried and their launch interval. The reply was "about one hundred" and "like a six-shooter." The tour continued and as the Soviets

Members of the international press can be seen in the foreground as they watch preparations for the first launch of a Regulus I missile from Helena *during a WestPac deployment on 17 June 1957 near Okinawa. (Courtesy of Brisco Collection)*

TM-1170 is successfully launched. After guidance by both Trounce and radio command control, the missile was guided to impact on Okino Diato Shima. (Courtesy of Brisco Collection)

Macon *in Boston Harbor with two Regulus I missiles on her fantail. (Courtesy of Fortson Collection)*

were leaving the ship, the senior officer turned and remarked to his guide in perfect English, "We know that you don't carry one hundred of those missiles!" and then left. This contact was duly reported to the Director of Naval Intelligence, US Navy.[26]

Captain Harry Hull, *Macon*'s Commanding Officer at the time, remembers that there were always questions about the security classification of the Regulus program. Upon return to Boston, he called the Office of Chief of Naval Operations and asked if it was possible to display the Regulus I missile on the launcher while in port. The answer was yes, and so from then on, whenever possible, *Macon* would display a missile on the launcher. Hull remembers really stopping traffic in *New York* City during a long weekend visit. *Macon* had been assigned a berth of the west side of Manhattan in the North Hudson River. *Macon* backed into the berth so that the fan tail was close to the Westside Highway but not underneath. This made the highway a perfect spot for viewing the missile and the ship. Once the ship had tied up, Lieutenant Thomas Fortson, the Missile Officer, rigged out a missile, bringing traffic to nearly a complete halt.[27]

Macon's last Regulus launch took place on 29 September 1958, with the flight serving as a test of the US Air Force Air Defense Zone radar system for the Northeastern United States. *Macon* launched the missile 50 nautical miles off of Boston and then guided it on a 100 nautical mile flight to Naval Air Station, New Brunswick, Maine, where a recovery team from Guided Missile Service Squadron TWO recovered the missile successfully. During her three year career as a Regulus missile cruiser, *Macon* had a perfect launch record with twenty missiles successfully launched. This is a reflection of the professionalism of the missile operations teams as well as the enthusiastic embracement of the concept by all of her commanding officers.

Endnotes

1 *Project RAM; Shipboard Installation. Bureau of Ships conf ltr ser 440-015 of 3 February 1953. Naval Historical Center, Operational Archives.*
2 *Ibid.*
3 *Captain Outerbridge was the skipper of the USS* Ward *(DD 139) and sank a Japanese midget submarine off Pearl Harbor moments before the Japanese attack on 7 December 1941.*
4 *Personal interview, Captain Robert Munroe, USN (Ret.), February, 1990.*
5 *Personal communication with Captain Larry Kurtz, USN (Ret.).*
6 *Regulus I Flight Test Record, 1961, LVSCA A50-24, Box 7.*
7 *All war reserve OST flights used a depleted uranium core.*
8 *Personal interviews with Captain Larry Kurtz, USN (Ret.) and Lieutenant Commander Len Plog, USN (Ret.).*
9 *Ibid.*
10 *Final Report of Project OP/S317/X11, Evaluate the REGULUS Guided Missile for Service Use, COMOPDEVFOR ser 00262 of 16 August 1957, page 1-13 and 1-14. Naval Historical Center, Operational Archives.*
11 *Personal interview with Lieutenant Commander Joseph Zelibor, USN (Ret.). Personal papers verify the dates and activities of this deployment.*
12 *Dictionary of American Naval Fighting Ships, US Naval History Division, 1959. Volume IV, page 144-145.*
13 *Personal interviews with Jim Whitt, a Signalman 3rd Class at the time and Lieutant Commander S.C. Ager, USN (Ret.), Missile Warhead Officer.*
14 *Regulus I Flight Test Record, 1961, LVSCA A50-24, Box 7*
15 *Personal communication and interviews with Lieutenant Commander Ken Brisco, USN (Ret.).*
16 *Regulus I Flight Test Record, 1961, LVSCA A50-24 Box 7*
17 *Dictionary of American Naval Fighting Ships, Vol III, pages 289-291.*
18 *Regulus I Flight Test Index, 1961, LVSCA A50-24, Box 7.*
19 *Regulus I Progress Report #14, 1 January to 31 March 1956, page 81. LVSCA, A50-22, Box 5.*
20 *Ibid.*
21 *Personal communication with Vice Admiral Vernon L. Lowrance, USN (Ret.).*
22 *Post Exercise Report of Regulus Training Operation 23 April-18 May 1956. CA132/13:cn A9-8 Ser 0129 of 22 June 1956. Naval Historical Center, Operational Archives.*
23 *Personal communication with Rear Admiral Vernon L. Lowrance, December, 1992.*
24 *Final Report on Project OP/S317/X11. Evaluate Combinations of Surface Ship Launching of Regulus I with Final Control Exercised by Units Other Than the Launching Ship. 16 August 1957. Naval Historical Center, Operational Archives.*
25 *Personal communications with Captain Thomas E. Fortson, USN (Ret.).*
26 *Report of Contact with Three Russian Naval Officers During the Visit of the Ship to Oslo, Norway. CA-132 ser 0115A of 6 August 1958. Operational Archives, Naval Historical Center.*
27 *Personal communications with Rear Admiral Harry Hall, USN (Ret.) and Captain Thomas Fortson, USN (Ret.).*

First launch of a Regulus missile, FTM 1276, from Macon *on 8 May 1956. This was the first East Coast Trounce guidance mission. (Courtesy of Fortsar Collection)*

Regulus I tactical missile ready for launch from Macon. *The nose boom carried barometric sensors for use in detonating the warhead. (Courtesy of the Fortson Collection)*

The major change made to Tunny *during her modification to an SSG was the addition of a large deck mounted hangar, constructed of 16 foot diameter hull sections similar to the hangars of the troop carrying submarines USS* Perch *(SSP 313) and USS* Sealion *(SSP 315). Unlike the hangars used in the Loon program.* Tunny's *hangar door opened vertically to prevent listing while ramming out the missile. (Courtesy of Blount Collection)*

Regulus I Deployment Aboard Submarines

Chapter Ten: USS Tunny (SSG 282)

The decision to convert *Tunny* from a fleet submarine to the new guided missile submarine configuration was based on two factors. One was that the two submarines involved in guided missile launch operations during Project DERBY (See Chapter Two), USS *Cusk* (SS 348) and USS *Carbonero* (SS 337) were still in use and hence unavailable for conversion. Secondly, the *Cusk*'s missile hangar could neither be reached from the inside the hull nor was it large enough to hold two Regulus I missiles. *Carbonero* did not have a missile hangar installed.

Tunny, after conversion and reclassification as an SSG, was recommissioned at Mare Island Naval Shipyard (Mare Island) on 6 March 1953, with Lieutenant Commander James Osborn, Commanding Officer. Many of her crew were selected from Project DERBY personnel. Commander William "Pappy" Sims, Officer-in-Charge of GMTU-5, the Navy unit training to prepare and launch Regulus missiles, had been offered command of *Tunny*. Sims felt that Osborn, his assistant officer-in-charge, was more than qualified for the job and strongly recommended him instead.[1] Osborn had previously been the DERBY Project Officer and was well acquainted with the idiosyncracies of submarine launched guided missile operations. Thus began the Regulus program's association with two dynamic and creative officers. Sims and Osborn were instrumental in the growth of Regulus into an operational system with Sims working from within the Bureau of Aeronautics at OP-51 and Osborn on board *Tunny* during the crucial first years of submarine Regulus I operations.

During the nine month conversion process, *Tunny* was fitted with a missile hangar aft of the conning tower with hangar access from within the hull; a launcher that could be elevated and a snorkel mast and a streamlined sail that housed the conning tower, both periscopes and other retractable masts. One of the four main engines and the auxiliary engine were removed. The missile hangar was designed to house two

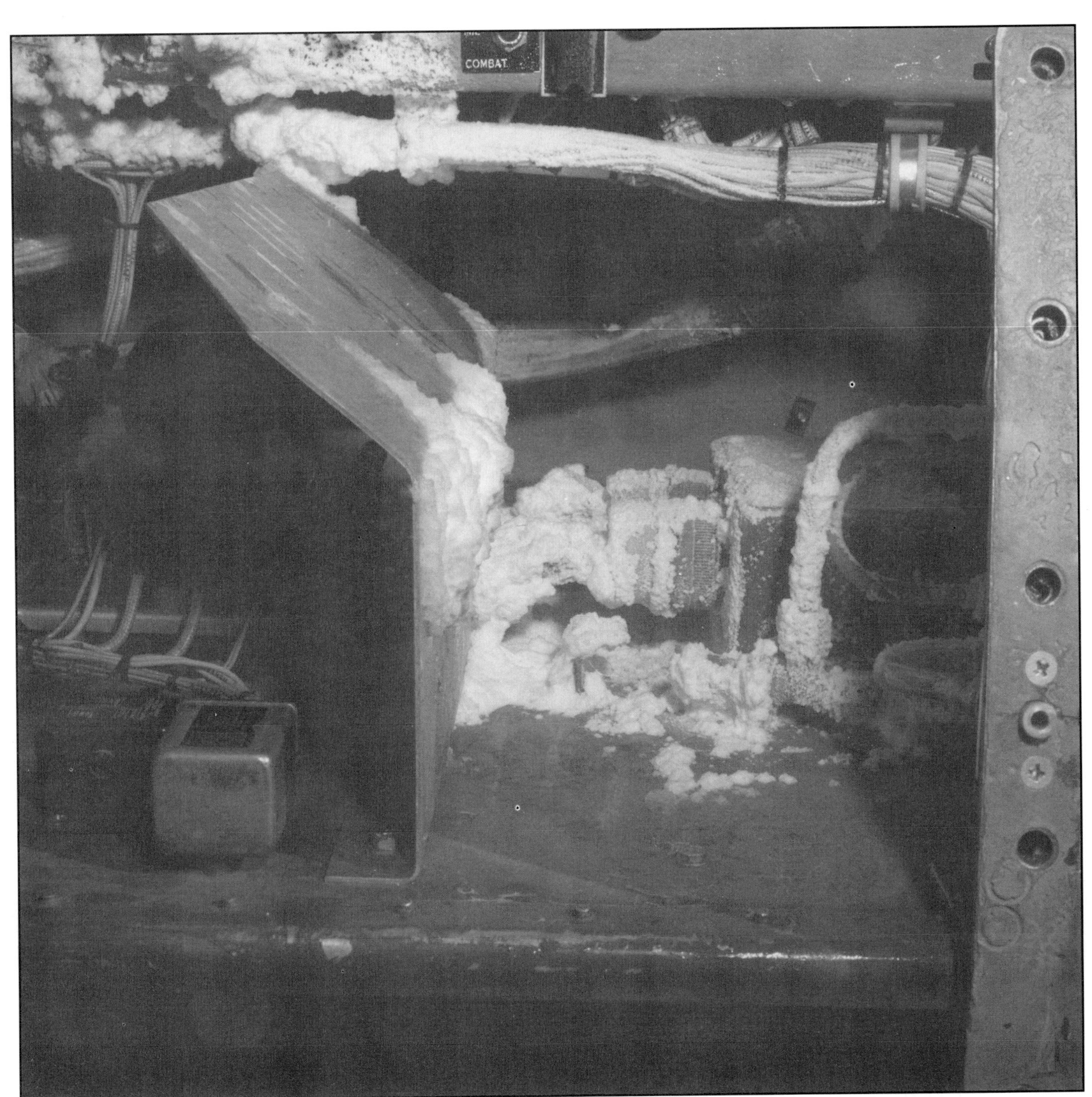

Salt deposits on the electric equipment inside FTM-1018 engine compartment from ingested saltwater spray. The saltwater spray ring was replaced by a jet exhaust deflector after just this one use. (Courtesy of Albrecht Collection)

FTM 1018 has been rammed out onto the launcher on 15 July 1953. The apparatus on the hangar door was salt water spray ring to keep the hangar door cool and prevent warpage during engine run up. (Courtesy of Albrecht Collection)

Regulus I missiles in a rotating cage. One missile was carried right-side up, ready for ramming out onto the launcher, while the second missile hung by its launch slippers upside down directly above the first. The circular cage which held the missiles rotated like a revolver cylinder to bring the upper missile into an upright position, ready to ram out after the first missile had been launched. Experience gained in planning the *Cusk* and *Carbonero* missile check out, launch and guidance centers was instrumental in planning *Tunny*'s installations.[2]

Tunny departed Mare Island on 15 May 1953 for San Diego and a three week shakedown cruise. She then proceeded to Port Hueneme for Regulus equipment checkout and dummy missile launch operations as well as training with drone aircraft during the third week of the shakedown. After a two week post-shakedown upkeep in San Diego, *Tunny* reported back to Port Hueneme on 9 June 1953 to commence Regulus operations. During the week of 6 July 1953, *Tunny* loaded her first production Regulus missile, FTM-1018, and commenced a complete checkout of the missile center pre-launch consoles and guidance stations. On completion of the pre-launch checkout, the missile was rammed out on the launcher, the engine started and the missile elevated. The engine was run up to 100% power and a new hangar door cooling system was turned on for the first time with the missile engine running. A spray ring using salt water was used to cool the hangar door during prolonged exposure to the hot jet engine exhaust gases. Cooling prevented deterioration of the door gasket material as well as minimized warping of the hangar door. Much to the consternation of the *Tunny* and Chance Vought engineers, the salt water ladened engine exhaust was sucked back into the missile engine intake. Salt deposits soon formed outside and inside the missile. The spray ring was quickly replaced by a slatted exhaust blast deflector that folded down between the launcher rails when not in use.[3]

The first Regulus I launch by a submarine took place 15 July 1953 and was the 58th launch of the Regulus program. Launched by Lieutenant John Duff and the *Tunny* missile team, FTM-1018 was successfully recovered at San Nicholas Island. For the next three months, *Tunny* participated in SSG-type standardization trials as well as conventional submarine operations. Regulus launches resumed in November 1953 after installation of the CP-98 Trounce Guidance Computer. *Tunny* launched three more missiles, all successfully, but did not take part in guidance since her P-1X Trounce guidance equipment was not completely installed. With a four-for-four launch record, the missile team, now headed by Lieutenant Sam Bussey, was anxious to work with the Trounce guidance system and be capable of independent operations.

1954

Tunny reported to Mare Island on 1 February 1954 for restricted availability to complete installation of the P-1X Trounce radar; modifiy the missile electrical umbilical cord installation, making it accessible for use inside the hangar while submerged; add a permanent external JP-4 missile fuel tank; and install a new internal missile communication system. In mid-April the work was completed.[4]

The first Trounce guidance operation by *Tunny* took place 1 June 1954 with the first submarine launch of a production tactical missile, TM-1040. The Trounce guidance system worked well during the twenty minute flight over the Sea Test Range off of Pt. Mugu.[5]

While Regulus launch operations appeared to be routine, this was far from true. Operational solutions were being found for various persistent missile and guidance system problems. Critical engineering problems were being solved at sea by the same personnel that had made Project DERBY successful. While these fixes temporarily solved the problem at hand, submariners had begun to modify each of the P-1X Trounce guidance radar course directing centrals to the point that each of the four sets, three on submarines and one onshore at Pt. Mugu, were literally custom-made installations. Only the electronic technicians working on these modifications really understood the intricacy of each of the myriad changes. These glaring deficiencies were not fully rectified until the introduction of the Trounce BPQ-2 system five years later.

In August 1954, Operational Development Force (OPDEVFOR) evaluation of the Regulus I program began (See Chapter Four). Phase I was "The Single Submarine Trounce Method of Tactical Delivery."[6] *Tunny*'s first OPDEVFOR operation on 26 August 1954 ended shortly after launch, when the missile parabrake prematurely deployed at 1,700 feet, causing the missile to splash. One week later, on her eighth

Rear view of FTM 1018 on the launcher with wings folded. The top of the hangar door spray ring can be seen at the bottom of the picture. (Courtesy of Albrecht Collection)

Moments after lift-off. This first Regulus I launch from a submarine was a complete success. FTM 1018 was launched three times by Tunny *and three times by Chance Vought during Phase A operations. (Courtesy of Albrecht Collection)*

launch, *Tunny* successfully completed a Trounce operation with a range of 113 nautical miles. Although the missile was lost on recovery landing approach to San Nicolas Island, the operation was considered a success because in a tactical situation the recovery would not have taken place. *Tunny* conducted four more OPDEVFOR launches in 1954 with a record of four successful flights in six attempts. Considering that these were the ninth through thirteenth Regulus launches for the *Tunny* missile team, this record was no small achievement.[7]

1955

The submarine OPDEVFOR program continued with 14 flight operations during the first five months of 1955 with only one failure at launch and three in flight. The culmination of this phase of the OPDEVFOR effort came in May 1955, with the first submarine operational suitability test (OST), where the weapon system was tested under tactical conditions without assistance from Chance Vought personnel. The OST was the first trip from the West Coast for *Tunny* in her new role as an SSG. The submarines of Submarine Division FIFTY-THREE, *Tunny*, *Cusk* and *Carbonero*, departed Port Hueneme, California for Pearl Harbor, under simulated wartime conditions.[8]

Tunny arrived in Pearl Harbor with two tactical missile airframes but no warheads. She docked at West Loch, the Naval Ordnance Depot in Pearl Harbor, where war reserve W-5 atomic warheads were assembled by conversion of two Mk-5 atomic bombs. Lieutenant Marvin Blair, the Warhead Officer for Guided Missile Unit FIFTY (GMU-50), was temporarily assigned to *Tunny* along with several warhead technicians. Blair remembers that an entire building had been set aside for his warhead team to use. It took four hours to convert from a bomb to a warhead and then another four hours to place the warhead in the missile and run mating checks. These warheads used training components (depleted uranium-238) and contained the equivalent of approximately 3,000 pounds of high explosives. The *Tunny* then took both missiles back on board and stayed in West Loch to perform missile and warhead acceptance checks.[9]

The first OST launch was 3 May 1955. Lieutenant George Clegg was the *Tunny* Missile Officer and recalls that *Cusk* and *Carbonero* took positions between the target and the launch position of *Tunny* west of French Frigate Shoals in the Hawaiian Islands chain. Once the missile was launched, the two guidance boats were to take control from the *Tunny* in sequence and then *Carbonero* would command the missile to begin the terminal dive maneuver. The destroyer USS *Cowell* (DD 574) was stationed off the target track in the vicinity of the target to obtain radar confirmation of the terminal dive.[10,11]

The launch was uneventful but when *Tunny* tried to take Trounce control, she was unable to establish either radar contact or control, and the missile continued down range at full throttle. *Cusk* was also unable to establish control. Cowell's air search radar obtained what appeared to be several "skin track" returns (later found not to be the case) and cleared *Carbonero* to send the missile its terminal dive signal. *Carbonero* transmitted the terminal dive command with no response and the missile continued flying to fuel exhaustion approximately 100 miles further down range. The flight was counted as a partial success. Clegg remembers that everyone was pleased, considering that this operation was the first time that two Regulus I's had been carried for an extended period in a realistic test of tactical conditions of prolonged submergence and approach to target. The missile that was launched had been carried in the upright position. Booster alignment had been maintained; and, except for the radar guidance failure, the missile had been launched successfully and had flown the correct flight profile.[12]

The second launch was at night, not by choice but due to delays caused by problems during missile checkout. The second missile had been in the inverted position in *Tunny*'s hangar since loading at Port Hueneme fifteen days earlier. A check of the boosters indicated they were out of alignment. Clegg found that after realignment the boosters would not hold the correct position. Clegg, Osborn and Lieutenant Commander Walt Dedrick, on board as Prospective Commanding Officer, debated whether or not to launch the missile. Osborn finally decided that if this was to be a true test of the missile, as it should be during an OST, they must align the boosters once more and carry out the launch. In

FTM 1018 with engine at full power and launcher fully elevated. Jack Welch, a missile operations engineer with Chance Vought is standing off to the left. (Courtesy of Loral Vought Systems Archives, Albrecht Collection)

this manner they could more completely identify any problems the system had and the capability of fixing them in an operational setting.[13]

Cusk was stationed 1000 yards off the *Tunny*'s port quarter to observe the missile launch. Having immediately dived after the launch according to operational doctrine, *Tunny* quickly resurfaced. Osborn, Clegg and Blair scrambled up to the bridge, expecting to watch the warhead's detonation since this was a short-range, low-altitude shot. They saw nothing in the night sky. A terse radio telephone message from *Cusk*'s skipper, Commander Eugene Pridinoff, was one word: "Splash." Pridinoff did not want to elaborate for security reasons. The submarines pulled close to one another and talked over the situation on the more secure short-range underwater telephone. Pridinoff told Osborn what had happened: the missile had splashed astern of *Tunny* immediately after launch. Pridinoff's description of the launch and subsequent crash made it obvious that booster alignment was the culprit. The thrust from the boosters had to be directed within 0.5 inches of the center of gravity of the missile/booster combination or stability problems would result that could not be countered by the control surfaces on the missile.[14]

Tunny headed back to Pearl Harbor, and experts from Chance Vought and the Navy worked for two weeks to develop a solution to the alignment correction procedures. On 20 May 1955, *Tunny* launched a short range, low altitude flight, 13 minutes long, and dumped the missile on target - a small success. Commander R.G. Anderson, Commander Submarine Division FIVE, summarized the exercise saying "...this was the first extended operation and provided negative but excellent information on the system's capabilities."[15] Lieutenant Commander Walter Dedrick relieved Osborn as Commanding Officer of *Tunny* on 6 June 1955 at Pearl Harbor and two months later, *Tunny* entered Mare Island for an extensive overhaul period.

1956

Launch operations resumed in June 1956 off the coast of Southern California after the Mare Island overhaul. As part of proficiency training, Clegg recalls *Tunny* made an undetected transit from Port Hueneme to the Hawaiian Islands. Approximately 150 miles north of the island of Oahu, they surfaced for 10-12 minutes for a simulated missile ram out and launch. Anti-submarine forces had been alerted and were searching for *Tunny* but the electronic countermeasures team did not detect the slightest hint of search radar activity.

On 22 July 1955, *Tunny*, positioned northwest of the Hawaiian Islands near Gardner Pinnacles, launched TM-1184, controlling the missile for the first 12.5 nautical miles before passing control to *Cusk* located north of La Perouse Pinnacle. *Cusk* controlled the missile for the remainder of the 185 nautical mile flight to impact. The next day *Tunny* launched and guided TM-1180 for a flight of 85 nautical miles to impact with excellent accuracy. [16] These launch operations demonstrated the maturity of the Regulus I program on the West Coast.

In late December 1956, the Phase I, submarine-Trounce guidance phase of the OPDEVFOR evaluation program was completed. *Tunny*, with *Cusk* and *Carbonero* had conducted 20 manned drone flights and 15 Regulus I missile flights during this period. High- and low-altitude flight profiles had been used to evaluate options for warhead delivery. The OPDEVFOR report concluded that both high- and low-altitude delivery profiles were acceptable, but that the accuracy of guidance equipment installed on board *Tunny* was marginal,

"The P-1X radar and CP-48 computer were unreliable, inaccurate and contained certain design deficiencies. The Mk 7 Mod 3 gyro was inaccurate. The missile launching and handling equipment was satisfactory but the environmental conditions in the missile hangar did not meet the specifications for Regulus I warhead storage. The shipboard checkout equipment was satisfactory. Sufficient, and well-trained personnel were available to operate the system and, missile maintenance (when spare parts were readily available), was satisfactory. Guidance equipment maintenance was a continuing major effort on all three submarines.[17]

While hardly a ringing endorsement of the program, the missile had performed well. Guidance equipment problems continued to plague the system and were finally resolved three years later with the installation of the Trounce BPQ-2 radar course directing central equipment (See Appendix I).

1957

The year began with the announcement of the impending move of GMU-50 and Submarine Division FIFTY-ONE (SubDiv 51) from Port Hueneme to Pearl Harbor in the late Spring. The move to Hawaii was undertaken to break away from the test and evaluation atmosphere at Pt. Mugu and begin to really shake down the system in an operational setting. Logistic facilities and training in Hawaii would permit a much more realistic appraisal of system deficiencies. This was a natural evolution of any weapon system being brought into the fleet, submarine or otherwise. Regulus assault missile operations with 7th Fleet carriers had moved to Hawaii a year earlier. Perhaps as important, the move to Hawaii would also put the Regulus submarine division that much closer to the Western Pacific theater of operations.

On 25 February 1957 the submarines of SubDiv 51 deployed to the Gulf of Alaska for the first OST operations under anticipated tactical weather conditions. The USS *Rasher* (SSR 269), *Tunny*, *Carbonero* and *Cusk* deployed separately. Bill Cannon, a Chance Vought Missile Operations Engineer assigned to GMU-50, was on board *Tunny* to help evaluate the effects of the artic weather on missile storage in the hangar as well as operations on deck. The Atomic Energy Commission had expressed concern about temperature and humidity conditions in the missile hangar, so one of his collateral duties during the cruise was to monitor the hangar environment. Cannon recalls that the weather conditions were really miserable and certainly were going to be a stressful test on the missile, warhead and launch equipment.

After preliminary exercises, all four submarines rendezvoused at Dutch Harbor, Unalaska Island, on 8 March 1957. Remaining operational details for the upcoming missile launch were worked out. After thoroughly checking out the missile systems and doing guidance loop checks with both guidance submarines, *Tunny* and the other boats departed on 10 March 1957 for the launch area near the Pribilof Islands, 200 nautical miles northeast of Unimak Island in the Aleutian Island Chain.[18]

Commander Richard Clark was Commanding Officer of *Cusk*. Clark recalls that *Cusk* arrived on station between *Tunny* and *Carbonera*. *Tunny* launched TM-1172 from a point just off the Pribilof Island, towards *Cusk* and *Carbonero*. The launch was normal but soon afterwards the transponder beacon return signal became intermittent. *Cusk* had acquired the beacon before it became erratic but *Carbonero* could not acquire track on the missile, so *Cusk* continued to control; and, as the range opened, it became more and more difficult to maintain contact. The tracking plot began to resemble a sine wave and finally Clark ordered the missile dumped. On board *Rasher* all eyes were trained to port, the expected direction of explosion. All were quite surprised when the OST warhead high explosives detonated to starboard.

On 30 April 1957, Lieutenant Commander Marvin Blair relieved Dedrick as Commanding Officer of *Tunny* in Port Hueneme, California. One week later, GMU-50 and SubDiv 51 were transferred permanently to Pearl Harbor, arriving 13 May 1957. The submarines were assigned to Submarine Squadron ONE and commenced training operations in the HAWAIIan area. The three Regulus submarines were reassigned to Submarine Division NINETY-ONE, which was significant, because there was no Submarine Squadron NINE. This division was therefore singled out to be designated by a number that did not correspond to an operational squadron so that it could operate as a Submarine Launched Attack Mission unit. With a Division Commander embarked in the launch submarine, the division would deploy as a tactical group.[19]

1958

In late January 1958, *Tunny*, *Cusk* and *Carbonero* again headed towards the Northern Pacific to participate in Operation SLAMEX, a Regulus tactical missile launch, and Operation TRANSITEX, an opposed transit to Vancouver Island from Alaskan waters with anti-submarine forces searching for *Tunny*. Transit to the launch area was also opposed but all three boats made it undetected, probably due in part to the truly adverse weather enroute. At one point *Tunny* was experiencing 45 degree rolls. An important part of this operation was another detailed study of the environmental conditions to which the W-5 warhead was exposed. With the storms *Tunny* encountered, Sandia Corporation weapons engineers again had a perfect test for the components.[20] When *Tunny* reached her operating area in the Bering Sea, all three engines and the Klineschmidt fresh water distiller/evaporator broke down. *Tunny* drifted for nearly three days with steadily diminishing battery power available. The missile launch was scheduled to take place but a few days later. Using patches to seal the outside of the cracked cylinder liners, the diesels were on line in time for the launch.[21,22]

Tunny launched the missile on 30 January 1958, again near the Pribilof Islands. The two guidance boats maneuvered the missile to the target area where the war reserve warhead detonated in a contact burst on a remote island. Two destroyers had been stationed 20,000 yards off the island to provide radar coverage of the terminal dive and impact. Target accuracy was within 1100 yards, more then sufficient for the W-5 warhead.

Having completed the SLAMEX portion of the exercise, *Tunny* began the opposed transit to Vancouver Island. Her track took *Tunny* across the eastern part of the north Pacific to the northern tip of Vancouver Island. As on previous exercises, *Tunny*'s speed of advance was high, requiring her to run on the surface or snorkel. *Tunny* was not allowed to use Loran A navigation in order to simulate wartime conditions of the system being out of operation. One alternative was to use star-sights but this would leave *Tunny* vulnerable again to the anti-submarine forces. Blair and his Executive Officer, Lieutenant Doug Stahl, conferred and decided to use the available, though very sketchy, bathymetric (sea floor contour) charts and attempt bathymetric navigation to reach the Vancouver area. A number of "guyots" or flat-topped underwater mountains were present a few hundred miles from Vancouver Island. One such guyot was sufficiently isolated from the others to permit its identification, hence to provide *Tunny* with navigational accuracy. They plotted a course to intercept this isolated guyot and when they reached what was thought to be the correct location, the boat made several figure eight passes using the fathometer and confirmed the guyout's identification. They then set course for Vancouver Island and again, without surfacing, were able to make landfall, all of this while remaining undetected. This transit demonstrated that the use of bathymetric charts might be practical for use in the future and also made the anti-submarine warfare forces' task much more difficult.[23]

Soon after her return to Pearl Harbor, *Tunny* participated in the first public launch demonstration of her missile system in the Hawaiian area. On 20 March 1958, the press and other invited guests of the Navy on board USS *McCain* (DL-3), watched as *Tunny* surfaced 500 yards away and launched a bright red training missile. After a successful launch, the 22nd consecutive one by the *Tunny* missile team, FTM-1267 was controlled by *Tunny* and *Carbonero*, using Trounce guidance, to a simulated terminal dive on a target off of Kauai with subsequent recovery at Bonham ALF.[24]

On 16 April 1958, *Tunny* launched her 51st Regulus I missile. One of the JATO boosters malfunctioned, causing the missile to skid to the left off the launcher, turn 180 degrees and head straight back at the submarine. Blair recalls watching through the periscope as the missile neared *Tunny*. Luckily it crashed 1000 yards away. Another problem that had cropped up during this launch was a malfunction of the hangar door hydraulics system. The hangar door was 13 feet in diameter and hinged on top. Two huge hydraulic rams opened and closed the door. The door was locked shut using a bayonet-type locking ring, both shutting and locking sequences being controlled with a single control lever. To open the hangar door, the ring rotated 15 degrees, releasing all the locking lugs. In the fully unlocked position the ring tripped a microswitch that diverted hydraulic pressure to the door opening rams. The valves controlling these functions were never very reliable and lasted a year at best. Once submerged, the enormous pressure on the hangar door tended to shut the door tighter than the bayonet locking ring and at deeper depths, the locking ring would actually be loose. If the valves were leaky, the locking ring would slowly rotate itself to the open position. This would trigger the microswitch and then the rams would try to open the door while submerged. While the outside pressure prevented the door from opening, it was nevertheless quite disconcerting to suddenly hear the hydraulic rams shuddering as they tried to open the door while *Tunny* was underwater.[25] Mare Island sent out their top mechanical engineers but they couldn't locate the problem. With time becoming a critical factor due to her upcoming deployment, and with no option of postponing or canceling *Tunny*'s departure, the Pearl Harbor Naval Shipyard came up with a makeshift solution that no one on *Tunny* liked at all. A large A-frame was mounted aft of the door and two chainfalls were used to raise and lower the door manually from the outside. These chainfalls would have to be mounted each time *Tunny* surfaced for a launch. This was

Tunny, *minutes after surfacing near the SEATO task force on 22 May 1958. The hangar door is nearly shut, and the launcher is moving into the elevated position. The apparatus between the missile tail, and the hangar door is the chain hoist A-frame installation. The* Shangri-La *is in the background. (Courtesy of U.S. Naval Institute Collection)*

Successful launch of TM-1198. The paint scheme on this missile was unusual. This is the only photograph of a tactical missile with a gray ventral and blue dorsal color scheme. The flight was routine, and the missile detonated 3,000 yards off the port side of the task force column. (Courtesy of Author's Collection)

clumsy, marginally effective, and certainly not a tolerable operational solution.[26]

Tunny departed Pearl Harbor in mid-May 1958 to participate in the annual SEATO demonstration exercise with the 7th Fleet. While this would not be the first Regulus I demonstration launch in WestPac, cruisers having done so since 1955, it would be the first submarine launch of the missile during the SEATO exercises and would be a dramatic demonstration of a new and powerful addition to the Pacific Fleet submarine force. Two Regulus launches would be the highlight of the exercise, the first from the *Tunny* and the second from the heavy cruiser USS *Toledo* (CA 133). This would also mark the first time that an OST was conducted using a W-27 war reserve thermonuclear warhead with depleted nuclear components.

Four days into the trip to the Okinawa rendezvous with the SEATO task force, Lieutenant George Skirm, the Missile Officer, finally convinced Blair and Commander Archer Gordon, Commander, Submarine Division NINETY, to let him try his solution to the hydraulic problem. Lieutenant Skirm had surmised that the fault was located in an hydraulic check valve. He and one of the submarine's enginemen found the necessary spring steel wire as well as a fountain pen barrel of suitable size around which to wrap the wire to form a spring. They installed the fix, *Tunny* surfaced and everyone held their breaths as the door was activated. To Lieutenant Skirm's credit, everything worked beautifully, giving *Tunny* much greater confidence that the upcoming launch would be infinitely more realistic and impressive. The chainfalls were removed and stored in the bilges.[27]

A perfect day dawned on 22 May 1958, with scattered clouds and a slight wind from the north. Immediately after *Tunny* contacted the task force visually, Vice Admiral Wallace M. Beakley, Commander 7th Fleet, ordered Blair to take station ahead as formation guide. The entire task force of 15 vessels in one long column formation was to follow *Tunny*. Having no previous experience at keeping formation in a submarine with surface ships the size of aircraft carriers, much less act as task force guide, Blair turned to the Gordon, and asked "What should I do first?" Gordon quietly answered "I suggest you get the hell out of their way before they run you over!"[28]

Tunny was 500 yards off to port of the aircraft carrier USS *Shangri-La* (CV 38) when she launched TM-1198. The missile followed the pre-set 60 degree dead reckoning turn and was picked up by two FJ-3D airborne guidance planes from the RAM Detachment aboard *Shangri-La*. After a pass down the starboard side of the column and out of sight, they turned the missile back toward the Task Force, crossing in front of the column just ahead of *Shangri-La* and well within sight of the SEATO senior officers. The missile war reserve warhead with dummy components was detonated approximately 3,000 yards off of the port side of the column. *Tunny* received a "Bravo Zulu" (Well Done) by the Commander-in-Chief, Pacific and was granted 30 days leave in WestPac at three ports of her choice. During this period the A-frame chain fall system was converted into a Tori Shrine by shipyard personnel and painted brilliant Oriental red.[29]

On the return trip, one day out of Pearl Harbor, Commander Submarine Forces, Pacific Fleet ordered Blair to turn off running lights, start zig-zagging and make best speed home. *Tunny* arrived at West Loch on 16 July 1958, picked up two war reserve W-27 warhead missiles, her war shot torpedoes and quickly reprovisioned. The A-frame assembly was removed and thirty-six hours later *Tunny* began the first submarine Regulus I deterrent mission. The crew had very little time with their families. Captain Blair recalls he had only a few hours with his wife before they left. Lieutenant Stahl remembers calling his wife and having her get their children out of school so that he could say goodbye. Everyone felt that this might be the start of something much bigger.[30]

The Lebanon situation was moving all too quickly towards a possible confrontation with the Soviet Union as it made motions towards sending troops into the Middle East. United States military bases worldwide had been put on heightened alert. Aircraft carriers were sorely needed in the Indian Ocean area of operations, so *Tunny* was directed to proceed to the Northern Pacific to relieve the sole carrier on station in this region. Navy RC-121 aircraft, forming

Tunny *moored across from the Japanese Maritime Self Defense Force patrol frigate* Nara *(282). (Courtesy of Christensen Collection)*

the radar barrier from Kodiak to Midway, were not informed of her passage so she had to transit undetected under essentially wartime conditions. This was not difficult, since *Tunny*'s electronic countermeasures equipment picked up the barrier aircraft radar well before she reached Midway. Five days of continuous snorkeling were required to transit through the Midway to Alaska radar barrier patrols. As the crisis abated, *Tunny* returned to Pearl Harbor in mid-August.

Regulus Deterrent Patrols 1959-1964

Tunny went into a restricted availability overhaul in early 1959. Lieutenant Commander Morris A. Christensen relieved Blair as Commanding Officer of *Tunny* on 1 July 1959. Christensen had been executive officer of *Cusk* during the research and development phase of Regulus I on the West Coast. He came to *Tunny* from the Bureau of Aeronautics where he had been instrumental in coordinating designs for submarine installations of Regulus I. *Tunny* had just come out of overhaul during which the new and greatly improved Trounce BPQ-2 guidance system had been installed.

Within three weeks of Christensen's arrival two unusual events took place. The first was the *Tunny* and a missile launch were featured as part of the popular comic strip "Buzz Sawyer." This was probably not the first time a highly classified weapons system was featured in some way in the Sunday morning paper but this exposure was certainly a change of pace for the "Silent Service."[31] The second event was even more unusual. A six-ship squadron of Japanese patrol frigates had arrived in Pearl Harbor a week earlier. The crew of *Tunny* was amazed to find berthed across the pier from them the Japanese Maritime Self Defense Force patrol frigate *Nara* (282). They calculated the chances of two ships with the same hull number being berthed this close together was on the order of a million to one. Gifts were exchanged and photographs taken to commemorate the event.[32]

On 23 October 1959 *Tunny* deployed on the first scheduled submarine strategic strike deterrent mission in the Regulus program. Although under the operational control of Commander Submarine Forces, Pacific, all Regulus submarines were assigned strike targets by the Commander-in-Chief, Pacific Fleet. The Single Integrated Operations Plan, the coordinated targeting of all U.S. military force's nuclear weapons, was still one year away from implementation. Upon departing the local operating area near Hawaii, the Operation Order directed SSG's to remain undetected at all times while in transit to and from their assigned patrol area. This directive required the boat to submerge even if friendly anti-submarine forces, such as P2V "Neptune" anti-submarine warfare patrol aircraft, RC-121 "Constellation" radar picket aircraft or surface ships on radar picket duty were detected.[33]

On 30 October 1959, while enroute to station, *Tunny* experienced a fire in the control cubicle. This was located in the Maneuvering Room, and was the heart of a diesel-electric submarine's propulsion regulation and control system. Here the propulsion power source was shifted from the main generators to the battery upon diving. The damage control procedure was to isolate the compartment, extinguish the fire and then effect emergency repairs. Due to the very rough seas, Christensen submerged to begin repairs which turned out to be mainly the replacement of burned contacts. *Tunny* proceeded to her patrol area on schedule.

On 27 November 1959, *Tunny* was on station, snorkeling to charge batteries and preparing for a Thanksgiving Day dinner. Captain Christensen was just starting to slice into a large turkey on the Wardroom table when everyone felt, then heard, a loud bump forward. With the Collision alarming sounding, Christensen ran to the Control Room while Lieutenant Commander Peter Fullinwider, the Executive Officer, rushed forward to assess the damage. Fullinwider reported from the Forward Torpedo Room that the sonar dome shaft had been shoved upward into the compartment and water was spraying in through the damaged shaft seals. All compartments were isolated and compressed air was bled into the Forward Torpedo Room to reduce flooding. The sonar dome shaft was lowered and locked in the down position. The leakage stopped when additional packing was placed around the shaft. Meanwhile the sonar watch had been searching for contacts but found nothing except the usual whale noises which had been present prior to the collision. *Tunny* secured from Collision Quarters and resumed her patrol routine with the forward sonar equipment inoperative. With no contacts other than the whales, it appeared that a whale had mistaken *Tunny* for a companion. *Tunny* returned to Pearl Harbor on 16 December 1959 and immediately went into drydock to have the damage inspected. The sonar dome was missing and the interior sonar head array damaged beyond repair so it was replaced.[34]

1960

On 12 February 1960, *Tunny* launched FTM-1436 north of Oahu on a routine exercise after the refit period. The launch was normal and the chase aircraft, two FJ-3D's from Guided Missile Group ONE took over airborne control for landing at Bonham ALF. Lieutenant Don Edgren, one of the safety-chase pilots, remembers that several minutes after they picked up the missile, he watched as the missile took a right hand turn on its own accord, heading for Oahu instead of Kauai as planned:

"I passed control to the second pilot and he couldn't control the missile either. The missile was heading right over the mountains towards downtown Honolulu. Then it climbed over the mountains on its own and let down again, heading towards Schofield barracks, Honolulu and Diamond Head. I was on the guard channel warning everyone that the missile was out of control. Just as I finished the communication, the missile went into a vertical climb, stalled at 20,000 feet and came down fluttering like a leaf. I remember looking at the possible crash sites and realized that my house was not too far away!"[35]

FTM-1436 ended up crashing in rugged jungle terrain approximately a quarter mile above the Camp Smith headquarters of the Commander-in-Chief, Pacific. Christensen remembers listening to the chase pilots on the Conning Tower speakers. He had observed the launch from the periscope with everything appearing normal until suddenly the missile veered to the right, definitely headed in the wrong direction. He immediately surfaced and headed toward port, manning all channels to hear the outcome. The Division Commander, Commander Pridinoff, was on board and briefed a somber wardroom on what to expect upon arrival in port. Once the gangplank was over, the first visitor was the Public Information Officer from Commander Submarine Forces, Pacific. He presented Christensen and Pridinoff with copies of the evening edition of the Honolulu newspaper with the first red headlines since 7 December 1941. It had a picture of *Tunny* on the front page and a brief caption describing the malfunction of a drone which had to be destroyed in a remote area of Oahu.[36,37] The crew was quite relieved, to say the least since at the time everyone had visions of the missile crashing in downtown Honolulu.

Tunny began her second deterrent patrol on 22 April 1960. Five days out of Pearl Harbor, a crew member came down with appendicitis. Christensen followed the recommendation of Chief Moon, his Hospital Corpsman, and *Tunny* changed course for Adak, making an unscheduled stop to drop off the crewman. This had a side benefit of allowing the fuel tanks to be topped off and additional fresh food taken on board. *Tunny* proceeded to her patrol area, arriving several days later. Lieutenant Forney H. Ingram, Assistant Missile Officer, remembers

Tunny *caught on the surface by a P2V patrol aircraft on 24 August 1960. A standing order for all Regulus submarines were to remain undetected during their patrols, including friendly forces on anti-submarine patrol. (Courtesy of Fullinwider Collection)*

this, his first deterrent patrol, as one of little restful sleep. This was due mainly to his berthing area, a folding bench seat for the Wardroom table. With 11 officers and 10 bunks, he was the junior officer on board, hence his makeshift bed.[38] The seat cover was made of slippery naugahyde. In a rolling sea, an all too frequent occurrence in the Northern Pacific, he remembers many a time picking himself off the deck and climbing back onto the bench.

In the days before the Submarine Force's attenna improvent program, communication attennas were continually flooding (grounding due to salt water intrusion). Enroute to the patrol area, *Tunny*'s crew discovered the solution. The whip antennas which were tight when they left the warm weather of Hawaii, were loose in their sockets in the cold northern climate. The officers and petty officers conferred and performed a test by placing an antenna assembly in the freezer box. It was a tight fit at the start and after an hour in the freezer had become loose due to the differing coefficients of expansion of the different metals in the assembly. With this information, the routine was initiated where as they proceeded north, the antennas were periodically tightened. This continued until the problem was resolved by redesign.[39]

Christensen recalls that this was the longest period anyone on board had remained submerged without a break. After departing from station and enroute to the forward refit base, Christensen invited all hands, in groups of four, to the bridge to enjoy the fresh air and a distant horizon. *Tunny* arrived at the forward refit base 17 June 1960. Remembering the earlier appendicitis problem, Christensen insisted all of the crew be checked, including himself, prior to leaving on the third patrol. The base doctor assured Christensen that he had nothing to fear at his age.

Tunny began her third patrol on 14 July 1960. For the first two weeks all was normal. Then, one night, after starting the snorkeling routine, a loud bang rang out close to the Conning Tower. Excessive back pressure immediately built up in the engine exhaust system. Since it was dark, they surfaced to find out what had happened. They found a six-inch hole in the snorkel exhaust mast. Christensen withdrew to the eastern most edge of the patrol area to attempt repairs. Floating ice in the area had confirmed their suspicions; the damage was due to ice. The air temperature was five degrees above zero, the water temperature thirty-one degrees. Lieutenant Ed Turner, the Chief Engineer, surveyed the damage and reported, "I can fix it if I can have the After Battery CRS (Corrosion Resistant Steel) hatch lining and the Engine Room cable covers." Christensen recalls asking how long and remembers Turner saying two, maybe three hours. "Get with it," he said and everyone hoped for the best while a much needed battery charge began.[40]

About midnight Turner reported that the snorkel mast was repaired as best it could be and ready for test. "What did you do?" asked Christensen. With a laugh, Turner replied, "We installed the biggest band-it patch in history. We layered gasket material, engine room cable covers, more gasket material, and the CRS after battery hatch liner and then tightened it all up with twenty band-it clamps." The battery charge completed, Captain Christensen dove the boat and rigged for snorkeling to test the patch. Everything worked fine and the patrol routine resumed.[41]

One week later Christensen found himself in a most embarrassing position. He had acute appendicitis. Chief Moon had correctly diagnosed the problem and put him quietly to bed with heavy doses of penicillin. Only Chief Moon, Fullinwider, and Chief-of-the-Boat Mathis knew of the dilemma. Fullinwider wanted to leave the patrol area immediately to get Christensen to a doctor. Christensen, however, was adamant about staying on patrol. Since the patrol plan already had Fullinwider acting as Command Duty Officer (CDO) during the daylight hours and Christensen as CDO during snorkeling at night, they developed a cover story that Christensen was deeply involved in studying for the nuclear power program and couldn't be interrupted except in extreme emergency. He answered the phone from his bunk and ran the ship through the senses of his conning officers. Each watch officer stopped by his stateroom for "going-on-watch-chats." His wardroom absence was explained by his studies and a strict diet that Chief Moon had put him on. Moon had him on intravenous feeding for five or six days. About a week after the onset of the attack, Christensen was up and about, with the crew commenting on the great effect the diet had and pestering Chief Moon for the secret.[42] The remainder of the patrol was without further incident except as they neared Adak enroute to Pearl Harbor. On 24 August 1960, thirty miles from port they were on the surface when the Officer-of-the-Deck was caught "napping." A P2V patrol aircraft from Adak photographed them on the surface. It was the only time during Christensen's two years as captain of the *Tunny* that this happened.

Tunny returned to Pearl Harbor from this back-to-back patrol on 12 September 1960. The *Tunny* wives had prepared an enormous lei of flowers for the Conning Tower and *Tunny* draped a huge sign across the bridge reading, "Welcome Blue Crew, She's All Yours," as she returned to the pier. This was a joke, referring to the Blue and Gold crews of the Polaris Program. There was no Blue Crew. The "Black and Blue Crews" of the Regulus program referred to the "bruising" routine imposed by the lack of relief personnel.[43] Patsy Christensen was of course there to greet her husband at the pier. She recalls "I couldn't figure out why all the senior medical officers from the squadron were on the pier too, surrounding Chris. They wouldn't let him free until he promised to make an appointment to have his appendix removed!" Christensen was taken to Tripler Hospital were he was operated on for a very infected appendix. *Tunny* immediately went into the shipyard for a three month overhaul.

Lieutenant Commander Christensen received a Commander Submarine Force Pacific Fleet Citation for his leadership during these patrols and *Tunny* received a Commander Submarine Forces, Pacific, Commendation as the outstanding ship in Submarine Division ELEVEN for 1960. Lieutenant Turner and two enlisted men received ComSubPac commendations for their actions in creating the patch for the ice damaged snorkel. In fact, the shipyard declared the patch the strongest part of the snorkel mast. Chief Corpsman Moon received a ComSubPac commendation for his handling of Christensen's appendicitis emergency.[44]

1961

On 30 June 1961, Christensen was relieved by Lieutenant Commander Douglas Stahl. They had previously served together on board *Cusk* during the early days of Regulus I development. Stahl had been Executive Officer of *Tunny* during the Lebanon Crisis patrol and was reporting back to her after a tour of duty as Executive Officer of Guided Missile Unit TEN, the submarine Regulus I support unit.

On 23 July 1961, *Tunny* left Pearl Harbor for her fourth deterrent patrol. Lieutenant Allen Shinn, Jr., recalls that at this point in the Regulus deterrent mission program the typical patrol left Pearl Harbor and refueled either in Adak or Midway Island. The stop was supposed to be for just a few hours but Stahl was usually able to time it such that they got an evening's liberty. This time it was Adak and Stahl decided to try sailing through Kagalaska Pass, just east of Adak Island, so that they could cut five hours or so of steaming off the trip and pick up a night's liberty in Adak. The usual and somewhat safer way was to go west around Adak and then into Sweeper's Cove.

It was daylight as *Tunny* neared Adak but the shore was shrouded in heavy fog. Stahl was on the bridge with Lieutenant Charlie Roberts, the Engineering Officer. Stahl recalls that Roberts called out "Rocks off to starboard." Stahl replied, "I have rocks off to port, Navigator, what course do you recommend?" Lieutenant James Burgess, the Navigator, promptly responded " I recommended a 180 degree course change!" Unknown to both Roberts and Burgess, the Executive Officer, Lieutenant Melim, was feeding Stahl course corrections based on a very clear radar picture. Not only was the entrance easily discerned, the radar showed this was the right opening. Just as *Tunny* reached the opening to the passage, she broke out of the fog bank into a stunningly beautiful scene that Shinn, Burgess, Ingram and Stahl still remembered clearly thirty-two years later. While the entrance was probably one-half mile across, at times it narrowed to 80 yards between islands with deep water beneath and sheer cliffs on each side. Seals were diving off the rocks all around, birds wheeling overhead and fish everywhere, deer by the streams and killer whales in abundance.[45] *Tunny* reached Adak early and all the crew enjoyed the chance at a brief liberty.

After leaving Adak they ran on the surface, diving only for trim purposes and to avoid radar contacts or electronic countermeasure alerts from the P2V search aircraft stationed at Adak. These aircraft overflew the operating area, so diving to avoid them was almost as commonplace as submerging for marine traffic. Spending this much time on the surface was contrary to normal doctrine. Since the orders were to remain undetected and did not explicitly say that snorkeling was mandatory, Stahl elected to try the more comfortable route. He maintained that snorkeling was hard on his sinuses, the engines, and the crew, not necessarily in that order, so if they didn't have to, with prudence, they wouldn't. Stahl's routine was to run submerged on the battery during daylight and then surface after dark to charge the battery and ventilate the boat. Once the battery was recharged, they would stay on the surface with the forward group of ballast tanks partially flooded down if the weather permitted. This provided better passive sonar reception and if something sounded like it was getting close, they could submerge and come back up when all was clear. In winter this meant 14 hours a day on the surface, but in summer they did not surface until 2300 and needed to get back down by 0500 or even earlier. Those six hours were just enough time to get the batteries fully charged and recharge the high-pressure air banks. Because of the length of time spent submerged during the summer, they would usually snorkel for 20 minutes or so about mid-day to ventilate the boat.[46]

This tactic of running on the surface resulted in a couple of close scrapes. During this patrol Shinn had the deck at dawn and Stahl came up as he usually did upon arising to take a look. It was foggy and perhaps because of this, Stahl decided to stay up a little longer. As far as they could tell there were no visible contacts. Suddenly, the electronic countermeasures operator reported a close radar contact, so close that if *Tunny* submerged she would have been an obvious "sinker" and thus alert the contact. By this time it was full light and if the fog lifted they would be readily identified as a U.S. submarine. Soon after the electronic countermeasures alert, the vessel sounded a fog horn and Stahl realized that between the radar frequency and the fog horn, the ship was a merchant vessel that had lit off its radar to make a navigational check. Seeing the small and somewhat erratic blip of the *Tunny* on his radar, the merchantman had sounded his fog horn to prevent a collision with what he apparently took to be a small fishing vessel. Stahl maneuvered around a bit and the merchant ship's captain then shut down his radar, satisfied that a collision was not imminent. *Tunny* promptly submerged with a collective sigh of relief from the crew and the captain.[47]

At the end of the patrol, *Tunny* headed to the forward refit base, arriving 28 September 1961. After a five week refit, she left for her fifth patrol on 4 November 1961. During this transit she was caught on the surface by a U.S. patrol aircraft. A new member to the electronic counter measures intelligence watch, a team of specialists that rotated between the five Regulus submarines, had recognized the radar signal as being that of a U.S. aircraft and had not warned the bridge even though *Tunny*'s instructions were to dive on all contacts. The pilot of the P2V had turned off his radar after making contact and descended through the overcast on the last radar bearing, catching *Tunny* flat-footed. Since the P2V's camera had jammed and according to their plotted "moving haven," *Tunny* was not expected at this location, the intelligence group at ComSubPac alerted *Tunny* to the probable presence of a Soviet submarine nearby. With Stahl's usual unorthodox procedures, they had been making 15 knots and not the 7 knots speed of advance in the operational orders, thus as far as the aircraft was concerned, *Tunny* could not have been the submarine sighted. Due to the imposed radio silence restrictions, *Tunny* couldn't respond to the message.[48]

Christmas, 1961, was miserable for the crew of the *Tunny*. Lieutenant Gene Heckathorn, Communications and Sonar Officer, recalled that two weeks of state 6/7 seas (22-33 knot winds) made keeping the boat at periscope depth a challenge. Snorkeling was almost impossible since they were always "flaming out", a term for a shut down due to excessive back pressure due to the snorkel being covered by a wave. On one occasion, one of the lifting rods to the Number 2 periscope broke while lowering the periscope. The whole assembly crashed into the bottom of the periscope well and the optics shattered, leaving only one periscope for the remainder of the patrol. As this were not enough, the heavy seas caused large pieces of the superstructure aft of the missile launcher to tear loose and clang against the remaining superstructure and piping. The noise was so loud that they couldn't hear anything on the sonar aft of the beam unless proceeding at a dead slow speed. On two occassions, Heckathorn and an auxiliaryman donned artic weather clothing and went out on deck to rip off or cut off large metal pieces. It was a dangerous operation and while they were clipped onto safety lines, only parts of the track for the clips remained intact. Captain Stahl improved the safety by travelling with the seas at the same speed as the waves. As pieces of metal were tossed overboard, the propellar on that side was stopped to reduce the chances of fouling the blades.[49]

A new morale program began at this time, the concept of "family-grams." Signal traffic during the week was busy and full but during the weekend the number of messages tapered off to the point that the communications personnel at ComSubPac were encoding magazine and newspaper articles in order to keep the message traffic pattern full. The idea that 100 word messages from wives and girlfriends could be used instead, on the weekends, was experimented with and became a key morale factor in the program. Stahl remembers that a suggestion was made to ensure only good news would be transmitted. He strongly urged that both good and bad news be sent since the best way to ruin time in port would be to hit a sailor with terrible news right off the boat. Stahl ended up experiencing this personally when he learned of his six-month old daughter's death while on patrol. A month later, when he returned to Pearl, he was much better able to cope with the tragedy and felt vindicated for his earlier insistence concerning the message content.[50]

1962

Tunny returned to Pearl Harbor on 12 January 1962. After a four month regular overhaul and three months of training, including firing nine missiles, on 24 August 1962, Stahl took *Tunny* on her sixth deterrent patrol. After topping off her fuel tanks at Midway, she turned north and reached the patrol area eight days later. One morning Shinn came up to relieve the morning watch only to find *Tunny* on the surface, steaming in front of and away from a group of ships. The annual Soviet resupply convoy from the White Sea to Petropavlovsk, a large Soviet naval base on the Kamchatka peninsula, had arrived in *Tunny*'s operating area and she was under orders to photograph as many of the vessels as possible. Once they made contact, they ran ahead of the convoy on the surface until first light, then dove and let them steam by on both sides while taking periscope photographs. Shinn remembers both periscopes being up and one view in particular of a Whiskey class submarine running on the surface with only a single crewman on the bridge, head lying on the rail, apparently asleep. Clearly they had no idea that any American submarine was in the vicinity.[51]

Early one morning in October 1962, in transit back to the forward refit base, Lieutenant Ingram had the 0400-0800 watch. After he was relieved, he made his rounds, reported to Stahl and sat down for breakfast. Ingram turned on the radio in the wardroom, tuned to Armed Forces Radio and began to eat. The radio announcer said that President Kennedy would shortly make an important announcement. Ingram alerted Stahl and they both listened to President Kennedy tell the nation about the Cuban Missile Crisis. Shortly thereafter, *Tunny* went to "Battle Stations Missile." Shinn was asleep and awoke not knowing if this was a drill or the real thing. On the previous patrols he had been the Communications Officer, and as such, would decrypt the encoded radio message ordering a launch or a test exercise. He now had no such role or information. The Communications Officer was not saying a word and Captain Stahl went through the entire countdown, an hour or so of procedures, without saying anything not in the script and with an unusual tension that most could sense. Shinn remembers thinking that this was the real thing, but was afraid to ask for fear of the answer. Finally Roberts, the Battle Stations Diving Officer, reported, "Ready to surface", meaning that the missile and all systems were ready at any time the Captain decided to surface. Shinn was watching the Skipper quite closely. If he ordered "Surface", it would be the real thing. At this point, however, the tension drained out of Stahl and he ordered the crew to secure.[52] Placed in the context of the ongoing Cuban Missile Crisis, this test of the system was all the more emotionally draining. The crew as a whole had very little information to go on since they relied on the ComSubPac news reports and these were rather sketchy at the time.

Tunny returned from her sixth patrol on 29 October 1962. On 5 December 1962 *Tunny* was awarded the first Missile Efficiency "E" award for overall excellence in guided missile operations. The award had been authorized by the Chief of Naval Operations, Admiral D.L. McDonald, earlier in the year.

1963-64

After a short refit and training period *Tunny* departed Pearl Harbor on her seventh patrol on 12 January 1963. When relieved on station on 15 March 1963, she proceeded to Naha, Okinawa, off-loaded the two tactical Regulus I missiles and departed for seven days liberty in Hong Kong and a seven day upkeep in Subic Bay, Philippines. Lieutenant Commander Byron Ruble relieved Stahl on 15 May 1963. In June 1963, *Tunny* received her second Missile Attack

Tunny's *last hoorah as a target for MK 37 acoustic homing torpedo tests off of San Diego. With compartment doors welded shut and the hangar giving her extra buoyancy,* Tunny *proved to be a difficut target to sink. The torpedo homed in on the propellers as expected, but* Tunny *refused to sink. Several hours later she was sunk by gunfire. (Courtesy of Blair Collection)*

"E" and Battle Efficiency "E" for Submarine Division ELEVEN. On 13 July 1963 *Tunny* began her eighth deterrent patrol. *Tunny* returned to Pearl Harbor on 3 October 1963. On 10 February 1964, *Tunny* began her 9th and last deterrent mission. Ruble remembers that morale remained high with support from the squadron and ComSubPac excellent through to the end. The Regulus I missile system was working fine at this point, the bugs had been worked out and even though it was an old system, it was well maintained. *Tunny* returned to Pearl Harbor on 11 April 1964 after an uneventful patrol. *Tunny* received a commendation from Commander-in-Chief, Pacific Fleet, for outstanding performance in carrying out successful Regulus patrols during the period 1 November 1961 to 27 June 1963.[53]

Post Regulus Operations

Lieutenant Commander Ruble was relieved as Commanding Officer of *Tunny* by Lieutenant Commander Robert Melim on 6 June 1964. Oddly enough, even though the program was winding down, *Tunny* fired her 100th Regulus I missile, and the last launched from a submarine, on 27 October 1964. This was a fleet training missile and was used for anti-aircraft fire practice by destroyers. All Regulus missiles were being expended at this time and so it was not recovered. After a limited overhaul from 26 April 1965 to 5 August 1965, *Tunny* was redesignated SS-282 and served as a troop carrier until her decommissioning on 28 June 1969 at Mare Island.[54]

In 1970, the Navy was evaluating a new model of the Mk-37 acoustic homing torpedo. Ironically, *Tunny* was being readied for sale to a salvage yard. It was decided to bring her out of mothballs and use her as the target for this new torpedo. They thoroughly stripped the ship and *Tunny* was sunk 19 June 1970 as a torpedo target by USS *Volador* (SS 490). The rudder was locked at five degrees to port and the skeleton crew left the submarine. *Tunny* circled slowly and the torpedo hit directly in her propellers. Since the watertight doors on all the compartments and the large missile hangar were tightly sealed, *Tunny* refused to sink and a destroyer had to complete the job using five-inch shells, finally sinking her near dusk.[55]

Endnotes

[1] *Personal interview with Captain William Sims, USN (Ret.).*
[2] Tunny-*Regulus, by Lieutenant Samuel T. Bussey, USN. Prospective Commanding Officer thesis. Provided by Commander Robert Owens, USN (Ret.).*
[3] *Project SLAM - Submarine Launched Attack Missile, Semi-Annual Report, 1 April to 30 September 1953, page III-2. Naval Historical Center, Operational Archives, ComSubDiv 51.*
[4] *Semi-Annual Report on Submarine Launched Attack Missile for the Period 1 October 1953 to 31 March 1954, page I-1. Naval Historical Center, Operational Archives, ComSubDiv 51.*
[5] *Regulus I Flight Test Index, 1961, LVSCA A50-24, Box 7.*
[6] *Evaluate Combinations of Surface Ship Launching of Regulus I with Final Control Exercised by Units other than the Launching Ship, Final Report on Project Op/S317/X11, 16 August 1957, page 5. Naval Historical Center, Operational Archives.*
[7] *Regulus I Flight Index, 1961, LVSCA A50-24 Box 7*
[8] *Personal interview with Commander George Clegg, USN (Ret.).*
[9] *Personal interview with Captain Marvin S. Blair, USN (Ret.).*
[10] *Personal interview with Commander George Clegg, USN (Ret.).*
[11] *FC4-11/5 A1-Project/X11 ser 0021 25 June 1955, Missile Operations in the Hawaii Area, from ComSubDiv 51. National Archives, Bureau of Aeronautics, Guided Missile Division, Record Group 38 Box 70.*
[12] *Ibid.*
[13] *Ibid.*
[14] *Personal interviews with Commander George Clegg, USN (Ret.); Captain Eugene Pridinoff, USN (Ret.) and Captain Marvin Blair, USN (Ret.). Personal communication with Admiral James Osborn, USN (Ret.).*
[15] *FC4-11/5 A1-Project/X11 ser 0021 25 June 1955, Missile Operations in the Hawaii Area, from ComSubDiv 51. National Archives, Bureau of Aeronautics, Guided Missile Division, Record Group 38 Box 70.*
[16] *Regulus I Flight Index, 1961, LVSCA A50-24, Box 7.*
[17] *Evaluate Combinations of Surface Ship Launching of Regulus I with Final Control Exercised by Units other than the Launching Ship, Final Report on Project Op/S317/X11, 16 August 1957, pages 1-9 to 1-12. Naval Historical Center, Operational Archives.*
[18] *Personal communications and interview with Bill Cannon. His personal papers contain the dates and incident descriptions.*
[19] *Personal correspondence with Captain Archer Gordon, USN (Ret.).*
[20] *Personal interview with Captain Marvin S. Blair, USN (Ret.).*
[21] *Ibid.*
[22] *Personal communications with Captain Richard B. Gilchrist, USN (Ret.).*
[23] *Personal interviews with Commander Douglas Stahl, USN (Ret.) and Captain Marvin S. Blair, USN (Ret.).*
[24] *"Schedule of Events USS McCain (DL 3) 20 March 1958." Personal records of Captain Marvin S. Blair, USN (Ret.).*
[25] *Ibid.*
[26] *Ibid.*
[27] *Personal interview with Captain Marvin Blair, USN (Ret.); personal correspondence with Captain Archer Gordon, USN (Ret.).*
[28] *Ibid.*
[29] *Personal interview with Captain Marvin S. Blair, USN (Ret.). Blair recalls that later he learned that what really impressed the SEATO dignitaries was that* Tunny *"disappeared" immediately after launch, leaving only her rotating radar antenna and a periscope to mark her position.*
[30] *Captain Blair still has the letter that Rear Admiral E.W. "Joe" Grenfell wrote to his wife that explained, in general, this sudden need for the redeployment of* Tunny *immediately upon her return from the Far East. It is a poignant reminder of the world tensions of this time.*
[31] *Buz Sawyer comic strip in 10 July 1959 Honolulu Advertiser. Pacific Fleet Submarine Memorial Association Archives.*
[32] *Personal records of Commander Morris Christensen, USN (Ret.).*
[33] *Personal interview with Commander Morris Christensen, USN (Ret.). There was some controversy with this data but it has been verified by reviewing the unclassified patrol logs of the two submarines involved,* Tunny *and* Grayback. *A complete list of tentative patrol dates was supplied by Rear Admiral Daniel Richardson and this permitted comparison of possible patrol dates with the actual ships logs at the National Archives.*
[34] *Personal interviews with Captain Peter Fullinwider, USN (Ret.) and Commander Morris Christensen, USN (Ret.).*
[35] *Personal interview with Lieutenant Commander Don Edgren, USN (Ret.).*
[36] *Personal communication with Commander James Burgess, USN (Ret.). Personal interview with Commander Morris Christensen, USN, (Ret.) and Captain Eugene Pridinoff, USN (Ret.).*
[37] *Pacific Fleet Guided Missile Crashes in Canyon Near Aiea, 16 February 1960, page 1, Honolulu Advertiser. Pacific Fleet Submarine Memorial Association Archives, Pearl Harbor.*
[38] *The increased manning for both officers and enlisted personnel was due to the need to produce qualified submariners for use in the Polaris program.*
[39] *Personal interview and communication with Captain Gene Heckathorn, USN (Ret.), who served as Communications and Sonar Officer on board Tunny from Jan. 60 to Jan. 62.*
[40] *Personal interview with Commander Morris A. Christensen, USN (Ret.).*
[41]
[42] *Personal interviews with Commander Morris Christensen, USN (Ret.) and Captain Peter Fullinwider, USN (Ret.).*
[43] *The reference to "Black and Blue" is something all of the Regulus submarine crews joked about, not just* Tunny.
[44] *Pacific Fleet Submarine Memorial Association Archives.*
[45] *Personal interview with Commander Douglas Stahl, USN (Ret.). Personal communications with Commander Forney Ingram, USN (Ret.) and Commander James Burgess, USN (Ret.) and Commander Robert Melim, USN (Ret.)*
[46] *Personal interview with Commander Douglas Stahl, USN (Ret.). Personal communication with Allen Shinn, Jr.*
[47] *Personal interview with Commander Douglas Stahl, USN (Ret.).*
[48] *Personal interview with Commander Douglas Stahl, USN (Ret.). Personal communication with Commander James Burgess, USN (Ret.).*
[49] *Personal communications with Captain Gene Heckathorn, USN (Ret.).*
[50] *Personal interview with Commander Douglas Stahl, USN (Ret.).*
[51] *Personal interview with Commander Douglas Stahl, USN (Ret.). Personal communication with Allen Shinn, Jr.*
[52] *Personal communications with Commander Forney Ingram, USN (Ret.) and Allen Shinn, Jr. Personal interview with Commander Douglas Stahl, USN (Ret.).*
[53] *Personal communication with Captain Byron Ruble, USN (Ret.).*
[54] *Navy News Release 26 June 1969. Pacific Fleet Submarine Memorial Association Archives.*
[55] *Personal interview with Captain Marvin S. Blair, USN (Ret.).*

Chapter Eleven: USS Barbero (SSG 317)

Barbero was chosen as the second Regulus launch submarine due to previous alterations during an earlier conversion to a cargo submarine. Two of her four main engines and two generators had been removed from the Forward Engine Room and the After Torpedo Room had been converted into crew quarters. This left enough space for installation of missile checkout and guidance equipment as well as the hydraulic equipment necessary to operate launcher and missile hangar systems.

On 1 February 1955 *Barbero* entered Mare Island Naval Shipyard (Mare Island), California, for conversion to her new role as a guided missile submarine. Lieutenant Commander Sam Bussey reported to the *Barbero* at Mare Island as Prospective Commanding Officer on 29 August 1955. He was intimately familiar with the Regulus due to previous duty as Missile Officer and then Executive Officer of USS *Tunny* (SSG 282), the first Regulus launch submarine, in service since 1953.

Bussey took command on 28 October 1955 upon formal recommissioning of the boat. Sea trials commenced on 12 December 1955 with the usual trip to the standard test diving area adjacent to the Farallon Islands, several nautical miles west of Golden Gate and the entrance to San Francisco Bay. This was the nearest area that had waters shallow enough to safely test dive a new or converted submarine and still be able to save the crew and salvage the boat in case of mishap.

At some point after conversion to a cargo submarine the weight and distribution records of *Barbero* were misplaced at Mare Island. Consequently the information necessary to compensate the boat to a neutral buoyancy condition after the addition of the heavy missile hangar and launcher assembly was unavailable. *Barbero* and *Tunny* both drew two to three feet more water then the average fleet submarine due to a 2 foot square box keel that ran two thirds of the length of the boat. It was full of lead ballast of approximately 200 tons weight to compensate for the added buoyancy of the missile hangar. For the always critical first dive, Bussey instructed the Diving Officer and Chief Engineer, Lieutenant Vogele, to cycle the ballast tank vents open and then shut since it was unclear as to whether the boat was too light or too heavy overall. The boat would not submerge. Bussey ordered a slower cycling, which was done, again with no perceivable downward motion. Still another cycle had the same results. Convinced that the compensation was on the very light side, Bussey and Vogele agreed to open the vents as on a normal dive.

The results were spectacular. *Barbero* immediately assumed a sharp down angle and plunged headlong towards the bottom more than 200 feet below. Officers and crew were thrown in every direction, but mostly forward. Both Bussey and Vogele quickly reacted with orders to shut all vents and blow all tanks. This succeeded in first slowing and then leveling the boat, but not before it had glanced off the sand and rock bottom, wiping off its brand new bottom sonar dome in the process. After surfacing, trim tanks were adjusted and the next attempt to submerge was successful. Upon return to port, *Barbero* entered drydock for inspection of the hull. No serious damage was revealed except for the sonar dome which was replaced.[1]

After completion of sea trials and acceptance by the Navy, *Barbero* sailed to Port Hueneme, adjacent to the Naval Air Missile Test Center, Pt. Mugu, California. The ship's missile checkout, launch and guidance equipment were fully tested using aircraft drone services provided by Guided Missile Group One (GMGRU-1. *Barbero* had the first submarine installation of the production version of P-1X Trounce radar guidance system, designated BPQ-1. Lieutenant Julian K. Morrison III, *Barbero*'s first Missile Guidance Officer, remembers that the BPQ-1 system and CP-98 missile navigation computer were more often disassembled than functioning properly. Much of this difficulty stemmed from the fact that many components were of World War II vintage and not designed for the exacting specifications that the Trounce system required. *Barbero* experienced two weeks of mechanical and electronic difficulties with the newly installed BPQ-1 Trounce system. The guidance radar sector scan reversing relays malfunctioned with consequent loss of antenna train. After replacing these with more rugged components, the drone training flights, conducted with the ship alongside the pier, proceeded smoothly.

The first Regulus I launch by *Barbero* was carried out on 14 March 1956 off Pt. Mugu. The missile crashed spectacularly almost immediately after launch due to misalignment of the boosters. This problem was rectified by the next launch two weeks later. The launch and flight were perfect with a direct hit indicated by the photo-theodolites on San Nicolas Island. This was not a tactical missile but rather a fleet training missile which was routinely guided to the target, given a "dump" signal to record location and then recovered.

Barbero launched only two missiles during operations on the West Coast. On 10 February 1956, *Barbero* had received official notice to proceed to her new home port, Norfolk, Virginia, where she would become the first Regulus I launch submarine in the Altantic Fleet, assigned to Submarine Division SIXTY-THREE. On 14 April 1956 *Barbero* began the transit to Norfolk. After a brief visit to Acapulco, Mexico, she approached the Panama Canal and in accordance with Operations Orders, conducted an undetected simulated missile attack on the Canal with Pedro Miguel Locks as the target. After visiting a number of ports in the Caribbean, *Barbero* arrived in Norfolk on 19 May 1956.[2]

Within Submarine Squadron SIX, five submarines formed Submarine Division SIXTY-THREE: *Barbero* was the Regulus I launch and guidance submarine while USS ***Torsk*** (SS 423), USS *Argonaut* (SS 475) and USS ***Runner*** (SS 476) were Regulus I guidance only; USS *Sea Lion II* (ASSP 315) was a specially configured special operations/troops submarine. Commander Walter "Pat" Murphy was Commander Submarine Division SIXTY-THREE. Murphy had been actively involved in both the Loon and early Regulus programs and was an avid supporter of the guided missile submarine concept. Missile maintenance support was provided by Guided Missile Unit FIFTY-ONE (GMU-51), the submarine Regulus I support unit located at the Naval Ammunition Depot (NAD), Yorktown, Virginia (see Appendix II).[3]

The schedule for *Barbero* evolved into operations for 6 weeks followed by 2 weeks upkeep alongside the Norfolk based submarine tender USS *Orion* (AS 18). Operations were usually conducted a week at a time since it took the submarines the better part of a day to run from the Destroyer/Submarine Piers at Norfolk to the 100 fathom curve 60 nautical miles away in the Virginia Capes Operating Area. Surface transit was four to five hours for the "normal" fleet boat. Due to having but two main engines, *Barbero* took closer to six or seven hours to transit from Yorktown to the operational area.

Barbero was often requested for publicity cruises or alongside demonstrations for cabinet level and military officials or the news media. While Bussey was Commanding Officer of *Barbero*, she made publicity visits to Washington, D.C., *New York*, Submarine Base New London, Submarine Base Key West, and the Naval Base at Mayport, Florida. At New London *Barbero* conducted several demonstration Regulus I ram-outs and simulated launches for the Commander Submarines, Atlantic staff and local interested submariners as well as serving as motivation for students from Submarine School. Through repeated practice the missile crew significantly reduced the launch time down to 5-6 minutes from "Battle Surface Missile" (submerged at 100 feet depth) to submerging after launch.

1957

In February of 1957, *Barbero* went south to the Puerto Rican operating areas as part of the general "Springboard" annual Atlantic Fleet submarine exercises away from the frigid waters of the wintery North Atlantic. A detachment from GMU-51 was sent to Naval Air Station Roosevelt Roads, Puerto Rico, to support the Regulus I missile operations. At the beginning of the exercise, *Barbero*, *Runner* and *Argonaut* transited submerged from Mona Passage, west of Puerto Rico, to the exercise area in the middle of the Caribbean. On the morning of 28 February 1957, *Barbero* was still about 20 miles west of Aves Island, almost in the geographic center of the Caribbean and several miles from the missile launch point. At this point the undersea Aves Ridge shelves up steeply from 2000 fathoms to 1000 and then 100 fathoms west of the island. Due to the distinctive undersea contours, Bussey knew exactly where the boat was even though they were still submerged. The decision was made by the Division Commander, Commander Murphy, embarked in *Barbero*, to launch from that known position. Final guidance was to be assumed by Runner and Argonaut, and thus a mile or so position error was not critical for the launch submarine. Bussey made one last 360 degree air search on the periscope and, with no visual or electronic countermeasures contacts, battle surfaced to launch in the cover of a rain squall. Bussey remembers all too well surfacing and taking another quick look from the bridge. Just as he was going to order the hangar door opened and the missile rammed out, a Navy Martin P5M "Mariner" antisubmarine warfare patrol aircraft dropped out of the clouds about one mile directly ahead. He immediately sounded the diving alarm and down the boat went in one of its fastest dives on record. Considering that they were now detected and that the airborne opposition would have held them down for the remainder of the day, Bussey then proceeded with the launch after surfacing again and radioing the aircraft to stay clear of the launch. *Barbero* launched a tactical missile with an inert warhead from a point west of Aves Island toward the target area between St. Croix and Vieques (about 200 nautical miles) where the guidance boats took over final control to terminal dive.[4] At the post-operation debriefing, the search aircraft pilot indicated that he had been concentrating on the Aves Island area using visual search with only periodic radar sweeps. He had decided to swing over and check out the rain squall as possible concealment and just happened upon *Barbero*.

Three days later *Barbero* conducted another OST launch but this one splashed immediately after takeoff. Commander Murphy reported the loss and remembers how Captain Slade Cutter, Commander Submarine Squadron FIVE, had a great deal of difficulty understanding how a simple report was sufficient to explain the loss of a $250,000 missile when the loss of a $10,000 torpedo required a formal Bureau investigation and all that might follow.[5]

During the same exercises, *Barbero* operated out of Guantanamo Bay and conducted several launches along the southern coast of Cuba. The only East Coast heavy cruiser armed with Regulus I, the USS *Macon* (CA 132), launched four missiles during the exercise with terminal guidance provided by the submarines of Submarine Division SIXTY-THREE. This was all part of the Operational Development Force evaluation of the missile system using the cruiser-submarine guidance combination.

Two views of Barbero *upon completion of her conversion to SSG configuration. Note the unstreamlined mast and snorkel arrangement clearly evident in the bottom photograph.* Barbero *had two main engines, and her stern torpedo tubes were removed during her earlier conversion to a cargo submarine. (Courtesy of Blount Collection)*

On 12 May 1957, *Barbero* carried out one of the more interesting public relations demonstrations on the East Coast. Always interested in good public relations, the Navy was featured on "Wide Wide World", a live national television show hosted by Dave Garroway. Originating from *New York*, the show was scheduled to carry live a launch of Regulus I from *Barbero* The USS *Boston* (CAG 1) was anchored and serving as the microwave relay to Dam Neck, Virginia. In addition, and within television range, was a destroyer-tanker refueling demonstration as well as another destroyer firing "Hedgehog" anti-submarine weapons. A crowded situation to say the least. After five rehearsals and numerous change of plans, the stage was set. Due to the strict time constraints of the program, *Barbero* had 40 seconds to launch the missile so no mishaps or holds could be tolerated. In addition, the weather was overcast at 600 feet so the missile was set to cruise at a constant altitude of 500 feet so it would not disappear into the overcast immediately after launch.

Commander Murphy was in the television studio in *New York* with a headset monitoring the operational frequency, advising the producers as to the progress of the countdown and providing live commentary during the operation. Due to the location of the ships in the demonstration, Bussey had *Barbero* heading towards the refueling group just as launch took place. The boosters ignited and before the JATO smoke cleared, Bussey called for right full rudder and all ahead standard to start *Barbero* turning to starboard. They then straightened out and passed just astern of the refueling group as it thundered past. *Barbero* simultaneously angled across the stern of the *Boston* and cleared the operating area. The missile was successfully recovered at Naval Air Station Chincoteague.[6,7]

In June 1957, Bussey was relieved by Lieutenant Carlos P. Dew. The new Division Commander, Submarine Division SIXTY-THREE, Commander Charles Styer, prepared at this time what he believes was the first Regulus deployment and launch operations plans for *Barbero* against Soviet targets. Commander Styer recognized the limitations and somewhat restricted capabilities of this submersible nuclear strike force, but the plan was approved and utilized nonetheless by the Commander Submarine Forces, Atlantic Fleet.[8]

In the Fall of 1957, units of Submarine Division SIXTY-THREE participated in Operation STRIKEBACK, a NATO joint services exercise simulating the Allied response to an attack by the Soviet Union on Europe. This was the largest multi-national NATO exercise to date and also included Mediterranean forces. Several aircraft carrier task groups and a Regulus I task unit comprised of *Barbero*, *Torsk* and *Runner* were engaged in the exercise. On the way to the launch position above the Artice Circle off the northern coastline of Norway, Commander Styer was embarked on board *Barbero* Several days prior to reaching the launch point, Lieutenant Morrison, Missile Guidance Officer, discovered that the Freiden calculator was not giving the same answer to the square root of nine twice in a row. This was a critical problem since the simulation was to be as real as possible. Morrison reluctantly reported this fact to Captain Dew and Commander Styer. Feeling that the simulation must assume wartime conditions and given the criticality of the mission to the Submarine Force, Styer decided that he and Morrison would attempt to repair the calculator. Commandeering half of the Wardroom table, they went to work, learning as they went since no manual was available. Forty-eight hours later the calculator was running like a charm, after having covered the table with gears, pinions, racks, and tiny screws. The two officers didn't find anything wrong that they could pinpoint and decided that the calculator probably had just needed cleaning.[9]

Sea conditions and wind force were extreme south of Iceland and they all had to surface and proceeded for four days to keep up with the operationally specified speed of advance. The *Torsk* lost its Officer-of-the-Deck, Lieutenant(jg) Bill Thompson, overboard with an extreme snap roll of the boat. Captain Dew immediately ordered the bridge personnel below, shut the Conning Tower hatch, and had the Officer-of-the-Deck stand periscope watch.

Barbero crossed the Arctic Circle on the way to the launch point, near the edge of the ice pack. For the simulated launch exercise the boat was within sight of good navigational landmarks on the northern Norwegian coast. Twice, at night, the missile and guidance crews got the missile checked out, on the launcher and simulated a launch, all without being discovered by the opposing forces. All concerned felt quite confident that the mission could have been successfully accomplished under actual wartime conditions.

On the way back to Norfolk, with each boat proceeding independently, Captain Dew decided one quiet Sunday morning that "Battle Stations Missile" would be a good drill, especially with two tactical missiles aboard. The air was very cold but there was only a slight wind. The seas were relatively calm but with a long, heavy swell from astern with deep troughs of 8-10 feet. The crew went through the drill with everything proceeding like clock work, the missile was in the raised launch position with engine at 100% thrust and the hangar door shut. With the drill completed, Lieutenant Robert Kutzleb, the Missile Officer, slowed the engine down to idle to let it cool down somewhat before shutdown. Kutzleb

Barbero's *missile team readies FTM-1255 for launch on 11 July 1956. This was the second launch of a Regulus missile from an East Coast submarine, and the missile was successfully recovered by the pilots of Guided Missile Group TWO after a 31 minute flight. FTM-1255 was successfully launched by* Barbero *seven times. (Courtesy of Blount Collection)*

Returning to Norfolk from Operation STRIKEBACK, the Barbero *launch team conducted a "Battle Stations Missile" drill with a tactical missile. A mishap near the end of the drill caused the missile, with engine at idle, to leave the launch rails and land on the after deck. Here the missile can be seen with its wings straddling the deck rails.*

recalls that he and the Chief Missileman "Red" Hager had discussed shutting the jet engine down with the missile still in the elevated position. Since there had been several reports of tailpipe fires due to fuel and lubrication oil draining back down into the hot tailpipe, they agreed to leave the engine running as the launcher was lowered.

At the forward end of each launcher rail was a mechanical stop that extended into the rail, preventing missile movement aft and possibly off the launch rails while the missile was in a horizontal position. The stops were automatically retracted as the launcher was elevated and then moved back into position by springs upon lowering the launcher. Both sets of stops had been well greased prior to the drill to insure proper movement. Normally this would not have been a problem but, unknown to Kutzleb and Hager, the springs malfunctioned and prevented the stops from extending into the locked position ahead of the missile launch rail slippers. As the launcher was being lowered, the launcher rams allowed the launcher to fall to the deck for the last foot or so of elevation. As it slammed down, the ship's stern dipped heavily into a deep trough and with the missile engine still running, the missile started aft off the launcher. As the missile moved aft, the launch umbilical was pulled away. As far as the missile was concerned, this was just like a launch and the engine went to full power. Lacking the thrust of the JATO bottles, it failed to fly, coming to rest supported by its wings on the port and starboard deck railings and running at 103% engine thrust.

Captain Dew was observing all of this through the periscope and his expression was one of total disbelief while the rest of the crew inside *Barbero* heard a horrendous crash when the missile hit the deck. Morrison was in the Conning Tower and Kutzleb quickly joined him from the Missile Center. The tactical missile had a war reserve W-5 warhead with enough high explosives to blow up everything aft of the conning tower. Morrison and Kutzleb quickly volunteered to go out on deck, through the Conning Tower hatch, to shut down the engine and disarm the warhead. Since the umbilical had disconnected during the missile's motion aft, the only alternative was to wait for the missile to exhaust its fuel supply and this meant waiting at least 40 minutes. They remember getting out on deck and racing through the jet blast to reach the access compartments on the side of the missile. They then sheepishly realized that neither had brought any tools. A quick check of the contents of their pockets revealed a jackknife and several coins. The cam-lock aircraft fasteners for the throttle and warhead compartments could be opened using a coin and shortly thereafter Morrison had the engine shut down while Kutzleb, also the Warhead Officer, quickly verified that the warhead was safe. There was never any danger of a nuclear explosion but the "pucker factor" was tight nonetheless.

Inspection revealed that the missile had crushed about 18 inches deep into the light steel superstructure aft of the launcher. No one on board had any idea what even the short fall from the launcher might have done to the warhead components. Furthermore, the missile was still full of jet fuel and had the two boosters armed with the igniters in place. It seemed the danger had lessened but the potential for disaster was still very real. With jacks and chain-falls best suited for other tasks, the missile crew and other volunteers managed to drag the missile back up onto the rails of the launcher. This permitted them to use the normal missile launcher equipment to stow the missile in the hangar. The salvage effort took the better part of eight hours and it was dark when the task was completed. The remainder of the voyage home was decidedly uneventful.[10]

1958

The annual Springboard exercise for 1958 was another full test of Regulus submarine operations on the East Coast. GMU-51 deployed to the Naval Air Station, Roosevelt Roads, Puerto Rico, with Guided Missile Group TWO, Detachment "Alpha" providing the recovery aircraft for fleet training missile operations. SubDiv 63 and *Orion* deployed to St. Thomas, Virgin Islands, with *Barbero, Runner, Argonaut, Torsk* and *Medregal* operating from the St. Thomas base for almost two months. *Barbero* picked up missiles from GMU-51 at Roosevelt Roads and then operated from St. Thomas until the missiles had been expended. GMU-51 also launched a large number of missiles in support of training operations for the guidance submarines. *Barbero* launch operations commenced 14 February 1958 with a successful Trounce guidance operation. Over the next three weeks, *Barbero* launched nine missiles, including one tactical missile. Eight of nine flights were successful with the only failure occurring during the OST on 11 March when a tactical missile failed to respond to Trounce guidance or chase aircraft commands and eventually splashed due to fuel exhaustion.[11]

Barbero made an emergency war footing deployment in July 1958 in response to the Lebanon Crisis. Alongside the *Orion* in Norfolk, *Barbero* was fully stripped to depart for a major overhaul at Philadelphia Naval Ship Yard. Urgent messages were received for deployment of the Atlantic Fleet, including the Regulus I submarines of Submarine Division SIXTY-THREE. *Barbero* was completely resupplied for war patrol in an around-the-clock operation. After reloading spare parts and total reprovisioning in 24 hours, *Barbero* left for Yorktown to pick up two tactical missiles and live warheads for the torpedoes. All were checked out in the middle of the night and then *Barbero* sailed for the launch station they had trained for during the NATO exercises ten months previously. After two days the emergency condition of military readiness was reduced and *Barbero* was ordered to return.[12]

In September 1958, the home port of Submarine Division SIXTY-THREE was changed from Norfolk, Virginia to San Juan, Puerto Rico. The primary consideration in the move was the weather. Training operations with missiles would be the rule rather then the exception because of the good weather in the Caribbean, while just the opposite was true at Naval Air Station Chincoteague and the Norfolk area. *Barbero* remained home-ported in Norfolk for another six months because she was still undergoing a major overhaul at Philadelphia Naval Shipyard.

1959

Early in the Spring of 1959, the Postmaster General of the United States, Arthur Summerfield, expressed interest in demonstrating the use of a guided missile to deliver official U.S. mail. Secretary of Defense MacElroy and Secretary of the Navy Thomas Gates agreed and *Barbero* was chosen for the task. Earlier Regulus I and Regulus II launches had carried private letters and the first Regulus II launch from a submarine had carried letters that were delivered to the President and other high officials but these were not considered official U.S. Mail by the Post Master General. On 8 June 1959, *Barbero* launched FTM 1269 one hundred nautical miles off the coast of Florida. After a flight of 21 minutes the missile was recovered at Naval Air Station Mayport, Florida; the 3,000 letters were suitably hand canceled and sent to high ranking government officials, including President Eisenhower.[13] After the mail flight, *Barbero* reported to Philadelphia Naval Shipyard for a restricted

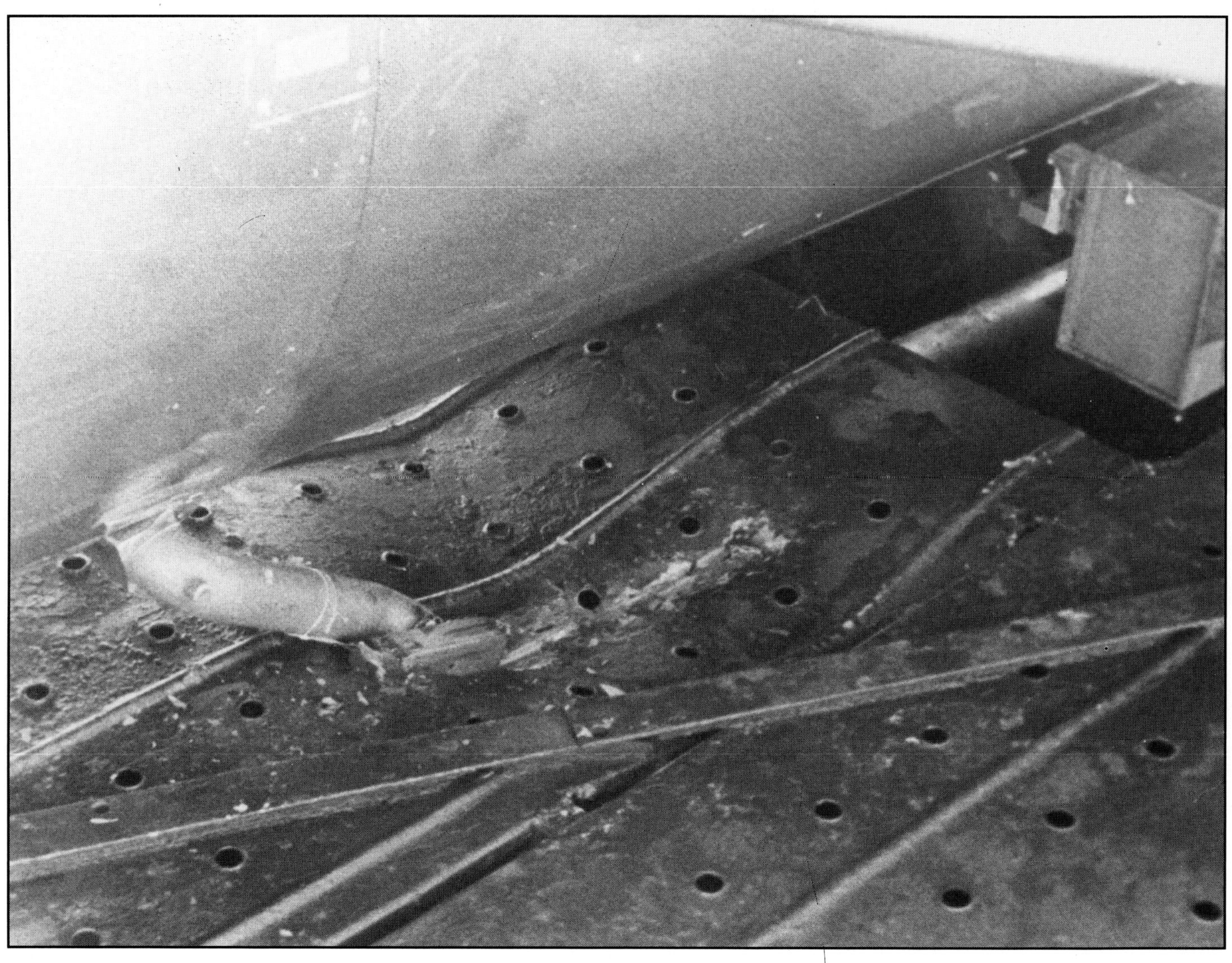

Bottom: The forward launch slippers dented and gouged the deck plating. Barbero's *missile team had to rig a hoist to raise the missile sufficiently to place it back into the rail guides and then pull the missile back into the hangar. (Courtesy of Morrison Collection)*

12 August 1959 and Barbero *is transiting the Panama Canal enroute to Pearl Harbor, and the consolidation of the Regulus submarine missile forces in the Pacific. (Courtesy of Blount Collection)*

availability period. Lieutenant James Watkins, a future Chief of Naval Operations and Secretary of the Department of Energy, was Executive Officer at this time.

Lieutenant Commander Robert Blount, USN, relieved Lieutenant Commander Dew in early July, 1959. Blount was very familiar with the Regulus program, having served as GMU-50 Warhead Officer, Submarine Division FIFTY-ONE Operations Officer, Engineering Officer on *Tunny* and Executive Officer on *Cusk* during Regulus I operations. *Barbero* was now scheduled to move to her new home port at Roosevelt Roads. All of the household belongings of the crew had been shipped and families were making final arrangements for quarters. All the plans were for not when the decision was made to consolidate all Regulus submarine assets in the Pacific because the first Polaris boats would be deploying in the Atlantic. *Barbero* and *Medregal* were therefore transferred to Pearl Harbor, Submarine Squadron ONE, and all of the crew had to wait for their household belongings to catch up with them in Pearl.[14]

On 12 August 1959, *Barbero* cleared the Panama Canal and rendezvoused with *Medregal* at Mazatlan, Mexico. They continued to Pearl on the surface, conducting torpedo and missile training operations on the way. The two submarines reached Hawaii on 24 August 1959. Submarine Squadron ONE was reorganized with Submarine Division Eleven consisting of *Tunny*, *Barbero* and the nuclear-powered guided missile submarine USS *Halibut* (SSGN 587), which was due to arrive 21 March 1960, as the launch submarines with *Cusk* as the division's only guidance submarine. Submarine Division TWELVE consisted of USS *Grayback* (SSG 574) and USS *Growler* (SSG 577) as the launch submarines and *Carbonero* and *Medregal* were the guidance submarines.

In early November *Barbero* made a deployment to the Western Pacific. This was a familiarization cruise and was to include a tactical missile launch during a SEATO demonstration. Three weeks out of Pearl Harbor *Barbero* stopped at Chi Chi Jima, 150 nautical miles north of Iwo Jima. Six hundred miles due south of Tokyo Bay, Chi Chi Jima had a large protected anchorage surrounded by mountains. In the mid-1930's the Japanese had honeycombed the island with fortifications. Blount remembers being very impressed with the amazing facilities that had been placed inside the mountain and understood why the Commander Submarine Forces, Pacific, planned to use Chi Chi Jima as a replenishment site for Regulus I submarines.[15] The launch for SEATO was scrubbed due to a tropical storm in the Okinawa area and *Barbero* returned to Pearl Harbor.

Regulus Deterrent Patrols 1960-1964

Barbero conducted her 46th missile launch on 4 August 1960. This launch also marked a major milestone in the program, the 1000th launch of Regulus I. The weather was foul and the successful launch took place in heavy seas, a true test of how far the program had come in nine years. Rear Admiral Roy S. Benson, Commander Submarine Forces, Pacific Fleet, was on board to observe and pushed the button for the launch off of Oahu. Accuracy was excellent on target but the missile was lost on recovery due to radio control interference with the recovery aircraft by a newly installed US Army high-powered tropospheric scattering radio transmitter located on Kauai.[16]

Barbero departed 30 September 1960 for her first deterrent patrol. She was on station in her patrol area in November 1960 when USS *George Washington* (SSBN-598) sailed from New London for the first Polaris deterrent patrol in the Atlantic. The first Commanding Officer of the *George Washington*, Commander Jim Osborn, had been the first Commanding Officer of *Tunny* after her Regulus I conversion as well as the second Officer-in-Charge of GMU-50. The Missile Officer on *George Washington*, Lieutenant Bernie Botula, had been the third Missile Officer of *Barbero*.

After a stop at Adak, to top off fuel, *Barbero* proceeded to the patrol area and commenced a slow race track pattern. During this patrol the submarine was required to be at periscope depth 24 hours a day so that a continuous listening watch on the very low frequency (VLF) loop antenna could be maintained. This dangerous practice was due to a misunderstanding with the ComSubPac staff on the continuity of the very low frequency broadcast. This procedure, coupled with nighttime snorkeling, was required in order to receive missile launch instructions at any time of the day or night. A later installation of the floating wire antenna permitted considerably greater latitude in patrol depth.[17]

Lieutenant Eugene Lindsey, Assistant Engineering Officer, clearly recalls that shortly after arriving on station for this first patrol, the "normal" day would begin when the sun went down and the snorkel went up. Four meals a day were served with a heavy meal during the mid-watch. If one had a stuffy head and was sleeping during snorkeling periods while the atmospheric pressure in the boat rapidly fluctuated from sea level to 5500 feet in heavy seas, one awoke with considerable pain, a silent scream and a curse. After a few days of this it was amazing how almost all hands could sleep soundly while the snorkel head valve opened and shut almost constantly in a heavy sea.[18]

Top surface speed of *Barbero* was 13 knots in a flat sea. In order to account for heavier than normal seas and periodic battery charging, transit speed of advance was usually given as eight knots. For fuel conservation purposes, it was sometimes limited to six knots. With the long distances to be transited and the need to dive upon sighting any ships, aircraft or with an electronic countermeasures warning, this combination made for some long voyages. The ship's officers were convinced, however, that if *Barbero* was discovered by the Soviets, it would undoubtedly lead to a most "unfortunate accident" and the possible disappearance of *Barbero* without a trace. Hence, great emphasis was placed upon alert watch standing and top performance of radar, sonar and electronic countermeasures equipment.[19]

When submerged, the boat's performance was less than something to behold. In a calm sea she submerged to periscope depth quite a bit more quickly than the usual fleet boat due to the missile hangar and the lead ballast in the keel. In a heavy sea, the missile hangar caused her to "hang up" due to surface effect and submerging took considerably longer. Top submerged speed was 6 knots, but only for half an hour. The usual submerged speed was 1.5 knots at 1/3 speed, 2.5 knots at 2/3rd speed and in a crisis "Standard" might produce as much as 4 to 4.5 knots. These speeds represent slightly half of what the submarine was originally designed for.

There was one occasion during the first patrol when *Barbero* went for 17 days without a navigational fix. It was typhoon season and the skies were always overcast. LORAN A radio beacon navigation coverage was spotty at night, usually with a single line that was of little value. Add to this two underwater logs for use in dead reckoning and a gyrocompass that was constantly

Change of command ceremony 29 April 1961. LCDR Blount was relieved by LCDR Carlton A.K. McDonald after two Regulus patrols on board Barbero. *Installation of the BPQ-2 Trounce guidance system and a floating wire low frequency antenna was part of the refit. (Courtesy of Blount Collection)*

"tumbling" due to the foul weather and the dilemma of knowing just exactly where the *Barbero* was located became apparent. Lindsey recalls one wardroom discussion during the 17 day wandering period. There seemed to be a consensus that if they had to launch they would find a specific small island in the patrol area and use it as the navigational point of reference. Lieutenant Douglas Murray, the Executive Officer and Navigator, was a master at the art of navigation. At the end of the 17 days the boat was determined to be less then ten miles from the dead reckoning position.[20]

The charts available for the patrol areas were circa the 1930's and filled with inaccuracies. The mountain peaks were perfect navigation landmarks except that many of their locations on the charts had no connection with reality. Murray took on the task of making his own navigation charts. This involved taking a series of running fixes coordinated with panoramic photographs through the periscope. The developed photographs, section of the charts and the bearings were then fitted together on the wardroom table. A new chart was thus created and the process repeated until the geography and navigation were correlated. A large, loosely bound volume was made for distribution to the other Regulus submarines for further updates and refinements.[21]

The first patrol lasted 60 days and *Barbero* then transited to the forward refit base, arriving on 2 December 1960. After a three week refit she left on 23 December 1960 to return to her patrol station. It took 10 days to get to Adak where several hours were spent refueling and taking on engine lube oil. This was critical on *Barbero* since even though she had two reliable engines, they leaked oil quite freely. The covers on the sump, over the pistons, weren't tight and the oil was lost to the bilges. In the cold Aleutian water temperature, the oil readily coagulated and was not easily recoverable. The engine room gang, assisted by others not on watch, had to form a bucket brigade to move the solidified oil in the bilges to the Officers' Head in the After Battery compartment to flush it into the sanitary tank and eventually overboard.

1961

The second patrol lasted 71 days. As they were returning to Pearl, *Barbero* received a radio message that from this point forward, the equivalent of four Regulus tactical missiles were to be on station at all times. Therefore, *Barbero* and *Tunny* would patrol as a four missile unit, while *Growler*, *Grayback* and *Halibut*, with four missiles (*Growler* and *Grayback*) or five missiles (*Halibut*), would patrol separately. This meant a more arduous patrol schedule since four patrol units would be rotating instead of five. Captain Blount recalls that the crew reaction was very positive. The crew took pride in being part of the elite of the Submarine Force and knew that their mission was critically important to the defense of the nation.

Several days south of Adak, *Barbero* was in heavy seas and routinely rolling 20-30 degrees. The boat was running with the main induction valve shut, snorkel mast raised and the upper conning tower hatch shut. Normally when transiting on the surface, the very low open bridge, the result of SSG conversion economy, was manned with two lookouts on the periscope shears and the Officer-of-the-Deck on the forward bridge structure. When the weather worsened, the lookouts first were ordered out of the shears onto the bridge deck and harnessed to the bridge railing. As the green water started rolling up to their feet and the bridge area filled with water, one lookout would be sent below. If conditions worsened, the officer would inform the Captain and obtain his permission to stand watch in the conning tower. On this particular day the waves were periodically drenching the bridge watch. Lindsey recalls that he asked permission to go below with the remaining lookout. He

Three photo sequence of the demise of Barbero *on 7 October 1964 during live fire tests of the new MK-37 acoustical torpedo. Top: Torpedo hits aft. Bottom: Rudder flying through the air.* *Continuation on page 132.* *(Courtesy of Richardson Collection)*

Continued from page 131: Down by the stern, Barbero *refused to sink until P2V "Neptune" patrol aircraft attacked her with depth charges. The missile hangar was removed for use as a hyperbaric chamber, but the launcher was left intact. (Courtesy of Richardson Collection)*

quickly opened the hatch to the conning tower and dropped below with the lookout, swinging hard on the lanyard to shut the hatch. The quartermaster had started to dog it shut when a monster wave hit. The wave smashed in the thin metal shield surrounding the bridge, the shield to which Lindsey had been attached just moments before. *Barbero* rolled at least 65 degrees since the Arma Mk 19 gyrocompass had tumbled at its limit of 65 degrees. It was most sobering to examine the damage several days later when the storm abated and allowed access to the bridge.[22]

Upon arrival at Pearl Harbor, 4 March 1961, *Barbero* went into restricted shipyard availability for installation of the new very low frequency floating antenna system, the new Trounce BPQ-2 Radar Course-Directing Central guidance system, overhaul of missile launcher and hangar equipment and an additional cold storage box for provisions. One of the officer staterooms was converted to include the launch and checkout console for the new guidance system.

During this shipyard availability, Blount was relieved by Lieutenant Commander Carlton A.K. McDonald on 29 April 1961. After eight weeks of around-the-clock work in the yard, they had almost two weeks of refresher training, shooting torpedoes and launching missiles. During this upkeep, the SSG personnel allowances were increased to 135% so that an officer and a number of enlisted men could remain behind during each patrol to attend special training courses, and take leave so that continuous patrol duty wouldn't become too demoralizing. Compared to the two complete crews available for each Polaris submarine, this wasn't much but it provided some relief from the "back-to-back" patrol schedules of the SSGs. In addition, this policy permitted an increased number of qualified officers and enlisted men for the rapidly expanding Polaris program.

The third patrol for the *Barbero* was in company with *Tunny*.[23] It began 23 July 1961, stopping enroute at Midway Island before heading north to relieve *Grayback*. Lieutenant Jack O'Connell, the Supply Officer, recalls that on this patrol the electronic intelligence team leader was a Marine captain, quite affable and popular with the wardroom. He used a single expression for all events, saying softly, reverently, and with laughter or anger, "Shit man, f_ _ _" After a while everyone started using it, shortening the phrase to "SMF". This is when O'Connell and Lindsey designed the insignia for the "North Pacific Yacht Club" with the nautical signal flags for SMF positioned diagonally across the middle. They had patches made for their submarine jackets and lapel pins fashioned for more formal attire. Months later at a cocktail party the innocent question was asked by one of the wives what SMF stood for and the quick reply was "Submarine Missile Force" of course.[24] Both *Barbero* and *Tunny* went to the forward deployment base for refit on 28 September 1961, returning to their patrol areas on 4 November 1961. The fifth patrol ended on 12 January 1962 upon return to Pearl Harbor.

1962-64

Lieutenant Commander George Mueller relieved McDonald in June 1962 and made two patrols which lasted from 24 August 1962 to 29 October 1962 and 12 January 1963 to 15 March 1963. A story that has proved hard to verify entails the "borrowing" of the Adak totem pole by the *Barbero* during this time frame. Lieutenant Commander James Greer relieved Mueller on 15 May 1963. Greer recalls that when he took command of *Barbero* she was in the best material condition of any of the submarines he served on. *Barbero* made her last two patrols from 10 July 1963 to 28 September 1963 and 5 January 1964 to 13 March 1964.

After they returned from the last patrol in late March 1964 the ship and crew had 30 days to recoup. Then they had the month of May to provide services in the Pearl Harbor area until the 1st of June when decommissioning procedures were started. After the last deterrent patrol, her wardroom and crew complement was rapidly reduced until Lieutenant Dan Richardson became, simultaneously, the Executive Officer, the Operations Officer, the Supply Officer, the Chief Engineer, Navigator, and almost every other officer billet aboard the boat. Richardson recalls that the boat was gradually stripped of almost any equipment that could be returned to the Navy supply system (or moonlight requisitioning by other enterprising submarine crews). During drydocking, the lead ballast in the keel and the missile hangar were both removed, the former because of the great value of lead and the latter to be used as a pressure chamber at a Honolulu hospital. The final decommissioning of *Barbero* was on 30 June 1964. With the boat decommissioned, Lieutenant Commander Greer moved on and Lieutenant Richardson became Officer-in-Charge of the skeletonized *Barbero* crew.[25]

New acoustical torpedoes were then entering the Submarine Force weapons inventory, so the plan was to utilize *Barbero* as a live target to test the capabilities of these new weapons against a surfaced submarine. Only one of the main engines remained operational and the batteries had been removed. Thus the boat rode extremely high in the water and was very unstable, with a minimum of diesel fuel, no spare parts, and no rations on board. The skeleton crew was composed of a few former *Barbero* men, with the majority from other ships at the waterfront in Pearl. The torpedo tests consisted of taking *Barbero* to sea, setting her on a straight course, and everyone on board retreating to a torpedo retriever boat which came alongside for that purpose. Lieutenant Nate Newman controlled *Barbero* using radio control equipment in an especially equipped helicopter.[26] For almost two months and seventeen attempts, firing runs with torpedoes carrying inert warheads were made by various submarines, but none were considered totally satisfactory. Finally the decision was made to fire a war shot. On 7 October 1964, *Barbero* went to sea and, as usual, the crew was evacuated to a torpedo retriever vessel. *Barbero*'s skeleton crew, alerted to the ultimate and final test, had worked overtime to ensure that all compartments were as airtight as possible, making it more difficult to send "their boat" to the bottom. This time the retriever headed towards Pearl Harbor, rather than positioning herself to reboard Richardson and the crew. By the time the torpedo was fired by the USS *Greenfish* (SS 351), the retriever was hull down towards home, and Richardson didn't even see the *Barbero*'s last minutes. In a twist of fate, the Greenfish's Executive Officer, Lieutenant John McDonnell, was given the honor of firing the final torpedo. McDonnell had been Missile Officer on *Barbero* during her first Pacific deterrent patrol. McDonnell later told Richardson that *Barbero* was struck in the stern, the most acoustically active portion of the ship and he was sure that the rudder had flown through the air. After this hit, *Barbero* was down by the stern in the late morning, but still adamantly refused to sink. After five depth charges were dropped by P2V aircraft, *Barbero* sank gracefully into the deep water off Oahu, Hawaii.[27]

Endnotes

1 Personal interviews with Lieutenant J.K. Morrison III, USN (Ret.) and Commander Samuel T. Bussey, USN (Ret.).

2 Personal interview with Commander Samuel Bussey, USN (Ret.).

3 Personal interview with Captain Walter P. Murphy, USN (Ret.).

4 Personal communications with Captain Walter P. Murphy, USN (Ret.) and Commander Samuel Bussey, USN (Ret.).

5 Personal correspondence with Captain Walter Murphy, USN (Ret.).

6 Ibid.

7 Personal correspondence with Commander Samuel T. Bussey, USN (Ret.).

8 Personal communications with Captain Charles Styer, USN (Ret.).

9 Personal interview with Lieutenant Julian Morrison III, USN (Ret.) and personal communication with Captain Charles Styer, USN (Ret.).

10 Personal interview with Lieutenant Julian Morrison III, USN (Ret.) and personal communication with Lieutenant Commander Robert Kutzleb, USN (Ret.).

11 Regulus I Flight Test Index, 1961, LVSCA A50-24 Box 7. Personal communications with Captain Ralph Jackson, USN (Ret.).

12 Personal interview with Lieutenant Julian Morrison III, USN (Ret.).

13 Missile Delivers Submarine's Mail, 8 June 1959 New York Times.

14 Personal interview with Rear Admiral Robert Blount, USN (Ret.).

15 Ibid.

16 Pacific Fleet Submarine Memorial Archives, Pearl Harbor.

17 Personal interviews with Captain John O'Connell, USN (Ret.), Commander Eugene Lindsey, USN (Ret.) and Rear Admiral Robert Blount, USN (Ret.).

18 Ibid.

19 Personal interviews with Commander Eugene Lindsey, USN (Ret.) and Rear Admiral Robert Blount, USN (Ret.). Neither indicated that this had ever happened. Their comments are more a reflection of what each felt the Soviet Union was capable of doing. In light of the continued harassment and actual attacks on several Air Force reconnaissance aircraft in 1958-1959, this does not seem to be an implausible concern. For further information see "American Espionage and the Soviet Target," Jeffery Richelson, 1987, William Morrow and Company.

20 Personal interviews with Commander Eugene Lindsey, USN (Ret.), Captain John O'Connell, USN (Ret.) and Rear Admiral Robert Blount, USN (Ret.).

21 Ibid. All of the Regulus submarines created these types of charts and freely exchanged them.

22 Personal interview with Commander Eugene Lindsey, USN (Ret.) and Rear Admiral Robert Blount, USN (Ret.).

23 While Barbero and Tunny operated as a unit, they were not in direct communication with each other during these patrols.

24 Personal interviews with Captain Jack O'Connell, USN (Ret.) and Commander Eugene Lindsey, USN (Ret.).

25 Personal communications with Captain James Greer, USN (Ret.) and Rear Admiral Daniel Richardson, USN (Ret.).

26 Personal communication with Nate Newman.

27 Personal communication with Rear Admiral Daniel Richardson, USN (Ret.).

Chapter Twelve: USS Grayback (SSG 574)

Grayback was the first submarine to be completely designed and built at Mare Island Naval Shipyard. Her construction was funded under the 1953 Fiscal Year shipbuilding budget as an improved 563 class, the most modern U.S. Navy fast-attack diesel-electric submarine at the time. Her original design was to retain the desirable features of the Portsmouth Naval Shipyard and Electric Boat Tang class submarines, while simultaneously improving on several of the problems of this class as well as reducing the building costs as much as possible. Midway through construction, *Grayback* was cut in half for the insertion, between the Forward Torpedo Room and the Forward Battery Compartment, of a 40 foot hull section containing Regulus missile checkout and launching equipment as well as storage for four Regulus I or two Regulus II missiles.

1953

Lieutenant Commander William E. Heronemus, a naval engineer, reported to Mare Island in June 1953 to take charge of the design and construction of *Grayback*. One seemingly small change from the Tang design was to increase the hull diameter by six inches. While this seems minuscule in any ship this large, Heronemus realized early on how much better her internal layout became with this extra room. New features included a special torpedo stowage and handling system in the forward torpedo room, and a new arrangement for the periscope, ship control and attack center. There were many improved "habitability" features for the crew, including individual air conditioning and full crew bunk installations, meaning no more "hot" bunking. Equally important was the fact that *Grayback* was the first to have submarine acoustic engineering concepts applied from the very beginning of construction. Her snorkel design was copied in all subsequent snorkel installations. The superstructure of *Grayback* was also a new concept, being made of aluminum and hinged for ready access to air piping and valves as well as the engine mufflers.[1,2] An innovation introduced by Mare Island during the design and construction of *Grayback* was the extensive use of a one-quarter scale model of the entire ship and several full scale models of complex ship compartments. This eliminated many man-hours of complex mathematical calculations and permitted evaluation of design alternatives as construction progressed.

Grayback's keel was laid 1 July 1954. As the lead ship in this new fast attack class it was imperative to get her launched and commissioned as soon as feasible. By November 1955 *Grayback* was eighty percent complete and scheduled to be launched in February 1956. Heronemus heard of an impending Bureau of Ships decision curtailing diesel-electric submarine construction and quickly returned to an idea that he had conceived several months earlier, that of converting *Grayback* into a Regulus I launcher. He needed 30 feet of additional forward deck space to ram out and launch a Regulus I. His original design provided for four Regulus I's to be carried, extra torpedoes, or a mix of torpedoes and mines. *Grayback* would now have many of the characteristics that would later became part of USS *Halibut* (SSGN 587), another Heronemus design.

Heronemus brought this concept to the attention of both Chance Vought and BuShips. All parties were intrigued because at this point only USS *Tunny* (SSG 282) and USS *Barbero* (SSG 317) carried Regulus Is. On further study, it was suggested that instead of Regulus I, the *Grayback* carry Regulus IIs. This required a total of 40 feet of additional forward deck space so that the longer Regulus II could be loaded onto the launcher and trained to either side for launch. The major problem now was the size of the missile hangar. Heronemus opted for two side-by-side hangars at deck level so that flooding could be more easily controlled. By using HY-80 low carbon, high strength steel, the hangars would be relatively impervious to all but catastrophic collisions on the surface. The presence of a larger safety tank served to counteract initial hangar flooding.[3]

In January 1956, work stopped on *Grayback* and she was cut in two just aft of the Forward Torpedo Room to accomodate four additional 10 foot long hull sections for installation of Regulus I and II missile checkout and launch equipment as well as missile guidance consoles and ship's navigation systems. The added length also allowed the twin Regulus missile hangars to be faired into an efficient hydrodynamic form. Two hatches were installed which allowed access to the hangars from within the hull while underway. Compared to hull shapes of other submarines of the time, *Grayback* appeared quite awkward.

The missile launcher was mounted topside between the outer doors of the twin missile hangars and the forward end of the bridge fairwater. Unlike the launcher on *Barbero* and *Tunny*, *Grayback*'s could be traversed to either port or starboard to service either missile hangar and trained to either side for launch. Furthermore, it was originally designed to be almost fully automatic with limit switches controlling hydraulic systems and launcher position. Television cameras permitted the missile launch officer to observe the launcher position from within the missile control center. The final 10 foot section of the addition was welded into *Grayback* 13 March 1957 and on 2 July 1957, *Grayback* was finally launched.[4]

1958

Grayback was commissioned on 7 March 1958, with Lieutenant Commander Hugh G. Nott assuming command. The qualification tests for the Short Rail Mark 7 launcher began four months later using Regulus I and II dummy mass sleds of concrete and steel balanced to simulate the weight of a missile. The Regulus I launch tests using a six-ton sled were uneventful, with the first firing demonstrating that the launcher could easily withstand the booster rocket thrust. The more telling test was the use of a 13-ton Regulus II sled. On 25 July 1958 the Regulus II dummy was launched and the sled flew for 350 yards before nosing down, hitting the water and

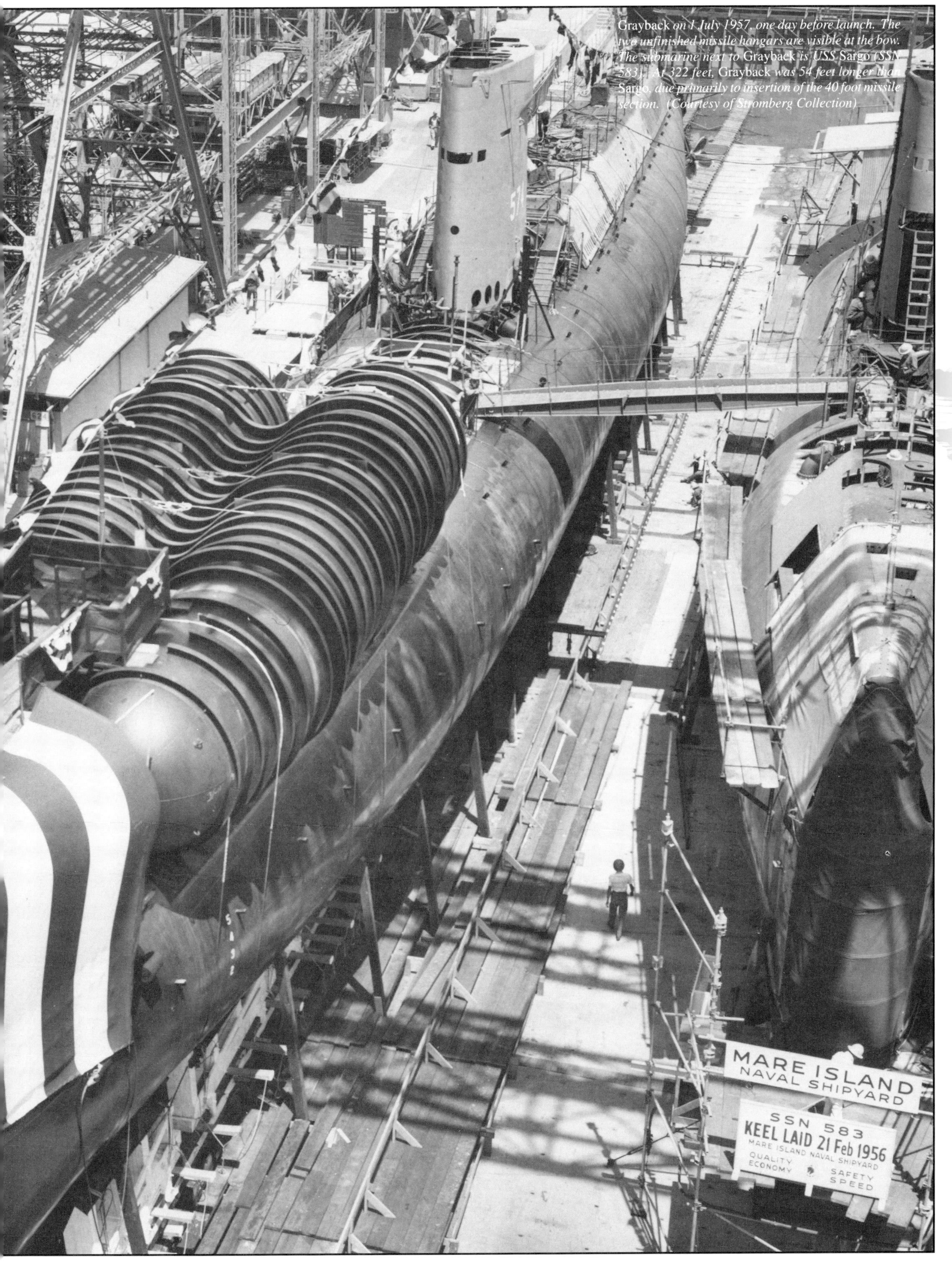

Grayback *on 1 July 1957, one day before launch. The two unfinished missile hangars are visible at the bow. The submarine next to* Grayback *is USS* Sargo *(SSN 583). At 322 feet,* Grayback *was 54 feet longer than* Sargo, *due primarily to insertion of the 40 foot missile section. (Courtesy of Stromberg Collection)*

Four months after commissioning, Grayback *tested the SR MK 7 Regulus I/II missile launcher in a series of tests conducted in the mudflats near Mare Island Naval Ship Yard. A concrete and steel dummy missile simulating a Regulus II was successfully boosted from this launcher. Notice the protective pod for the port side television camera just beneath the hatch on the sail. This camera and three others were removed early in the program due to continued flooding of the housings with subsequent camera failure. (Courtesy of Stromberg Collection)*

On 16 September 1958 Grayback *successfully conducted the only launch of a Regulus II missile from a submarine. GM-2016 leaps from the launcher less than a second after booster ignition. While is appears that the missile is launching directly over the bow, in reality it is already clear of the starboard side of* Grayback. *F7U "Crusader" chase aircraft could barely keep up with the missile as the booster dropped away. (Cannon Collection)*

cartwheeling several times before coming to rest in the mudflats.

Grayback had been the focus of considerable media attention during her construction at Mare Island. A review of *Grayback*'s scheduled launch of the first Regulus II from a ship of any kind was featured in the 3 August 1958 issue of "This Week," a nationally syndicated Sunday newspaper magazine. In mid-August *Grayback* hosted the printed press and television media for a review of the Regulus II system featuring a engine run up while moored to the pier at Mare Island.[5]

On 16 September 1958, *Grayback* surfaced approximately one mile off Port Hueneme. The port hangar door opened and a blue and white fleet training Regulus II missile, GM-2016, was rammed out, her wings, tail and noseboom folded back to facilitate hangar storage. The wings and tail unfolded automatically and the crew quickly placed the noseboom in the forward locked position. After attaching the fuel umbilical, the J79 engine was started and brought to full thrust. At 1112, GM-2016 was launched as the chase aircraft, three F8U-1s, swooped down into escort position. The single 115,000 lb thrust JATO bottle ignited and GM-2016 bounded into the sky, enveloping *Grayback* in dense white booster exhaust. After four seconds of thrust, the booster dropped away and GM-2016 was heading out into the Naval Air Missile Test Center Sea Test Range at 325 knots. A pre-programmed climbing turn was made and the missile began its 200 mile flight to Edwards Air Force Base. Missile guidance was by the out-of-sight control stations at Pt. Mugu and Edwards.[6] As preparations began for recovery at Edwards, the landing gear on the missile failed to respond to the gear down command from the chase aircraft. Despite repeated efforts to engage the gear release command, the missile stubbornly refused and the decision was made to make a belly landing on Rosemond Dry Lake. While this had been accomplished successfully before, the missile caught fire and was demolished. Ironically, the "missile mail" payload, a group of letters especially postmarked for the event, did survive and were mailed the next day to President Eisenhower and high ranking defense department officials. All of the Regulus II launch objectives for *Grayback* had been successfully met and since under wartime conditions this would have been a tactical missile without landing gear, the mishap on landing was seen as troublesome but not especially serious.

In early October 1958, *Grayback* left Port Hueneme for the Puget Sound operating area to conduct torpedo tube acceptance trials and ship's noise surveys. On the way to Puget Sound, cold weather and rough seas evaluation of the missile handling equipment also took place. She returned to Port Hueneme in mid-October and conducted her first Regulus I operation on 22 October 1958. Her missile team successfully launched and guided her first Regulus I under adverse weather conditions using the Trounce guidance system. On 29 October 1958 *Grayback* departed Port Hueneme for Pearl Harbor for further shakedown testing and refresher training. Four Regulus I launches in a two week period were successful with all four missiles recovered for re-use.[7] *Grayback* returned to the West Coast on 12 December 1958 for a week long visit to San Diego where she was again the subject of media attention as she conducted both military and press demonstrations of the Navy's newest and most powerful weapon system. One week later *Grayback* returned to Mare Island for her 10,000 mile checkup. Unwelcomed news of the official cancellation of the Regulus II program the day before was mitigated somewhat by the announcement that her new home port would be Pearl Harbor as she joined the rest of the submarine Regulus force.

1959

Grayback departed the West Coast on 23 February 1959 and arrived in Pearl Harbor on 7 March 1959 ready for duty, reporting to Submarine Squadron ONE. Over the next four months she conducted seven missile operations with six successful launches. Four of the operations were multiple guidance submarine tests with each successfully demonstrating the maturity of the Regulus I program.[8,9] On 3 July 1959 *Grayback* departed Hawaiian waters for "cold" water operations in the Gulf of Alaska operations area. These were specific tests of freezing temperature operations with war reserve W-27 thermonuclear warheads. One missile was launched, TM-1486, with dummy nuclear components, as a test of the W-27 warhead airburst fuzing system. The flight and terminal dive were executed properly but the airburst fuse failed. After brief stops at Dutch Harbor and Kodiak, she returned to Pearl Harbor on 31 July 1959.

Regulus Deterrent Patrols 1959-1964

After refit and extensive preparation, *Grayback* departed Pearl Harbor on 21 September 1959 for the her first Western Pacific cruise, the final evaluation run of the Regulus missile deterrent patrol concept. She arrived at the forward refit base 12 November 1959 for ten days of repairs and reprovisioning. On 16 November 1959 Nott was relieved by Lieutenant Commander John "Pete" C. Burkhart and *Grayback* was soon enroute back to Pearl Harbor for refit.

Grayback *as she appeared in March 1959 arriving in Hawaiian waters and her new home port of Pearl Harbor. Notice the streamline fairings for both hangar doors and the launcher stored in the flush position amidships. The hangar fairings were removed after the first patrol due to corrosion problems and the constant physical strain from the pounding North Pacific seas. (Courtesy of Author's Collection)*

Five months later *Grayback* began her first regularly scheduled Regulus deterrent patrol 31 May 1960. Completing seven weeks of patrol duties, she was relieved on station and arrived at the forward refit base, arriving 30 July 1960. After a four week refit and reprovisioning, she began her second deterrent patrol on 24 August 1960, returning to Pearl Harbor on 29 October 1960. After this patrol *Grayback* underwent restricted availability at Pearl Harbor Naval Shipyard.

1961

Normal requirements during pre-deployment training meant firing several exercise torpedoes. Lieutenant Donald Henderson, Engineering Officer, remembers that from 17 to 18 May 1961, just prior to the ship's third deterrent patrol, *Grayback* was scheduled for her torpedo firings with USS *Epperson* (DD 719) assigned as target vessel. During the two days various live firings were conducted. The target vessel routinely provided estimates of hit or miss distance with respect to the middle of the target. All firings during this exercise were reported to *Grayback* as misses. Since the *Epperson's* commanding officer had a reputation in Pearl Harbor of not caring very much for submariners, *Grayback* assessed this factor as possibly influencing the miss distance estimates. But all was not lost. One more torpedo shot was scheduled for 18 May 1961. *Grayback*'s torpedo fire control system malfunctioned and the running depth for the last shot was zero feet, meaning the torpedo had been incorrectly set to run on the surface. It hit the *Epperson* just slightly forward of the midship, fortunately more or less on the outboard end of a transverse stiffener, below the shipfitter's shop and above a fuel tank. Two or three feet more either way and it could have been pretty serious. Nonetheless, the 'dish' caused by the torpedo hit was indeed disfiguring and suffice it to say *Epperson's* skipper was enraged. On the other hand, *Grayback* did score a certain hit. Damage to the *Epperson* included a small leak to be repaired at the next regular drydocking and a 30-inch by 36-inch dent in her side.[10]

Grayback departed for her second set of back-to-back patrols on 5 June 1961. The third patrol ended on 13 August 1961 at the forward refit base. After four weeks of upkeep and replenishment she departed for her patrol station on 12 September 1961 and returned to Pearl Harbor on 13 November 1961. On 9 December 1961, Burkhart was relieved by Lieutenant Commander John J. Ekelund, previously the Executive Officer of *Growler*.[11]

1962

After a four month refit at Pearl Harbor Naval Shipyard, *Grayback* left on her fifth deterrent patrol on 2 April 1962, returning to Pearl Harbor on 3 June 1962. *Grayback* departed for her sixth deterrent patrol on 7 October 1962. The weather was even more miserable then usual. Ekelund vividly remembers one incident that happened while they were snorkeling at night in the patrol area. Lieutenant Frank Talbot was Conning Officer. The head valve of the snorkel was operating properly but what they did not know was that ice was building up on the seat of the valve. This was due to the combination of low temperature, the velocity of air flow and the sea-spray being sucked into the induction piping. Consequently, with each shutting of the head valve, *Grayback* started to ship more and more water. On one dunking this turned into a large and steady flow of ice cold water. The word was passed "Flooding in the Engine Room" and Captain Ekelund went immediately from the Wardroom to the Conning Station to tell Talbot to keep the up angle off. It was too late as the Diving Officer had already taken action to get back up to snorkel depth.

Ekelund ordered the After Group blown to take the up angle off. The problem with the high water in the Engine room and an up angle was that it almost guaranteed flooding the low centerline generator. By the time Ekelund got to the Control Room the threat of excessive flooding had passed and the ship was under control. The generator did flood as Ekelund had known it would and the ground reading went to zero, a dead short. They spent the rest of the patrol alternately washing it out with distilled water and ventilating it with a heater and blower but they were not able to return the unit to service.[12]

Grayback was in her patrol area during the October 1962 Cuban Missile Crisis. Jerry Beckley, a warhead technician, remembers being called to the wardroom along with the Missile Officer. The Captain and the Executive Officer, opened up the sealed Emergency War Order documents because a message had come through moving the defense condition to a higher state of alert. The envelope contained instructions for adjustments to the warhead. Beckley had to take out close to 60 cam-lock fasteners to remove the fuselage panel over the warhead compartment. The warhead had a high voltage thermal battery pack that was stored disconnected. He had to remove 4 bolts, turn the battery over

Grayback *pierside during* Tunny *Change-in-Command ceremony for* Tunny. *Note the missing superstructure covers aft of the sail.* Grayback *had returned 15 May 1963 from her seventh deterrent patrol just four days earlier.*

and reconnect to power up the warhead circuits. He then re-installed the panel and the warhead was almost ready. If launch was ordered, he would have had to insert whatever detonation settings the orders stated, such as contact or air burst. His last action would have been to flip the arming switch and the warhead would have been ready for launch. Beckley didn't have to go this far, but it was a very tense period of time and something he still clearly remembers. *Grayback* returned to Pearl Harbor 22 December 1962.[13]

1963

During the two months before *Grayback*'s next patrol, she received a new Ships Inertial Navigation System (SINS), the North American Aviation Autonetics N7, replacing the *Sperry* Gyroscope Mark I SINS equipment.[14] *Grayback* left on her seventh deterrent patrol on 20 February 1963. Just beyond the half-way point enroute to Adak, *Grayback* was running submerged and snorkeling. A particularly strong roll due to the normal inclement weather caused the After Main Battery circuit breaker to short, spraying sparks across the compartment. A fire started near the breaker in the After Battery Berthing compartment. Linoleum was burning, as well as mattresses, making a dense, choking, acrid smoke. Captain Ekelund went to the periscope stand and Lieutenant Bill Gunn, the Executive Officer, went to the fire scene to evaluate the damage and resolve the problem. Dense smoke had filled the berthing area before there was any idea of where the fire was located. The berthing space had been evacuated by the time Gunn arrived and so, with the smoke billowing out, the fire door between the After Battery Crew's Mess and the berthing spaces was shut and dogged tight. No sooner had they done this, then they heard banging on the door and muffled voices. The Hospital Corpsman and the Chief-of-the-Boat had gone into the compartment to clear it of all personnel. The After Battery water tight door was already shut. Normally the first rule in this situation is that once the door is shut, it is not open it until the problem has been solved on the other side. Gunn decided to open the fire door anyway and two crew members re-entered the smoke filled compartment, found the two trapped crewmen and dragged them out. They thought the fire was contained and no one else was left in the compartment. The fire door was then resealed. The emergency ventilation procedures were followed and ventilation of the compartment commenced. Gunn clearly remembers opening the fire door and entering the compartment. The mattresses were still smoldering and so they requested and got permission to open the After Battery Compartment deck hatch to topside to jettison the debris since the boat had surfaced by this time. As they removed the mattresses they came upon the body of Seaman James R. Jensen. He had been overcome by the fumes when he became trapped between the arcing breaker and the path to safety. Jensen had

Grayback *shown after her refit in 1967. August 1969 her designation was changed to LPSS 574, an amphibious troop and special operations transport submarine. The sail was raised during this refit and the hull lengthened 12 feet to accommodate new engines.*

crawled into the lowest bunk, apparently feeling safe since he was below the worst of the fumes and smoke. Several of the seamen that had evacuated recalled that he had tried to get past the arcing breaker, couldn't, and told them to go on, since he would be okay in the lowest bunk.[15] With the fire out, *Grayback* reported to Pearl Harbor that she was on the surface without propulsion. Within a few hours she regained propulsion and returned to Pearl for repairs. Two weeks later she was on her way back to her patrol station, much to the relief of the *Tunny* and *Barbero* whose patrols had been extended to ensure continuous target coverage. She returned to Pearl Harbor on 11 May 1963.[16]

Grayback departed Pearl Harbor on 7 September 1963 on what was to be her eighth and last deterrent patrol. A new battery had been installed prior to this patrol and the battery recharge was in progress as they patrolled submerged and snorkeling. Captain Ekelund was asleep and Gunn was Command Duty Officer. The boat had open tank battery ventilation, which meant that there was no ductwork between individual batteries and that the batteries vented through a ceramic dome into the battery well. Captain Ekelund recalls being awakened by Gunn who told him that they had just experienced a minor explosion in the Forward Battery compartment. As they talked, there was a considerable thump under Ekelund's bunk, located over the Forward Battery compartment and Ekelund needed no further explanations. The explosions were probably due to the ignition of small accumulations of hydrogen gas resulting from the charging process. Just what was causing the fire needed to be determined and resolved immediately.

Ekelund went to the Control Room and Gunn went to the accident scene, having already ordered the battery charge secured. Lieutenant (jg) Merle Pippin, Lieutenant(jg) John Regan and Gunn went into the battery spaces with fire extinguishers. The wooden wedges separating the individual cells as well as holding them in place were on fire. The battery spaces were cramped to begin with and the three men, encumbered with extinguishers, could barely maneuver but did succeed in putting out the fires only to hear a second set of muffled explosions. Gunn ordered everyone out. The fires were subsequently extinguished and *Grayback* retired to the eastern most part of her patrol area to examine the battery cells one by one. They found that ground paths had developed in the battery cooling system and the cell ceramic domes were also fouled, contributing to a heat build up that resulted in the fire. Once corrective fixes were applied they had no further problems from the batteries.[17] While they were resolving the battery problem, Ekelund also checked the stern planes since depth control had been a problem for most of the patrol. Everything appeared in order and the patrol was resumed.

At the end of the patrol 2 Nov 63, *Grayback* headed for Naha, Okinawa. As soon as *Grayback* was cleared for radio communications, Captain Ekelund requested a diver be ready in Naha to inspect the stern planes. Going into the channel, the tugboat carrying the harbor pilot stopped abruptly ahead of the *Grayback*, forcing her to stop. The bow started to fall off towards the nearby sea wall, so Ekelund backed the starboard engine full. There was an immediate loud thump and the ship shuddered. The Maneuvering Room reported a fire and that they were unable to answer bells. Ekelund remembers wondering what else could go wrong as he ordered "All Stop." Gunn went down to the Maneuvering Room to determine the damage, expecting the worst. He found no fire but plenty of smoke and hot electrical equipment.

After mooring, a diver inspected the stern diving planes and reported that there was no port stern plane and the trailing edge of the starboard plane was badly chewed up. Further tests revealed that though the hydraulic system functioned properly, it had no effect on stern plane motion. When they were able to inspect the mechanical linkage connecting the hydraulic stern plane ram to the stern plane shaft, they found all of the nuts had backed off of the bolts that served to clamp the operating mechanism to the shaft. The result was that the planes had been freewheeling for most of the patrol. These nuts were supposed to be safety wired to the bolts after being tightened to prevent just such an event. There was no evidence of safety wire. This explained the lack of depth control during the patrol, as well as the recent events in the channel. When the starboard screw was backed full, water flow pulled across the stern plane and flipped it over 180 degrees, allowing the trailing edge to jam the propeller, torquing the shaft and causing the propellor to drop off. This event was just one more example of how important seemingly small things could be in submarine operations. Several consequences came to mind once they saw what had happened. If they had run up "All Astern" in the patrol area, they could have lost one or possibly both propellers in very inhospitable territory, both politically and in terms of adverse weather conditions. The shaft could have warped or broken inside the stern shaft tube, flooding the motor room and motors, resulting in major electrical damage and no propulsion capability.[18]

On 13 August 1986, after 28 years of service, Grayback, *painted international orange, was towed to a live fire site in the Phillipine Sea and used as a target by units of the USS* New Jersey *Battleship Battle Group during HARPOONEX '86. (Courtesy of DeJesus Collection)*

Grayback proceeded to Subic Bay, Philippines, with a submarine rescue vessel escort, for drydock and repair after the damaged stern plane was removed at Naha. She was in drydock for three weeks during which time a complete inspection of the motors and generators indicated no permanent damage caused when the shaft came to a full stop from full speed ahead. With repairs complete, *Grayback* departed for Pearl Harbor, arriving 10 December 1963.[19] Lieutenant Commander Bill Gunn relieved Ekelund as Commanding Officer of *Grayback* on 11 January 1964, having received these orders earlier during the previous patrol. Gunn remembers that he was eager to get the ship ready for her next patrol. He and Lieutenant Commander Robert Owens, Commanding Officer of USS *Growler* (SSG 577), were summoned to the Commander Submarine Division ELEVEN office and told that which ever submarine was the first to be ready would go out next. Both captains wanted to be next so they intensified replenishment efforts. Soon it was obvious that *Grayback* would be ready first and Gunn was called again to the Division Commander's office, expecting to be told the date for the start of his next patrol. Instead he received a completely unexpected order: *Grayback* and *Growler* were to stop refit activities immediately pending further notification. After what seemed like weeks, they were told to prepare to return to Mare Island for decommissioning. Their roles as part of the submarine strategic deterrent force had come to an end.

After the decision to decommission had been made and during the standdown period, on 29 January 1964, *Grayback* launched FTM 1455, the last Regulus I missile from a 574/577 class guided missile submarine in the Pacific. *Growler* accompanied *Grayback* on the operation and hosted the dependents of both boats and GMU-10.[20] Commander Robert Hale, Commander Submarine Division ELEVEN had arranged this trip so that the girlfriends, wives and children had an opportunity to see what the crews of these submarines had been doing for the past five years.[21] *Grayback* and *Growler* sailed together for Mare Island. The crews' families had prepared enormous leis that were draped over the sail areas as they left Pearl Harbor. Both crews stored them during the transit and rigged them out as they entered Mare Island making for a memorable arrival at the submarine piers. *Grayback* was decommissioned on 25 May 1964.

Post Regulus

In early November 1967, *Grayback* began extensive conversion to remove missile checkout, launch and guidance equipment and replace it with amphibious operations equipment. The hangars were converted into wet and dry compartments for underwater operations with special operations teams. The conversion also included lengthening the ship 12 feet so that she could be re-engined with low speed and more reliable engines, and raising her sail 10 feet for better submerged handling.[22]

Grayback was recommissioned 9 May 1969. Commander Bill Gunn, then Chief Staff Officer of Submarine Squadron FIVE, rode *Grayback* as the Commander Submarine Forces, Pacific representative during *Grayback*'s acceptance trials at Mare Island and during her first underway submerged swimmer and swimmer delivery vehicle operations at San Diego. Shortly thereafter, *Grayback* deployed to her new home port, Subic Bay, Philippines. For 11 years *Grayback*, redesignated as LPSS 574, operated as a unit of the 7th Fleet both in the combat environment of Vietnam and in conducting special warfare training with U.S. and Allied Forces in Asia.[23]

On 13 August 1986, after 28 years of service, *Grayback*, painted international orange, was towed to a live fire site in the Phillipine Sea and used as a target by units of the USS New Jersey Battleship Battle Group during HARPOONEX '86.[24,25]

Endnotes

1 *Personal communication with Captain William E. Heronemus, USN (Ret.). Grayback Will Give Its Crew Full Bunks, 1 Jul. Heronemus, USN (Ret.).*

2 *Grayback Will Give Its Crew Full Bunks, 1 July 1954, Vallejo News-Chronicle.*

3 *Personal communications with Captain William Heronemus, USN (Ret.).*

4 *Personal communications with Lieutenant Commander Ham Stromberg, USN (Ret.).*

5 *Personal communications with Lieutenant Commander Louis A. "Pat" Kilpatrick, USN (Ret.).*

6 *Navy Jubilant Over Missile Shot, 17 September 1958,Oxnard Press-Courier.*

7 *Regulus I Flight Index, 1961, LVSCA A50-24 Box 7.*

8 *Personal communication with Captain Donald Henderson, USN (Ret.).*

9 *Regulus I Flight Index, 1961, LVSCA A50-24, Box 7.*

10 *Pow! Practice Fish Runs Silent — But Not Deep, 25 May 1961, The Honolulu Advertiser.*

11 *Ship's Log, USS GRAYBACK (SSG 574). Naval Historical Center, Ship's Histories.*

12 *Personal communication with Rear Admiral John J. Ekelund, USN (Ret.).*

13 *Personal communication with Chief Warrant Officer Jerry Beckley, USN (Ret.).*

14 *Personal interview with Captain William Gunn, USN, (Ret.).*

15 *Seaman Jensen was about to be released from active duty since he was in the Naval Reserve. He had gone home on emergency leave to attend his father's funeral and returned to Grayback for one last patrol.*

16 *Personal interview with Captain William Gunn, USN (Ret.); personal communication with Rear Admiral John J. Ekelund, USN (Ret.).*

17 *Ibid.*

18 *Personal interview with Rear Admiral John J. Ekelund, USN (Ret.).*

19 *Personal interview with Captain William Gunn, USN (Ret.).*

20 *GMU-90 had been redesignated as GMU-10 on 1 July 1959.*

21 *As luck would have it, a sequencing lanyard failed to work and the missile splashed 500 yards astern.*

22 *Navy Times 15 June 1981.*

23 *Ibid.*

24 *Mare Island Naval Shipyard "Grapevine," 19 September 1986.*

25 *Personal communication with Jim Christley of the Nautilus Museum, Groton, Connecticut.*

CHAPTER THIRTEEN: USS *Growler* (SSG 577)

Growler, like USS *Grayback* (SSG 574) was an improved 563 Class submarine. Built at Portsmouth Naval Shipyard, Kittery, Maine, *Growler* had a similar outward appearance to *Grayback* but was quite different in her internal layout; aft of the missile guidance center the layout was nearly identical to the smaller USS *Darter* (SS-576).[1] *Growler* was launched on 5 April 1958 at Portsmouth Naval Shipyard, Kittery, Maine and commissioned on 30 August 1958 with Lieutenant Commander Charles Priest, Jr., assuming command.[2]

Growler began her sea trials on 4 November 1958 in the traditional submarine test area off the Isle of Shoals. A successful first day was spent on the surface conducting full power runs, testing various ship systems and cycling all masts. At dawn on 5 November 1958, the *Growler* crew prepared to conduct the first test depth dive. After submerging to periscope depth, she then proceeded deeper, leveling off at 50 foot increments as the crew checked all systems and hull fittings subject to sea pressure. As *Growler* passed the fleet-type submarine test depth of 475 feet, the majority of her crew were in new territory, never having been this deep before.[3] Everything was fine until *Growler* reached 75 feet short of her test depth.

Radioman Leonard Powers was in the Radio Shack directly across the passage way from the Sonar Room. Powers remembers hearing a loud pop and looking across the passage way towards the source of the sound only to find a stream of water roaring down from an empty one-half inch cable fitting in the overhead of the Sonar Room. Captain Priest immediately ordered "Emergency Surface" while everyone nearby grabbed buckets and began collecting the water, passing it along to the galley for disposal. Most of the water was flowing into bilges or staying within the four-inch deck coaming that surrounded the Sonar Room. Unlike most of the crew's experience on the fleet-type submarines, where the compressed air rushed into the ballast tanks during an emergency surface evolution, at this much greater depth the air seemed to barely hiss. Lieutenant(jg) Robert Duke, the Communications Officer, was monitoring the depth gauge in the Chief Petty Officer's quarters and recalls the strange sensation of *Growler* slowly rising to the surface with a slight down angle due to the flooding. *Growler* surfaced with only superficial damage.[4,5] The Portsmouth Naval Shipyard Planning Superintendent, Lieutenant Commander Hank Hoffman, went topside and determined that an unused cable fitting opening had been plugged with a temporary blank for dockside tests which had not been replaced prior to sea trials. With all the time lost and additional costs if they returned to port, Hoffman suggested to Captain Priest, Jr. that a solution was readily available on board. The cable hole was slightly smaller than the diameter of a nickel and with two nickels sandwiching a rubber gasket, Hoffman was able to securely plug the hole. A compartment air pressure test indicated no leakage present and the trials resumed with torpedo firing and other ship's system tests. The temporary plug was removed in the shipyard, mounted on a plaque with the label "The Cheapest Repair in Shipyard History," and was the start of the ship's commemorative plaque collection.[6,7]

On 15 November 1958 *Growler* conducted her first missile operation test when she launched a 56 foot long, 13 ton dummy mass sled balanced to simulate a Regulus II missile. Much to the chagrin of shipyard officials, the first three attempts failed due to electrical problems. On the fourth try, the sled was successfully launched, splashing into the ocean 2,000 yards away as planned.[8]

1959

With acceptance trials completed, *Growler* headed south for her shakedown cruise. After successful completion of torpedo firing trials, *Growler* headed for Naval Air Station Roosevelt Roads, Puerto Rico and the start of her Regulus I launch operations. *Growler*'s first missile launch took place 24 March 1959. Since the BPQ-2 Trounce guidance equipment was not yet installed, USS *Runner* (SS 476), a Regulus guidance submarine, took control immediately after launch and guided the missile during the 30 minute flight. The next flight was a two-boat Trounce guidance operation in combination with USS *Argonaut* (SS 473) and *Runner* and was again successful.

Growler completed another three launches, all successful, over the next two weeks. Missile operations were then brought to an abrupt halt by a failure in the launcher elevation mechanism. The Short Rail Mark 7 launcher was overly complicated due to automatic sequencing and safety controls. Elevation was controlled by limit switches that were positioned to prevent the elevation screws from over extension. These switches failed and the launch rails were forced off the screws, stripping the top of the threads in the process. Repair was seemingly impossible since the boat did not have the necessary tools to re-cut the stripped threads. Priest remembers that, without being asked, off-duty crew members would come topside to take turns trying to repair the threads by filing them back into shape with hand files. He realized his efforts to bring to the crew the team spirit so necessary to successful operation of a submarine had been successful.[9]

Growler returned to Portsmouth for post-shakedown availability. The launcher was modified to prevent the recurrence of the limit switch failure. The BPQ-2 Trounce guidance radar and

Growler, *pictured here on 11 November 1958 during her sea trials.* Growler *differed from* Grayback *both internally and externally.* Growler's *pressure hull was 16 feet in diameter,* Grayback *was 18 feet;* Growler *had only four forward torpedo tubes,* Grayback *had six. (Courtesy of Author's Collection)*

Growler *conducts her third Regulus I (FTM-1433) in Puerto Rican waters. The missile was successfully recovered at NAS Roosevelt Roads, Puerto Rico. Note the two streamlined pods, one just below the open hatch on the sail and one just below the closed starboard hangar door. These were television cameras that were later removed due to unpredictable operation. The missile is shown just prior to engine ignition. (Courtesy of Kutzleb Collection)*

electronic equipment installation was also completed. During this time period *Growler* received orders to her new home port, Pearl Harbor. One guidance submarine, USS *Medregal* (SS 480) and the other East Coast Regulus I launch boat, USS *Barbero* (SSG 317), were also moving to Pearl Harbor as all Regulus I operations were being consolidated in the Pacific. *Growler* departed Norfolk 27 July 1959. After several days in Key West, Florida, where she put on several missile ram-out demonstrations, *Growler* left 14 August 1959 for transit to Pearl Harbor via the Panama Canal.

During the long and slow transit the crew had one memorable swim call. On 26 August 1959, Priest and the Executive Officer, Lieutenant Commander John C. "Pete" Burkhardt decided it would be appropriate to make a movie, from the surface, of *Growler* at periscope depth, snorkeling and then surfacing, ramming out a missile and running the missile engine up to full power. A life raft was inflated and a volunteer crew consisting of Lieutenant(jg) Robert Duke, Lieutenant(jg) William Lindeman, Torpedoman First Class John Haney and Commissary Steward Oscar Weigant, paddled 50 yards off to start filming. While submerged and circling the raft, Priest recalls observing the raft and seeing everyone waving quite energetically. He took this to mean that the filming was working out well. When they surfaced and recovered the raft, Priest learned the rest of the story. Duke recalls:

It was very, very quiet and actually pretty lonely in the raft, even with three fellow volunteers. After successfully filming *Growler* as she submerged, we were preoccupied with trying to ward off shark attacks. While we were watching for the periscope, I felt a heavy rippling along the bottom of the raft. After the second time, I asked Lindeman, Haney and Weigant if they felt it. They had and as we talked I looked over the side of the raft and saw a six-foot shark pass under the raft, turning to try to take a bite out of the raft's underside. I calmly asked for the shark repellent and received a 'There is no shark repellent, Sir.' I then asked for the flare gun and received a 'There is no flare gun, Sir.' We were completely ill-equipped and were about to face the consequences. I took an oar, ready to hit the shark the next time it made a pass. Meanwhile, Weigant was standing up, waving a shirt at the periscope he had just spotted. I felt sure we were all about to be dumped into the water. After I got Weigant to sit down; and, with Haney paddling like mad towards the periscope, the shark made another pass and this time I managed to give it a good rap on the nose. Much to my amazement, the shark disappeared for the next five minutes.

Meanwhile, *Growler* surfaced 100 yards off the raft and prepared to ram out the missile. The movie camera was on the floor of the raft, bouncing around in the salt water, useless. The shark returned but this time he had a friend which was quite a bit larger. The newcomer never made a run on the raft but the smaller one continued to worry us. As *Growler* approached to recover us, the sharks, of course, disappeared and everyone on board remained skeptical of our story.[10]

Growler arrived at Pearl Harbor 7 September 1959 and was assigned to Submarine Squadron ONE. Missile operations resumed on 2 October 1959 with the first Trounce guidance flight for the *Growler* guidance team. The operation was successful and the missile recovered at Bonham Auxiliary Landing Field on the Island of Kauai. *Growler*'s first tactical missile operations took place in late October with two highly successful and accurate terminal dives to impact. Her first unsuccessful launch occurred 8 December 1959 when the missile did not program over to cruise settings and splashed astern. Over the next three months she launched an additional three missiles, including two tactical missiles for warhead development testing. Prior to her first deterrent strike patrol, in nine launch operations *Growler* had lost one missile at launch and none while in flight.[11]

Regulus Deterrent Patrols 1960-1964

Growler's first deterrent patrol began on 12 March 1960. A major problem during transit to her assigned patrol station was the gradual loss of both aluminum sheet metal fairings around the missile hangar doors. Started by corrosion due to electrolysis between the aluminum and steel and exacerbated by the heavy seas encoun-

Missile engine exhaust throws up a rooster tail water spray just prior to booster ignition. (Courtesy of Kutzleb Collection)

Launch across the bow. Later modifications to the launcher permitted launch directly over the bow only. (Courtesy of Kutzleb Collection)

Bows on view of Growler *taken 1 January 1960. The Regulus I rigged out on the launch is a tactical missile. Note the bulged nose or chin on the missile, indicating either the W-5 or W-27 warhead could be carried. (Courtesy of Author's Collection)*

Growler *is shown just after getting underway from the submarine base in Pearl Harbor to take the King of Thailand to sea for a day. The king is on the left, Commanding Officer Lieutenant Commander Robert Crawford is facing the King and Executive Officer Lieutenant Ekelund is conning the ship. The view forward shows some of the structural loss atop the port missile hangar as a result of heavy weather on th preceding patrol. (Courtesy of Ekelund Collection)*

tered in the miserable North Pacific winter weather, the aluminum fairings disintegrated and were lost overboard.[12] During this first mission, Lieutenant John J. "Joe" Ekelund, Executive Officer and Navigator, developed an innovative method to determine the submarine's position in the assigned operating area. The technique was quite simple and similar to that used by submarines to determine the range of a target ship. Using navigation charts, Ekelund identified mountain peaks and their height as listed. He then observed the mountain through the periscope and, utilizing the built-in periscope stadimeter, he could superimpose the image of the base of the mountain on its peak. This double image and known peak height provided a good approximate range to the mountain that was read on the stadimeter dial. Using the range so determined, one can could calculate the amount of height which was not seen (was below the horizon) and correct the charted height to the observed height. Using the observable height a second, more accurate range could then be measured. Three iterations of this sequence would yield a navigationally useful range. Using more than one peak, he could accurately determine his position.

Ekelund remembers that the first "interesting" experience on this patrol involved the Number One periscope. *Growler* was snorkeling at night and the Conning Officer reported to Ekelund that he had sighted a white object. With no sonar contacts reported and no ice seen during the previous several hours, a complete sweep of the horizon revealed white objects completely surrounding the boat. They had sailed into an ice field. Immediately all masts were lowered but not before the periscope was hit by a large ice flow, damaging it enough to render it useless.

Priest and Ekelund both recall that from then on the mission was routine, except when it came time to head back to Pearl Harbor. On 2 May 1960 the mission was extended three days after Gary Powers' U-2 aircraft was shot down over the Soviet Union. Morale sagged temporarily when this announcement was made. After seven weeks on station in terrible weather, even three days was a major burden. *Growler* returned to Pearl Harbor on 12 May 1960.

Priest was relieved by Lieutenant Commander Robert Crawford on 7 June 1960. Crawford had served on Regulus guidance submarines on the West Coast and was returning to submarine duty after completing a tour in the Bureau of Aeronautics at the submarine-launched guided missile desk. The day Crawford reported for duty was the same day a catastrophic fire occurred on USS *Sargo* (SSN 583). Ekelund recalls that at about 1700 hours he heard a fire alarm sounding on the base. He went to the bridge and saw columns of smoke over the buildings in the direction of nearby piers. Sargo was on fire, with the flames being fueled by a break in the oxygen transfer line in the stern compartment. The fire was finally extinguished by flooding the stern compartment.

Growler and her crew became involved when Crawford was asked to be host of the King of Thailand during his State Visit since *Sargo* was now no longer available. A good part of the rest of the night was taken in making all of the myriad of preparations, including meals during the cruise, planning for proper honors, alerting all of the crew that the uniform would be Full

Dress Whites with swords. The day went perfectly and the crew and officers of *Growler* were justifiably proud that when COMSUBPAC needed something done well without prior planning, they had been selected.

One month later *Growler* was awarded the Battle Efficiency "E" for overall excellence in Submarine Squadron ONE during the previous year. Launch operations resumed in August with two fleet training missile flights and then a tactical missile low-level profile flight. This flight was somewhat different in that the *Growler* missile team launched the missile on shore at Bonham and transferred control to the *Growler* guidance team on board the submarine for the remainder of the flight. The missile was expended as planned.

Growler's second deterrent mission began 10 November 1960 and she returned to Pearl Harbor 18 January 1961. After two months upkeep and two successful missile launches, she left 18 March 1961 on her third mission.[13] Lieutenant Commander Robert Owens had reported to *Growler* as Prospective Executive Officer in February and was serving as Assistant Ordinance Officer. He recalls that the transit to Adak, Alaska for refueling and then to the assigned station was uneventful. One morning he went up to the bridge to shoot the morning star sight. Unfortunately, dense fog lay on the water surface and there was no discernible horizon. The bridge was above the fog layer while the deck, perhaps 20 feet below, was completely hidden. Suddenly the electronic countermeasures alarm began to blare from the speaker on the bridge. The operator recognized it as being transmitted from a Soviet ship. Due to the intensity of the transmission it was determined that the ship was close aboard. Crawford and Owens simultaneously observed a radar mast suddenly appear above the low lying fog. Apparently *Growler* was inside of possible radar detection range. Crawford

Growler *in drydock at Pearl Harbor, date unknown. Note that the streamline fairing for the port missile hangar door (and starboard door, not shown) has been removed. Clearly shown is the port saddle tank added during the Fall 1961 overhaul when* Growler's *sail was increased 7.5 feet in height to improve depth keeping capabilities. (Courtesy of Harmouth Collection)*

Grayback *(outboard) and* Growler *pierside at Mare Island Naval Shipyard, May 1964, shortly after returning from Pearl Harbor at the end of the Regulus Deterrent Patrol program. Note the much shorter sail on Grayback. The white line draped on her sail is the flower lei that crewmember's wives made for both boats when they left Pearl Harbor. (Courtesy of Priest Collection)*

made the decision not to dive in order to avoid possible sonar detection. *Growler* changed course to head directly away from the contact and escaped undetected.[14]

Growler returned to Pearl Harbor 12 May 1961. Lieutenant Commander Donald Henderson relieved Crawford 24 June 1961. During the change of command ceremonies *Growler* was awarded a Submarine Force Unit Citation by Rear Admiral Roy S. Benson, ComSubPac, for her previous mission. *Growler* immediately entered Pearl Harbor Navy Shipyard for overhaul. One addition was the installation of a *Sperry* Gyroscope Mark I Mod 0 Ships Inertial Navigation System (SINS) and the first LORAN C navigation system.[15] A second modification during overhaul was an attempt to improve the handling characteristics of *Growler* at periscope and snorkeling depth. The problem was one of fluid hydrodynamics. The top of the missile hangar fairings were nearly one-half the height of the sail. At periscope depth this made for some difficult handling and a roller coaster ride as the Bernoulli effect caused the hangar deck area to act like an airplane wing and make the boat move towards the surface. This was especially apparent in rough weather. While *Grayback* and *Growler* had nearly identical exteriors, *Grayback* had a slightly different shape to her missile hangars that lessened this unwanted Bernoulli effect. By adding 7.5 feet to the height of *Growler*'s sail, the hangar surfaces would be 7.5 feet deeper at periscope depth and in theory, depth keeping problems would be somewhat mitigated. This also meant adding 7.5 feet to each of the periscopes, communications and radar masts as well as the electronic countermeasures equipment and snorkel. This was not a small undertaking by any means. The additional height of the sail changed considerably the metacentric height, a measure of ship's stability. To prevent excessive rolling on the surface, additional saddle ballast tanks were added outboard of the main ballast tanks.[16]

A welcome modification was also made to the missile launching equipment. The original trainable and transversable launcher that had been designed to launch both Regulus I and II missiles was removed and replaced with one that simply transversed to either missile hangar for missile ram out. Launch was forward over the missile hangars. The removal of the myriad of microswitches and associated hydraulics greatly simplified launcher operation and made this launcher much more reliable. *Growler* completed her overhaul in early December 1961.

1962

After eight weeks of refresher training, *Growler* left Pearl Harbor on her fourth deterrent patrol on 11 February 1962, arriving at Midway Island five days later to disembark a sick crewman. Leaving Midway Island the next day, *Growler* arrived at the patrol area on 24 February 1962.[17] *Growler* departed for the forward refit base one month later, arriving 24 April 1962. After a four week repair and upkeep period, *Growler* departed 24 May 1962. Arriving on station in early June 1962, she commenced her fifth deterrent patrol. *Growler* returned to Adak on 23 July 1962, departing for Pearl Harbor the next day. Lieutenant Commander Gunn, now Executive Officer, had a battle flag made that read "Black and Blue Crew, No Relief Required!" They were flying this banner upon return to Pearl Harbor on 1 August 1962. Rear Admiral Bernard A. Clarey, ComSubPac, joined *Growler* as she entered Pearl Harbor and upon seeing the unfurled flag flying on the mast, put his hand on Henderson's shoulder and asked if they really meant it. Henderson responded that it was true, the Regulus submarine crews took great pride in the fact that they did not need the Blue and Gold two-crew system used in the Polaris submarines. *Growler* received a ComSubPac Unit Commendation for both the fourth and fifth patrols.[18]

After a 30 day upkeep, *Growler* began her customary refresher training with both torpedo and missile firing excercises. Submarine officers who aspire to command of a submarine must undergo a series of rigorous qualifying tests, exams and practical evaluations, all under the watchful eyes of the senior officers on board. Henderson remembers a most memorable prospective commanding officer evaluation that took place at this time. One of the steps in the evaluation process requires that the candidate personally prepare an exercise torpedo for firing. This meant supervising the loading of the torpedo on board, acting as the Approach Officer (assuming the position of the Commanding Officer during the attack) and upon gaining a satisfactory firing solution, fire the torpedo.

The operating area was off of Barbers Point, Oahu. By seagoing standards, the area was reasonably close inshore but not dangerously so. Areas such as this were frequently utilized to reduce the transit time for torpedo recovery vessels. The assigned target was a Pearl Harbor-based submarine rescue vessel. Lieutenant Gene Wells, the ship's Torpedo Officer, was being evaluated and had done very well up this particular day. His fire control party attained a firing solution on the target's speed course and range. Wells fired his personally prepared torpedo and just like in the movies, he started a stopwatch to time the period of the torpedo run to determine when it should intercept the targe; and in this case, locate the torpedo after the run. Exercise torpedos were set to run in one of two modes, either high speed, short range or low speed and long range. Usually one would select the high speed option to minize the opportunities for targets sighting the torpedo and manuevering to avoid being hit.

Wells selected the high speed option, but, due to equipment malfunction, it was not entered into the torpedo. For reasons that were never clear, the torpeod ran the low speed, long range run. Henderson recalls everyone counting down the time with no result, i.e., the torpedo could still be hearding whining away. It kept running and running and running and then the sound finally stopped. Both Wells and Henderson were at the periscopes and were astonished at what they saw. To their amazement, as the whining sound stopped, they saw the torpedo break the water surface and run up the beach, finally coming to rest between two large fuel storage tanks in the Barbers Point fuel farm!

One can only imagine the initial response of the torpedo retrieval team back at the base when *Growler* requested a cherry-picker retrieval crane to proceed to the middle of the naval air station fuel farm. Wells passed his torpedo firing test since on the balance, the shore-based fuel facility was considered a worthwhile target.

Growler's sixth deterrent patrol, the third with Henderson in command, began on 24 November 1962. Weather in the assigned station area was again miserable. For Christmas dinner Henderson decided to go deep so the crew could enjoy the meal in relatively stable conditions. A thousand foot floating wire antenna permitted *Growler* to submerge to three hundred feet and still receive messages. While wave motion could still be felt at 300 feet, the meal was really much more enjoyable. A novel relief during this patrol was contributed by a Quartermaster Second Class who had been on board *Growler* for all six patrols. Traditionally, daily routine reports are made to the Commanding Officer at 0800, 1200, 1600 and 2000 hours. The 1200 hours report consisted of fuel and water on board, magazine and missile hangar temperatures, average specific gravity of both the forward and aft battery cells, ship's position and that all chronometers (precision time pieces set to Greenwich Mean Time) had been wound and compared with each other. This report was normally made to the Commanding Officer during lunch. The other officers present paid little attention since it was usually so monotonous and routine. On this particular day this Quartermaster Second Class gained everyone's full attention when he recited the following poem in place of the routine report *(see poem below)*:

Needless to say, this received a lavish round of applause. *Growler* returned to Pearl Harbor on 11 February 1963 and received a COMSUBPAC Unit Commendation for this patrol. In addition, CINPACFLT issued a Unit Citation to all officers and men of Submarine Division ELEVEN for the period 1 November 1961 to 27 June 1963.[19]

Lieutenant Commander Robert Owens relieved Henderson on 1 June 1963. *Growler* conducted two more deterrent missions, 14 June 63 to 12 August 63 and 14 October 63 to 13 December 63. In early 1964 the decision was made to decommission *Growler* and *Grayback*. *Growler* and *Grayback* sailed for Mare Island Naval Shipyard, Vallejo, California together and were decommissioned in May 1964.

Post Regulus, the Growler Museum

After decommissioning on 25 May 1964, *Growler* was placed in the Inactive Reserve Fleet at the Puget Sound Naval Shipyard, Washington. Twenty-five years later it was decided that she was a burden to the annual budget and the Navy decided to use her as a torpedo test target for nuclear attack submarines. Fortunately these tests were never conducted. Instead, through the efforts of Mr. Zachary Fisher, of New York, and by an act of Congress, on 8 August 1988, *Growler* was assigned to become part of the Intrepid Sea-Air-Space Museum in New York City.

In early 1989, *Growler* departed Puget Sound under tow. Proceeding through the Straits of San Juan de Fuca, she began a journey of six thousand nautical miles. After transiting the Canal, *Growler* was towed to a civilian shipyard on the west coast of Florida. While in the shipyard, *Growler* received both exterior and interior hull repairs, most important of which were the changes made between the missile hangars and the hull. These changes were made to facilitate access for visitors at the museum. On 18 April 1989, *Growler* was moored to the north side of Pier 86 in the Hudson River, her final "Home Port." The entire cost of this operation was absorbed by Mr. Fisher, founder and chairman of the Intrepid Sea-Air-Space Museum. On 26 May 1989 *Growler* was "re-christened" at Pier 86 and is now one of the most popular exhibits of the Intrepid Museum complex.

Endnotes

[1] *The author has been unable to find interior plans for Grayback that would permit a detailed comparison.*
[2] *America's Fighting Ships, Vol III, 1968.*

Good afternoon Captain and the rest of you
Here's the good word from the O.D. and the crew.

The chronometers wound just about nine
Then checked and compared with Greenwich Mean Time.

1252 is the gravity now
And since we've submerged its bound to go down.

The magazines checked and found to be well
With temperature normal, 51 sounds swell.

Now I don't wear a mask and I don't hide my face
The noon reports lately have been a disgrace.

So I'll make this poetic
To keep up the pace.

Now thanks for your patience in hearing me out
I'll see you tomorrow, on that there's no doubt.

[3] *This is for the "thick" skinned fleet-type submarine. The "thin" skin fleet-type submarine had a test depth of 312 feet.*
[4] *Personal communication with Commander Robert Duke, USN (Ret.).*
[5] *Personal correspondence with Leonard E. Powers, RMC(SS) USN (Ret.).*
[6] *Personal interviews with Captain Charles Priest, USN (Ret.) and Rear Admiral John J. Ekelund, USN (Ret.).*
[7] *Growler Sea Trials, July 1960, page 128-131, U.S. Naval Institute Proceedings.*
[8] *Personal communications with Lieutenant Commander R. "Sam" Kutzt, USN (Ret.).*
[9] *Personal interview with Captain Charles Priest, Jr., USN (Ret.).*
[10] *Personal communications with Commander Robert Duke, USN (Ret.) and Torpedoman First Class John D. Haney, USN (Ret.).*
[11] *Regulus I Flight Index, 1961, LVSCA A50-24, Box 7.*
[12] *Personal interview with Captain Charles Priest, Jr., USN (Ret.)*
[13] *Ship's Logs, USS Growler (SSG 577), Naval Historical Center, Ship's Histories Branch.*
[14] *Personal communications with Commander Robert Owens, USN (Ret.)*
[15] *Personal interviews with Captain Donald Henderson, USN (Ret.) and Captain William Gunn, USN (Ret.).*
[16] *Ibid.*
[17] *Personal interview with Captain Donald Henderson, USN (Ret.).*
[18] *Ibid.*
[19] *Personal interview with Captain Donald Henderson, USN (Ret.).*

Not all recoveries were completely successful. On 26 March 1959, **Growler** ***launched FTM 1458 on a 31 minute Trounce operation. While the guidance portion of the flight was successful, the recovery was not. FTM 1458 did not fly again. (Courtesy of Hoffman Collection)***

Chapter Fourteen: USS Halibut (SSGN 587)

Halibut was the only U.S. Navy submarine built from the keel up as a Regulus guided missile submarine. The only submarine of her class, *Halibut* was built to test the concept of a nuclear-powered guided missile submarine with improvements and changes to then be incorporated into the Permit Class of nuclear-powered guided missile submarines. Cancellation of Regulus II and the success of the Fleet Ballistic Missile Program stopped the further design and construction of Regulus missile submarines with the result that the Permit Class boats were built as nuclear- powered attack submarines.

1955

The story of the somewhat torturous route of the design of *Halibut* starts in April 1955. The Bureau of Ships assigned Mare Island the project of determining the size and general features of a submarine missile hangar to house both the Regulus I and II guided missiles. W. J. Imboden was the Project Engineer and a quarter-scale wooden model was constructed to demonstrate the scheme. In August 1955, P.A. Green, H.G. Ulbrich and G.D. Childs were called to Washington to assist in the preparation of the preliminary specifications for a diesel-electric submarine utilizing their hangar concept. In December 1955, they returned to Mare Island and joined the contract team headed by Imboden, Lieutenant Commander C. Russell Bryan as Project Officer and 50 marine architects and engineers.[1]

Several months earlier, in the fall of 1955, the new Chief of Naval Operations, Admiral Arleigh Burke, had met with Rear Admiral Hyman Rickover and officers of the Bureau of Ships to discuss how nuclear propulsion concepts could be more quickly integrated into the fleet. With Fiscal Year 1956 funding providing for five conventional powered submarines, Admiral Burke wanted two of these to be converted to nuclear power. After the 1956 building program was approved, Burke made the far reaching decision that all future submarines would be nuclear powered.[2]

Lieutenant Commander Walter Dedrick, an officer who had been involved in the early Regulus I program on board *Cusk* and *Tunny*, was present at a briefing where Bureau of Ships staff had tried to persuade the Admiral Burke that *Halibut* would be a much better as a diesel submarine. Dedrick felt this presentation had been heavily slanted against nuclear propulsion and suggested to Commander William Heronemus (Mare Island Design Superintendent) that they write a counter-proposal. Dedrick and Lieutenant Commander Byran (Heronemus' Assistant Design Superintendent) sat down in Heronemus' office and wrote operational scenarios comparing nuclear and diesel-powered boats. They reached a conclusion completely opposite to that of the Bureau of Ships staff and convinced Heronemus to propose a nuclear plant instead.[3] Thus the unique design for *Halibut* was born. The design details were hammered out by Commander E.E. Kinter (Nuclear Power Superintendent), Commander W.E. Heronemus (Design Superintendent) and H.L. Graybeal (Head Nuclear Engineer).[4]

The preliminary design submitted to the Ship's Characteristics Board by Lieutenant Commander Charles Slonim called for four topside hangars of the new diesel-electric guided missile submarine design. These would be positioned forward and aft of the sail, permitting two separate launchers to be available for firing two missiles at one time. The final design eliminated three of the four hangars and placed a single large hangar in the bow with one launcher. Contract design resumed in June 1956, was submitted to BuShips in October and approved in December, 1956. With construction of the nuclear-powered *Sargo* well underway, a full scale mockup of the reactor spaces was already available, greatly facilitating the redesign of *Halibut's* stern spaces.[5] The keel of the nation's first nuclear-powered guided missile submarine was laid on Mare Island's Naval Shipyard (Mare Island) Ways One on 11 April 1957, the 57th birthday of the U.S. Navy's Submarine Force. *Halibut* was the second nuclear powered submarine to be built at Mare Island and the ninth nuclear-powered submarine constructed.

1959

Halibut slid down the ways on 6 January 1959 after two years of construction. After one year of equipment installation and power plant testing, *Halibut* was commissioned 4 January 1960, with Lieutenant Commander Walter Dedrick assuming command. On 17 July 1959, *Halibut* was placed "in service", meaning that Dedrick accepted responsibility for the operation of the submarine and custody of the reactor.[6] The missile system installation was not complete due to the removal of the Regulus II checkout and launch equipment.

During the first dive, Lieutenant Commander J. Taylor Rigsbee, the Engineering Officer, noticed that the auxilary seawater pumps were airbound. This was not uncommon on the surface in particularly heavy swells but *Halibut* was submerged. These pumps were required by several ship's systems, including the reactor. Rigsbee vented the pumps, only to have the problem reoccurr almost immediately. By now several alarms wree sounding and Rigsbee, Dedrick and the Shipyard Commander, Rear Admiral Leroy Hunsinger, were called into the Captain's stateroom and vigorously questioned by Rear Admiral Hyman G. Rickover as to why the problem had not been fixed. Once free of the interview, Rigsbee, Dedrick and Hunsinger decided that the problem must be due to a plugged pump intake. Compressed air was used to purge the intake and the problem was solved.[7]

The problem was not a flaw in the design of *Halibut*. A piece of the material that had plugged the intake was recovered, examined at Mare Island and found to contain seaweed and popcorn shrimp. A chain of events that no one could have foreseen had caused an abnormal growth of seaweed around the pump intake at dockside. *Halibut* was moored to a pier next to the shipyard power plant. The warm exhaust water from the plant had facilitated the growth of algae and seaweed in the nearby area. Intermittent use of the pumps dockside had prevented complete clogging of the intact screen but just barely. *Halibut* had apparently passed through a school of shrimp during the course of the trial and they had served to completely plug the intake. As a result of this incident, standard shipyard procedure for nuclear-powered submarines after prolonged shipyard periods required a diver to check the pump intakes for fouling.[8]

In mid-September 1959, engineering sea trials began. Missile system sea trials began in late November 1959 due to alterations caused by the removal of Regulus II equipment.[9] Lieutenant Commander Paul Early, Executive Officer of *Halibut*, recalls that Rigsbee and Lieutenant Howard Laron, Assistant Engineering Officer, had discovered a miscalculation in *Halibut*'s compensation, the weight and balance calculations necessary to place the submarine in neutral buoyancy. As a result, 50 tons of lead ballast were placed, temporarily, in the superstructure on top of the hull between the hangar door and the sail. The lead was in temporary steel bins lined up fore-and-aft outboard over the crown of the ballast tanks. These rows of bins, about 2.5 feet high, acted as longitudinal bulkheads.

Since *Halibut* was designed to surface and launch missiles in a State Four sea, one of the sea trial tests was to surface in such conditions and ram a dummy missile onto deck.[10] They surfaced heading into a state 4 sea with a light wind from dead ahead. They were at "Battle Stations Missile" with the inner hangar door open and the low pressure blower on the main ballast tanks. Dedrick ordered the hangar door opened and the launch sequence to begin. He increased speed slightly to improve stability since they were rolling gently. Early, who was at the periscope, noticed that the bow was just dipping into the tops of the Pacific rollers and the waves would flow smoothly up over the hangar hump. When they passed over the raised faring, the waves would pour into the inner hangar door like a waterfall, too thick to see through into the hangar. Soon a request was received from the missile hangar compartment to begin pumping the hangar room bilges with the drain pump. They had difficulty getting a good suction, so it was quickly reported from the hangar that the bilges were filling. Dedrick asked Early to go down and have a look at the problem.

Early discovered the problem was that a good deal of water was being trapped by the rows of ballast bins and this water was now up to the level of the outer door lip. As the boat pitched down into the swell, the free water would stream forward and cascade into the space between the two doors. With the available space between the two doors already filled with water up to the door lips, the trapped water would spill into the hangar. As the ship rolled slightly, more water would enter the area between the bins through the limber holes and the process repeated. The water level between the hangar doors was about at thigh level and when Early turned to look forward and saw the distance down to the keel and bilges, the water fall left a lasting impression. The shipyard fix to this problem was straight forward and the problem did not reoccur.[11]

Two weeks after commissioning, *Halibut* got underway on her maiden voyage to Puget Sound for comprehensive sound and maneuver-

Photograph of the interior of the 1/4 scale model of Halibut *illustrates the sheer size of the missile compartment. This model enabled several configurations to be studied. Above, Regulus II configuration with a missile rammed out on the launcher but none in the hangar. (Courtesy of Mare Island Naval Shipyard Collection)*

ing trials on the Navy's sound ranges located in Dabob Bay and Carr Inlet. During transit to and from the sound and torpedo trials, the missile team used a dummy missile for drills on missile handling, including transfer from the shock mitigation system, ramming the missile onto the launcher and elevating the launcher. During the return voyage to Mare Island, *Halibut* participated in a simulated missile attack exercise off of San Diego which went undetected. *Halibut* returned to Mare Island on 22 February 1960 for provisions and spares prior to her shakedown cruise to New Zealand and Australia.

When clear of the coast of California enroute to Hawaii, Dedrick held the usual missile drills. During this first drill of the voyage, the launch slippers broke off in the slipper sockets, dropping the missile onto the launcher, still unelevated, fouling the chains in the extended position and preventing the hangar doors from being closed. The dummy missile had been used so much in previous trials and training that fatigue cracks had gone unnoticed in launch slippers. The crew used portable jacks to bring the missile free of the chains, retracted the chains and then slid the missile handling skid up to the launcher. The missile was transferred to the skid and pulled back into the hangar and the hatches shut. The whole operation took four hours in a slight swell and made for an interesting beginning to the shakedown cruise.[12]

Halibut arrived at Pearl Harbor on 21 March 1960. Practice missiles were loaded and she immediately began missile operations. The first Regulus I launch from *Halibut* was on 25 March 1960 off Oahu. FTM 1456 was launched and completed a 30 minute flight before being recovered successfully at Bonham Auxiliary Landing Field. One week later *Halibut* conducted a maximum range mission with USS *Carbonero* (SS 337), serving at the down range guidance boat. After *Halibut* guided the missile 200 nautical miles, *Carbonero* took over for the 77 nautical miles to handoff for missile recovery. This was the longest two-submarine Trounce guidance flight of the program.[13]

Halibut returned to the West Coast in mid-July for post-shakedown refit. Her post-construction overhaul took place from 3 August to 5 November 1960. On 7 November 1960, *Halibut* departed Mare Island for Pearl Harbor to become part of Submarine Division ELEVEN, arriving 21 November 1960 after several days in San Diego. Lieutenant Charles Baron was relieved as Missile Officer by Lieutenant William Gunn. Gunn was fresh from USS *Grayback* (SSG 574), having made her first and second deterrent patrols as Weapons Officer and Navigator. Baron worked with Gunn to assure a smooth transfer of missile operations. Gunn and the *Halibut* missile team continued a string of successful launches with four flights in four weeks as they prepared for their first deterrent patrol. The final test of operational readiness would be the participation of *Halibut* in the annual SEATO exercises off Okinawa.[14]

Gunn recalls that they left Pearl Harbor with five tactical missiles, two of which carried war reserve warheads minus the nuclear components for use during the upcoming Operational Suitability Test (OST) at the SEATO exercise. *Halibut*'s orders were to demonstrate the stealthiness of this new weapon system by surfacing near the US carrier on a specified day, fire the OST missile and again submerge. At Subic Bay, Philippines, they rendezvoused with two guidance submarines and proceeded to join up with the task force.

Launch day, 8 February 1961, dawned with high overcast and a slight sea swell. The SEATO Task Force was steaming 65 nautical miles east of Okinawa waiting for the demonstration to begin. On cue, *Halibut* surfaced two hundred yards off the starboard beam of USS *Lexington* (CV 16), rammed out TM-1459 and conducted the 1025th launch of Regulus I. After a thirty minute flight, the W-27 war reserve warhead with dummy nuclear components was detonated as an airburst.[15,16] Instead of returning directly to Pearl Harbor, *Halibut* went to Naha, Okinawa, to replenish her Regulus I loadout, replacing the

Above, Regulus II configuration with a full load of four missiles. Note the scale-size 6 foot sailor mannequin on the top of the torpedo rack. The Regulus II nose boom was hinged to permit folding back during storage. (Courtesy of Mare Island Shipyard Collection)

expended OST missile and exchanging the remaining unused OST missile. GMU-10, the Regulus I submarine support unit, had flown out a detachment and three tactical missiles in an exercise to demonstrate the feasibility of replenishment of Regulus missile boats away from the facilities at Pearl Harbor. After loading the second replacement, Gunn's missile crew went through the acceptance checkout which included an engine run up on the launcher in the lowered position. Gunn recalls that the engine had started, but was still at low power, when the missile launch slippers slipped their locks and started moving toward the hangar door. First Class Electrician Wilbur Willis was in between the missile and the closed outer door of the missile hangar, standing between the launcher transfer rails. He crouched down, reached up and pulled an umbilical cable, shutting down the missile as it glided over him down the track and lightly impacted the outer hatch. The nose was slightly dented and the missile had to be replaced. Luckily, GMU-10 had three missiles available so the third was checked out, accepted and *Halibut* now had her full complement of missiles.[17]

Regulus Deterrent Patrols 1961-1964

Halibut began her first deterrent patrol the next day, returning to Pearl Harbor on 10 April 1961.[18] After a three week resupply period she departed for her second deterrent deployment on 1 May 1961, returning again to Pearl on 28 June 1961. On 29 June 1961 Dedrick was relieved by Lieutenant Commander W.R. "Bus" Cobean, Jr. *Halibut* was also awarded Pacific Submarine Force Unit Citation for her two previous deterrent patrols at this change of command ceremony. Lieutenant Commander Cobean, who had been the first Squadron Engineer of the Polaris Program, was pleased to be getting into the Regulus I program. *Halibut* then underwent her first yard availability period since joining the submarine force. *Halibut* successfully launched her 13th consecutive Regulus I shortly after leaving the yard.[19,20] On 21 October 1961, *Halibut* was awarded her second Pacific Submarine Force Unit Citation, making her the only unit in

Regulus I configuration without torpedo racks and with the #2 missile in the ram out position. While the shock mitigation gear looked complicated, missile movement to and from the storage positions was relatively easy. (Courtesy of Mare Island Naval Shipyard Collection)

The first launch of a Regulus I missile, FTM-1456, by Halibut's *missile team took place on 25 March 1960 off the coast of Oahu. The missile was recovered after a 30 minute flight, the beginning of a string of seven successful launches.* Halibut's *masts have been censored from this photograph. (Courtesy of Peed Collection)*

On 8 February 1961 Halibut *surfaced 200 yards off the starboard beam of USS* Lexington *(CV 16) and conducted the 1025th launch of a Regulus I missile. After a 30 minute flight, TM 1459 was expended in a terminal dive maneuver directly over the target. (Courtesy of Peed Collection)*

the Pacific Submarine Force to receive the award twice in the same year. Commander Dedrick was awarded the Navy Commendation Medal for his efforts during her second deployment. In December, Commander Submarine Division ELEVEN awarded *Halibut* the COMSUBPAC Fire Control Excellence Award for her torpedo operations from July to December.[21]

On 20 December 1961, *Halibut* left Pearl for her third deterrent patrol, returning on 31 March 1962. During May and June, *Halibut* conducted extensive torpedo and Regulus I missile firing operations, including eight successful missile launches. Prior to deployment on 9 July 1962, her fourth deterrent patrol, *Halibut* was awarded the Battle Efficiency "E" for fiscal year 1962 by Commander Submarine Division ELEVEN. She was also ComSubPac's candidate ship for the Arleigh Burke Fleet Trophy for the 1962 competitive year.[22] Returning to Pearl Harbor on 15 September 1962, *Halibut* commenced a short upkeep period prior to departure for Mare Island and a reactor core change. Enroute to Mare Island, *Halibut* became aware of the mounting tensions of the Cuban Missile Crisis. Cobean volunteered to proceed to San Diego, the Regulus I West Coast depot location, take on a load of missiles and proceed as needed. *Halibut* was directed to continue to Mare Island as scheduled, arriving on 23 October 1962, the third anniversary of the start of Regulus I submarine deterrent patrols.[23]

Lieutenant Commander John F. Mangold, Executive Officer of *Halibut*, relieved Cobean on 1 November 1962. For the next five months *Halibut* underwent core replacement and a limited overall, being ready for sea on 8 March 1963. Arriving at Submarine Base Pearl Harbor on 19 March 1963, *Halibut* immediately underwent a two week upkeep and participated in local area training exercises including two missile firings, both successful. On 29 April 1963, *Halibut* departed on her fifth deterrent patrol. During this time she steamed her 100,000th mile (of which 77,609 were submerged) and made her 444th dive.[24] Returning on 6 July 1963, *Halibut* entered Pearl Harbor Naval Shipyard for battery replacement. After five weeks of upkeep, *Halibut* returned to sea for local area training with torpedo and missile firings, successfully launching her 27th Regulus I. On 2 November 1963, Commander Mangold was relieved by Commander Harold S. Clay. During the change of command ceremonies, *Halibut* was presented with the missile "E" by Captain R.A. Kaufman, Commander Submarine Squadron ONE, in recognition for outstanding missile launch operations for the previous year. *Halibut* also received a letter of commendation from the Commander-in-Chief, Pacific Fleet, to units of Submarine Division ELEVEN for their meritorious service in the missile deterrent program.[25]

On 19 November 1963, *Halibut* began her sixth deterrent patrol. Notified during this patrol that she had received the Fire Control Excellence award for the third straight year, *Halibut* took permanent possession of the award. On 29 January 1964, *Halibut* returned from patrol and went directly into a five week upkeep at Pearl Harbor Naval Shipyard. *Halibut* participated in local training operations for next two months. On 7 May 1964, *Halibut* left on her seventh and the last Regulus submarine strategic deterrent patrol, returning on 14 July 1964. While deployed *Halibut* was awarded Submarine Division ELEVEN's Battle Efficiency "E" for fiscal year 1964.[26]

POST-REGULUS

Upon return to Pearl Harbor, *Halibut* had all missile handling, checkout and guidance equipment removed. Sonar test equipment was installed. During the Fall of 1964, *Halibut* participated in the PERMIT Class submarine Weapons System evaluation. This program involved eight conventional and nuclear-powered submarines in a thorough test program to provide information on the Permit class attack capabilities. Subsequent conversion in February 1965 during a major overhaul led to *Halibut*'s redesignation as SSN 587. After 11 years as a research submarine, *Halibut* was decommissioned on 30 June 1976. On 12 July 1993 *Halibut* entered the Navy Recycling Program and the process was completed on 9 September 1994.[27]

Endnotes

[1] *Foresight of M.I. Engineers Gave 587 Nuclear Power, The Mare Island Grapevine, 9 January 1959 page 9.*

[2] *The Nuclear Navy 1946-1962, Richard G. Hewlett and Francis Duncan, 1974, University of Chicago Press, page 265.*

[3] *Personal communications with Captain William Heronemus, USN (Ret.).*

[4] *Ibid.*

[5] *Ibid.*

[6] *Nuclear Power and Guided Missiles, by Lieutenant Commander J.T. Rigsbee, U.S. Naval Institute Proceedings, July 1960, page 118.*

[7] *The compressed air hose that was designed for this purpose was not on board due to the partially completed state of Halibut at this time. A substitute was rigged but after this incident, Rickover decided that all systems had to be fully installed, with complete spare parts, before sea trials of any kind could take*

place. Personal communications with Captain J. Taylor Rigsbee, USN (Ret.) and Rear Admiral Paulare parts, before sea trials of any kind could take place. Personal communications with Captain J. Taylor Rigsbee, USN (Ret.) and Rear Admiral Paul J. Early, USN (Ret.).

[8] Ibid.

[9] "Nuclear Power and Guided Missiles," by Lieutenant Commander J.T. Rigsbee, U.S. Naval Institute Proceedings, July 1960, pages 118.

[10] A State Four sea has a wind speed of 11-16 miles per hour and 5-8 foot height waves with whitecaps.

[11] Personal communications with Rear Admiral Paul J. Early, USN (Ret.) and Captain J. Taylor Rigsbee, USN (Ret.).

[12] Personal communications with Rear Admiral J. Paul Early, USN (Ret.) and Captain Charles Baron, USN (Ret.).

[13] Regulus I Flight Index, 1961, LVSCA A50-24, Box 7

[14] Personal communications with Captain William Gunn, USN (Ret.) and Captain Charles Baron, USN (Ret.).

[15] Personal communication with Captain William Gunn, USN (Ret.).

[16] Regulus I Flight Index, 1961, LVSCA A50-24, Box 7.

[17] Personal communications with Captain William Gunn, USN (Ret.)

[18] Details of Halibut's patrols are sketchy at best due, apparently, to their classified nature.

[19] Personal interview with Captain Warren R. Cobean, Jr., USN (Ret.).

[20] Regulus I Flight Index, 1961, LVSCA A50-24 Box 7.

[21] USS HALIBUT (SSGN 587), 15 September 1966, Ship's History, Pacific Fleet Submarine Memorial Association Archives.

[22] Ibid.

[23] Personal interview with Captain W.R. Cobean, USN (Ret.).

[24] USS HALIBUT (SSGN 587), 15 September 1966, Ship's History, Pacific Fleet Submarine Memorial Association Archives.

[25] Ibid.

[26] Halibut fired thirty-two missiles without a launch failure.

[27] Personal communication with Ms. Debbie Franz-Anderson, Editor, "Salute," the Puget Sound Naval Shipyard newspaper; Naval Institute Proceedings, June 1995, page 47.

Regulus

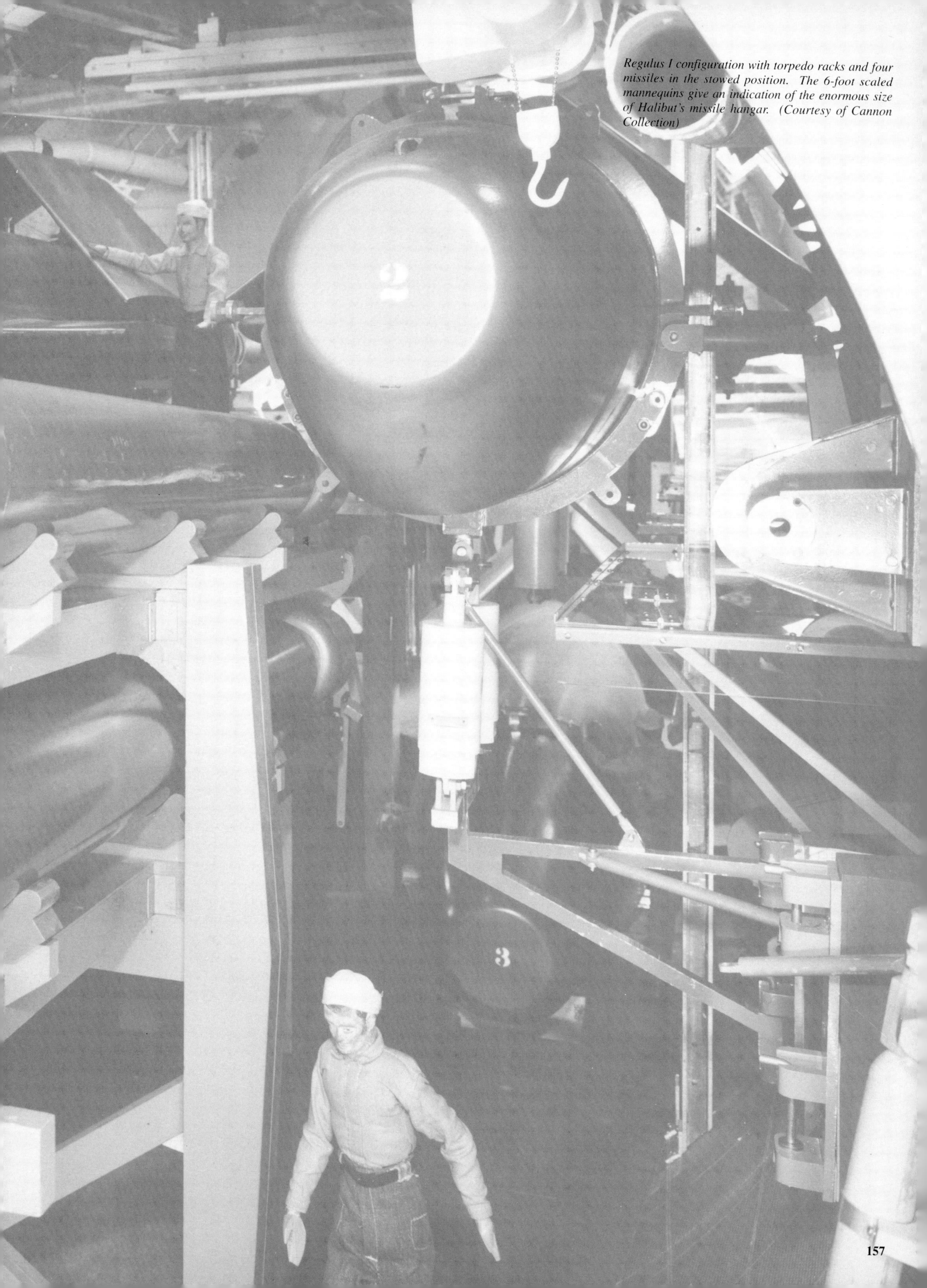

Regulus I configuration with torpedo racks and four missiles in the stowed position. The 6-foot scaled mannequins give an indication of the enormous size of Halibut's missile hangar. (Courtesy of Cannon Collection)

On 23 April 1959, FTM-1346's starboard booster did not fire. Four seconds after ignition the missile crashed in a ball of fire. This was the second to the last launch for GMU-51.

Part IV

Appendices and Glossary

Appendix I: Guidance Systems

Three remote control guidance systems were utilized in the Regulus I program: radio command control (RCC), bipolar navigation (BPN) and paired pulse radar (Trounce). A fourth system, designated SHORAN and similar to BPN, was initially to be the primary guidance system but was discarded prior to hardware development. While several mid-course guidance systems were considered for Regulus II, only one guidance system was flight tested, the Regulus Inertial Navigator (RIN). RIN was originally considered as a mid-course guidance system to replace Trounce in the Regulus I program but evolved into the main guidance system for Regulus II. Positive flight termination (PFT) was a range extension system that was installed in tactical missiles for use with both the Regulus Assault Missile and submarine Regulus I missions.

Radio Command Control

RCC was not a new concept when it was adopted for use during the flight test phase of the Regulus program. What was new, however, was the adaptation of controls meant for propeller-driven aircraft to the much different responses of jet-driven aircraft. Commands were transmitted by an FM radio link. A console, coder and transmitter made up the control station. The control station, housed either in the control aircraft or in the ground station, commanded three proportional channels: pitch, turn and throttle. These settings were proportional to dial or lever settings, which could be varied continuously over a limited range. In addition, a 16-position selector control, coupled with an "increase" or "decrease" momentary switch, permitted control of operations such as parabrake deployment, landing gear down, tracking smoke, etc. The nuclear warhead arming commands were sent through this selector switch system.

Bell Aircraft produced three RCC guidance system "lots." Lot I, installed in the first five flight test vehicles, was limited to a 50 mile range. Besides this short range limitation, radio frequency drift caused considerable difficulty. While lakebed operations were always planned to begin at dawn before high temperatures and unstable air developed, the Lot I equipment was still too susceptible to external temperature. Lot II equipment design began in June 1950 and was completed in December 1951. The Lot II transmitter was a modification by Bell Aircraft of the transmitter built by Collins Radio Company. Lot II had greater reliability but was still limited in range. In the Bell Aircraft Lot III RCC system, the transmitter and receiver were built by Collins. Shifting operation to the higher frequency of 400-500 megacycles resulted in greatly increased range. Reliability at high altitude was increased tremendously with pressurization of the RCC component packages. Flight testing of Lot III equipment began 3 April 1953 and it was accepted one week later.[1]

In the event that RCC carrier signal was lost, a destructor circuit provided a means of automatically destroying the missile. If carrier signal strength was below a preset minimum for more then six seconds, a steady stream of smoke was generated as a visual indicator to the chase pilots. The proportional controls then moved to zero pitch, zero bank and ninety-percent throttle. After 30 seconds elapsed, fuel cut-off occurred and the missile was programmed into a vertical dive.

Using the RCC system with Regulus I at take-off or landing usually involved a minimum of two pilots and one aircraft. The front cockpit was designated as ABLE, the rear cockpit was BAKER. The BAKER pilot flew close formation with the missile, responding to all attitude and altitude changes. The ABLE pilot controlled the missile by way of the RCC system. In practice, the ABLE pilot would make control changes for the missile that were reflected in the TV-2D's position due to the close formation flying of BAKER. In effect, ABLE was controlling the TV-2D via radio control. For the recovery sequence, ABLE would control the missile such that at the moment of touchdown, the missile was lined up on the runway centerline while the TV-2D would be to the right but would not land with the missile.[2] On more than one occasion a pilot in a single seat aircraft, the TV-1 or one of the Navy escort guidance aircraft, landed Regulus I single handedly.

Bipolar Navigation

The BPN system was conceived as a mid-course guidance system for the ramjet powered, supersonic Rigel surface-to-surface missile program.[3] Tactically, BPN consisted of two picket submarines whose geographical positions were accurately known and were within 150 nautical miles of the target. Once within 150 nautical miles of the guidance submarines, the missile guidance system would transit a steady radio signal to the picket stations. The pickets would then determine, via radio signals transmitted to, and returned from the missile, the range to the missile. After a series of range and time delay determination steps, the two stations would begin transmitting radio signals with time delays based on their range to the missile. The missile guidance system detected the signals from the picket submarines and directed the autopilot to maneuver so as to keep the signals from both stations arriving simultaneously.[4]

After several minutes for the missile flight path to stabilize, the picket submarines would begin to shift their timing delays by predetermined increments until the hyperbola representing simultaneous signals at the missile passed through the target. The hyperbola that passed through the target could be calculated in advance for most targets and a missile guidance acquisition point calculated for the launching submarine. The missile would be launched on a dead reckoning heading with a timer set to trigger the transmission of a picket interrogation signal by the missile. While the missile would be radiating a signal for a relatively short period of time, the guidance submarines would have to be on the air constantly. This was a major drawback in terms of susceptibility to electronic countermeasures and detection.[5]

Many constraints were apparent in this system. As just mentioned, the susceptibility to electronic countermeasures was significant. The earlier the missile could pick up the desired time delay hyperbola, the more accurate the guidance. This was offset by the increased vulnerability to the transmitting picket station submarines which would normally be within sight of geographical landmarks on shore to ensure accurate positioning. Once acquisition was achieved, both picket boats had limited maneuvering capability, and this was usually only along the target bearing. Each had to estimate its own speed and dial into the system a corrective factor for the change in signal delay.[6]

Between July 1952 and April 1955, 241 TV-2D drone and 13 missile flight operations were conducted at Pt. Mugu to evaluate the BPN guidance system.[7,8] The first phase utilized 200 TV-2D aircraft and eight missile flights. These tests utilized both picket submarine BPN stations and ground based stations. The results demonstrated the feasibility of using BPN as a mid-course guidance system for Regulus I, warranting the need to upgrade system components. In January 1955 an improved version of the BPN system, AN/DPW-3(XN-3), was utilized to test these changes.

The final BPN tests were conducted using two production tactical missiles to determine the accuracy of a radio terminal guidance system also based on BPN. Both flights were successful with the final flight on 29 April 1955 a particularly good evaluation of the system as the Point Vincente station disengaged guidance due to an equipment failure, repaired the problem and reestablished guidance without loss of control. The delivery of the missile to the dump point was highly accurate but the dive to impact was not. The radio terminal guidance system was too sensitive to a "jitter" in the master station circuitry, preventing accurate terminal dive control. The circular error probable for delivery to the terminal dive point using BPN guidance were incredibly accurate, on the order of 50 to 250 yards. Admittedly these were under ideal conditions that were nowhere near tactical situations, nonetheless, these were impressive results.[9]

The cancellation of BPN in December 1955 was not arbitrary. The Operational Development Force evaluation, in the middle of its program, did not conduct missile operations due to the cancellation but had conducted 15 manned drone flights. The conclusions were: removal of the guidance submarine's Number 2 periscope due to installation of the BPN antenna was undesirable; while the semi-automatic operation of the system reduced personnel training to a minimum, the reliability of the equipment on submarines and the missiles was poor; the system required two submarines and if one failed, there was no alternative guidance capability; and, lastly, the system was susceptible to electronic countermeasures to a larger extent than the alternative Trounce guidance system.[10]

TROUNCE

The Trounce guidance system originated in the Navy's Loon program under Project POUNCE (See Chapter One). Loon was the "Americanized" version of the German FZG-76, more commonly known as the V-1 or "buzz" bomb which the Navy had been launching from submarines since 1947. Results from Project POUNCE indicated the need for a more secure guidance system than the radio control system currently used on Loon.

In response to Project POUNCE recommendations, Project TROUNCE was established 17 May 1950.[11] TROUNCE personnel continued to launch Loon missiles and in addition studied improvement of submerged tracking and control techniques. Primary in this effort was the continued research and development of a pulsed radar guidance system that was more secure than the previous radio control system for Loon. This pulsed radar system was given the name Trounce and was to serve as a secondary guidance system for the Regulus missile since BPN, under development for the Rigel missile, was considered the primary Regulus guidance system. Trounce I was first test flown in a manned drone aircraft on 23 April 1951.[12]

Originally five commands were suggested for encoding into the Trounce guidance system; right, left, up, down and dump. The Loon missile had no automatic control functions but when the Trounce system was adapted to Regulus, the up and down control functions were removed since automatic altitude controls could be programmed into the autopilot package.

The Trounce installation was referred to as the radar course directing central. The first radar course directing central was designated as the P-1X. Initially installed on the Loon launch and guidance submarines USS *Cusk* (SS 348) and USS *Carbonero* (337) as well as at the flight test center at Pt. Mugu, P-1X systems were largely customized installations built at the Naval Electronics Laboratory. They differed slightly between installations and suffered from not having a reliable parts supply system or maintenance manuals. In all fairness, the systems were in many ways laboratory experimental models and certainly not regarded as production equipment. Upkeep was a direct function of the experience and skill of the system operators.

The P-1X consisted of a radar transmitter, pulse pair encoder and guidance computer designated CP-98(XN)/UPW, that automated course change calculations. The CP-98 accepted inputs for the range, altitude and bearing of the missile in flight, the ship's motion, target range and bearing. From these inputs the computer automatically calculated and transmitted the appropriate course corrections. Built by Ultrasonic Corporation, Cambridge, Massachusetts, a breadboard model was available for testing in December 1951.

The first Trounce guidance system was installed on *Cusk*, with the first submarine-based Trounce guidance flight using a Loon missile taking place 28 June 1951. Three weeks later, on 20 July 1951, *Cusk* demonstrated Trounce guidance of a Loon missile to a range of 118 nautical miles, a tremendous increase in range over the former radio command control system, due in large part to the radar beacon that vastly improved the ability to track the missile flight at this extended range. *Carbonero* received a P-1X installation during overhaul in late 1953 and began Loon Trounce guidance operations in January 1954.

With the installation of Trounce equipment in Regulus missiles came a new series of Trounce

Run-up shed, GMU-10, 1961. (Courtesy of Peed Collection)

Booster Alignment Hangar-Regulus I, 1961. (Courtesy of Peed Collection)

component designators. Trounce I was the first Regulus Trounce guidance system and was similar to the version installed in Loon missiles. Later in the Regulus program, Trounce IA, IB and IC were built and utilized. Trounce IA utilized several different components and had enhanced reliability over Trounce I. Likewise with Trounce IB, but its improvements focused on improved defense against electronic countermeasures. Trounce IC permitted selection of three combinations of pulse pairs and also had the latest beacon, antenna and encoder modifications.

The shipboard Trounce radar course directing central was also available in three modifications during the program. P-1X was the designation for the first version and the same equipment was used as the initial installation aboard *Tunny*, *Cusk* and *Carbonero* as well as the Pt. Mugu shore- based station. AN/BPQ-1 was the production version upgrade of P-1X produced by Stavid Engineering. AN/BPQ-1 was installed on the East Coast Regulus launch submarine,USS *Barbero* (SSG 317) and the three East Coast guidance submarines, USS *Torsk* (SS 423, USS *Argonaut* (475) and USS *Runner* (SS 476). AN/BPQ-1 was much more reliable then the P-1X and, as a result, the West Coast submarines were upgraded in the field with the AN/BPQ-1 system. In 1957 the AN/BPQ-2 version of the Trounce radar course directing central was designed and built by Stavid Engineering. The first submarine installation was on board USS *Medregal* (SS 480), an East Coast Regulus guidance submarine.

Radar tracking of Regulus using the original sector scan system of the SV-1 required a skilled operator since the narrow acquisition angle of the missile radar beacon antenna meant that the commands would be received only if sent while the missile was in the narrow angle seen by the receiving antenna. Since the SV-1 scanned a twenty degree sector while the beacon signal only sent or received in a 7 to 8 degree center portion of this scan, tracking and sending signals was as much an art as a skill. With rigorous training, operators were able to reach required proficiencies but this was seen from the start as an area for system improvement.

Both Chance Vought and BuAer realized the inherent shortcomings of the Trounce guidance system. While much less complex than the primary BPN guidance system, Trounce relied on a World War II era radar system that had not been built to generate the kind of bearing accuracy needed for Trounce guidance of Regulus. While Trounce had a theoretical range of 230 nautical miles if the missile were at 35,000 feet, until the bearing angle error found in the submarine radars could be solved, the missile was limited to a 125 nautical mile operational range.

The first Regulus I flight using Trounce guidance took place 29 May 1953 with the launch of FTM-1029B, a training missile reconfigured as a tactical missile prior to tactical missile production. Two minutes after launch from Pt. Mugu, *Carbonero* successfully assumed Trounce control. After controlling the missile in a race-track pattern for 27 minutes at an altitude of 35,000 feet, *Carbonero* directed the missile to the dump point only to find the target was obscured by clouds. Control was turned over to the chase aircraft and an alternate site for dump was selected, unfortunately without phototheodolite coverage.[13]

In December 1955, BPN was formally canceled and Trounce guidance became the primary submarine and cruiser guidance system. The AN/BPQ-1 production version of P-1X represented a significant improvement in reliability. Cruiser installations of the AN/SPQ-2 radar required a beacon different than the submarine system. While suffering many of the same reliability problems as submarine Trounce, the cruiser Trounce program developed into a deployable system much earlier due in large part to the improved bearing angle accuracy of the cruiser radar. Though initially unable to transfer control to Trounce guidance submarines, this compatibility problem was solved by 1957 and the first cruiser-Trounce, submarine-Trounce flight took place on 14 March 1957 with launch and control from USS *Macon* (CA 132) and transfer to *Barbero* and *Torsk*.[14]

Part of the reliability issue was training of maintenance and operations personnel. Until the P-1X was unitized into the BPQ-1 version to match the "production" quality units built by Stavid Engineering, little if any documentation was available other then the experience and notes of the individual operators on each submarine. While several electronics problems were responsible for poor reliability in the early Trounce system, one stood out above the rest: mismatched magnetrons on the submarines and missiles. Magnetrons were the radar signal source. If the submarine based guidance system was to maintain control of the missile in flight, the shipboard magnetron frequency had to be compatible with the signal decoders on board the missile. At the same time, the frequency could not shift even the slightest amount or missile control would be lost. Missile tracking would be lost if the magnetron in the radar beacon failed to transmit on the correct frequency. To complicate matters, if two or more submarines were to control the missile, the magnetron on each had to be carefully matched. In the Loon Trounce program it was not uncommon to procure a large number of magnetrons and then go through and test each one to find a pair that matched closely enough to provide two guidance stations. The tubes had to match within two megacycles. During the Regulus program, radar signal tuning capabilities were added to the shipboard receiver so that if the missile transmitter drifted, missile tracking would not be lost.

A major oversight by critics of the Trounce guidance system accuracy was that of contribution of ship navigation error in the open sea. Simply put, any missile guidance system can be only as accurate as the error in the final guidance station's calculated (navigational) launch point's true geographic relationship to the target. Chance Vought detailed this problem in a report to BuAer in 1954 when it first proposed the RIN System. While the Regulus program magnified the need for vastly improved navigational aids, only the advent of the ballistic missile submarine forced the design and acquisition of the first inertial navigation systems for shipboard use.

An example of navigation error can be readily illustrated by the Regulus I Trounce flight launched from Pt. Mugu on 26 March 1954. Lateral control was assumed by the CP-98 guidance computer on *Carbonero* and range control assumed by the Pt. Mugu Flight Test Center CP-98 computer. Both flight legs met all objectives with both the Pt. Mugu and *Carbonero* successfully switching CP-98 control functions during the flight. The dump signal was sent and phototheodolite coverage indicated a 10 yard lateral and 2,000 yard range error. Investigation revealed that most of the flight range did not have first order survey information on the location of Gull Island Light which was the reference used during the exercise. The standard hydrographic charts used by *Carbonero* were found to be in error by 2,000 yards.[15]

In 1956, a contract was signed with Stavid Engineering for the BPQ-2 version of Trounce. BPQ-2 incorporated all of the changes and improvements that evolved from the BPQ-1 system. These included an integral antenna for use with air search, surface search and IFF (automatic identification signals), a larger antenna dish and the incorporation of a lobe-switching scan technique rather than sector scan. This conversion was a critical improvement that eliminated many of the reliability problems.[16]

Sector scan refers to the movement of the entire antenna through a "sector" of the compass. This is a mechanical operation and, with the narrow acquisition angle necessitated by the missile beacon antenna, this was hard on the radar hardware since it was originally designed as an air search radar with a continuously rotating antenna. Lobe-scanning, developed at the Naval Electronics Laboratory, utilized radar signal phase shifts to electronically scan a sector, thus allowing the radar antenna to remain stationary. Lobe-scanning eliminated significant wear and tear on the equipment but magnified yet another deficiency in the search radars of this time period, bearing angle error. This error refers to the difference between the indicated bearing angle on the radar scope and the true angle of the antenna. Submarine search radars had not changed significantly since the World War II and as such, the search radar was more of an early warning device and did not require pinpoint accuracy. Trounce guidance did require pinpoint accuracy since at the longer ranges, misalignment of the antenna meant loss of missile track and signal reception. Aligning the radar system with the correct bearing was not a trivial task; and, in the case of the submarine Trounce BPQ-2 system, was directly addressed in an exhaustive research and evaluation program on the East Coast.

In June, 1957, Lieutenant Commander Myron "Max" Eckhart, working in the Electromagnetics Systems Division of the Naval Underwater Sound Laboratory, New London, Connecticut, was assigned the task of evaluating accuracy and reliability problems associated with the redesigned Trounce AN/BPQ-2 radar course directing central guidance system and to "make it work." Eckhart had extensive training in radar, fire control and guidance mechanization. His orders gave him wide ranging authority to muster resources. Eckhart was given the authority to operate independently, with BuShips providing the financial support. In addition, he had the complete support of Cletus M. Dunn, civilian head of the Electromagnetics System Division.

Regulus I was, by this time, a reliable missile. The problem was that none of the specifications for the missile: autopilot, guidance computer, AN/BPQ-2 radar, gyrocompass or submarine installation bore any reference to each other or to the total mission specifications. The broad problem was to integrate the dynamic relationships of the several independent spherical coordinate systems involved, and to establish measurement, control and inter-coordinate system alignment accuracies in order to meet the required mission accuracy. Eckhart had to make sure that each subsystem generated equally accurate information for use in the overall guidance problem.

From the beginning, Eckhart realized if his test program was to be valid, it would have to be conducted in routinely less-then-perfect weather. The success of the system to date probably stemmed from the fact that virtually all of the developmental work had been done in the relatively benign weather found off the coast of southern California. Operations out of New London, Norfolk or Chincoteague promised to give more realistic sea and weather environments for evaluation and illumination of specific problem areas. On the one hand, the frequent thunderstorms permitted evaluation in stormy weather; on the other hand, the presence of the jet stream at 35,000 feet allowed evaluation of missile response to cross-winds on the order of 100 to 150 knots, surely the most strenuous the missile would encounter anywhere it might be deployed.

Eckhart and his team spent the first eight months defining the problem qualitatively and quantitatively. One of the first concerns was the establishment of static and dynamic test ranges of sufficient size to truly evaluate the BPQ-2 system while at the same time permitting accurate tracking. The static test range would permit testing of both submarine and missile components under carefully controlled and instrumented conditions. The dynamic test range would permit flight tests using both missiles and Trounce guidance equipped manned drone aircraft.

The static test range for isolation of component problems was constructed on Fishers Island off New London by using the laboratory's optical tracking facilities and a missile transponder trailer housing a Trounce transponder on a pad on Block Island. The fixed optical tracking facilities were on a high bluff overlooking the submarine operating area. The second optical tracker was mounted on Race Point Lighthouse off the southwest tip of the island. These two positions gave nearly ideal conditions for triangulation. Verification testing indicated that measurement accuracy and repeatability were within inches. Eckhart recalls that the first-order survey was the first indication that he really did have the authority to conduct his program rapidly. First-order coastal survey lead time was on the order of years, yet he requested and received one of the Block Island Sound area within several weeks. Location of the large transponder van on a pad in-and-amongst the houses of the super-wealthy landowners was another indicator. No complaints were voiced.

The dynamic test range ran from a guidance area off Cape Hatteras Beach to the tracking theodolite target at Wallops Island. Guided Missile Group TWO (GMGRU-2) provided piloted F9F-5KD drone aircraft carrying a Trounce transponder and guidance receiver. When missile launches were required, *Barbero* and Guided Missile Unit FIFTY-ONE, the submarine guided missile unit, provided launch services, and GMGRU-2 provided missile recovery and safety chase aircraft.

USS *Medregal* (SS 480), Lieutenant Commander Jack Padgett Commanding Officer, reported to Submarine Squadron SIX, Submarine Division SIXTY-THREE, Norfolk, on 20 December 1957. *Medregal* entered Charleston Naval Shipyard on 1 March 1958 for restricted availability while the first BPQ-2 guidance system was installed. On 11 June 1958, *Medregal* left the shipyard to begin an eleven month BPQ-2 evaluation program under the direction of Eckhart.

Eckhart found that the developmental model of the BPQ-2, as installed on the *Medregal*, worked well as a basic radar system. They were the first to realistically test the system with aircraft and found that the generated guidance commands had random errors. Systematic investigation began at the beginning of the guidance problem, making the radar system lock onto a static target while the submarine was running submerged. Both the static and dynamic test ranges were used in this first test phase. *Medregal* operated close-in to Fishers Island so that the optical instrumentation could track the antenna as it in turn tracked the stationary trailer on Block Island. The static range had only one problem. Sheer granite cliff faces were at both the beginning and end of the submerged run for *Medregal*. To get a useful run, the *Medregal* had to submerge with little forward motion, make the run and then surface quickly and come to a complete stop before hitting the cliffs ahead. As the need for closer and ever more precise work developed, this became even more delicate. Evaluation of the optical tracking data indicated that as suspected, antenna alignment was a critical problem area. These first static test range evaluations did little to isolate the components of the alignment problem but were necessary to generate an error envelope for the submarine's guidance installation prior to use in evaluating airborne equipment error.

Critical to the successful use of the dynamic test range was the ability to precisely locate *Medregal* during the 30 minute manned drone Trounce guidance flights. Eckhart located a short range land-based precision navigation system in Oklahoma, previously used during airborne magnetic anomaly surveys in the oil fields. Placed on Hatteras Beach, this system permitted precise tracking of the submarine while submerged but it had one slight drawback, limited range. Due to the shallow waters, *Medregal* would run up, on the surface, to a buoy marking the starting point for the run, and then move out to run submerged for the guidance operations. She had only a few feet of water under her keel and remained in this shallow water for the length of the run.

Since missile flights were expensive and at best might be as frequent as 2 or 3 per week, Eckhart turned to the GMGRU-2 Trounce equipped F9F-5KD drone aircraft. With a capability of as many as 6 flights per day, the flight test work went much more quickly. As lessons were learned in the aircraft system they would

Guided missile Unit FIFTY ONE facilities under construction at NAS Roosevelt Roads, Puerto Rico. Lower center can be seen the engine run up shed where full flight profiles were conducted on Weber dollies as a final check after missile maintenance had been performed. (Courtesy of Jackson Collection)

be evaluated on a missile flight to confirm the solution, as well as tested using the static range.

Testing on the West Coast at Pt. Mugu demonstrated that the missile autopilot functioned well and reliably. Unlike the frequently foggy but otherwise reasonable weather of the West Coast test area, the Norfolk, Virginia Capes area frequently had what could best be described as adverse weather conditions. Testing of the missile guidance system under these more demanding conditions revealed control correction rate limitations that seriously affected guidance accuracy. Missile cruise-to-target control correction rates could be relatively slow since as long as the course excursions were not excessive, minor corrections generating a sinusoidal track would be acceptable. However, once the missile neared the terminal dive or "dump" point, tight control correction was necessary so that the missile entered the supersonic dive directly on the target bearing.

Innumerable drone flights were used to identify and correct this problem. Eckhart recalls how unpopular his group was with the GMGRU-2 drone pilots since the more adverse the weather, to an extent, the more important the test flights were. The drone aircraft cockpits had autopilot gain, including aircraft response rates, as selectable functions for the pilot. Mimicking the missile commands made for uncomfortable rides. While it was certainly understandable that pilots would want to mitigate the rough flight by changing the control rates, this did not aid the testing program. While there was no cockpit recorder to sense any pilot induced changes in the settings, the radar track gave it way. Thus the pilot was signalled to break off the run and start again or abort and head back to the field. It soon became easier to endure the rough ride than return to the field due to control rate adjustment aborts. Eckhart remembers that the operations officer for GMGRU-2, Lieutenant Commander Ralph Mattus, understood the problem and did his best to have the pilots understand the situation. In more then one instance Lieutenant Commander Mattus flew the missions in truly adverse weather in order to keep the program on track. Eckhart recalls one time in particular where his team was noting several odd control reversal outputs. When asked, Mattus calmly replied that the autopilot was having trouble keeping the aircraft upright in the storm. The BPQ-2 team had not realized that flight conditions were that bad and quickly suggested that he return to base. He replied that it was not that big a deal and that they should leave the flying to him!

Measurements afloat could not isolate a family of systematic errors due to the absence of specified alignment between the BPQ-2 antenna and the gyrocompass; nor resolve indications of significant variability in the vertical reference output of the gyrocompass. The BPQ-2 antenna, mounted on a retractable mast, defined the submarine's horizontal reference plane for radar measurement of the missile's position. The gyrocompass, deep in the submarine, provided a vertical earth reference for transformation of missile tracking measurements into earth spherical coordinates for computation of missile guidance commands. The gyrocompass bedplate was its reference for the submarine horizontal plane. Optical alignment equipment was not equal to the task of relating the antenna and gyro horizontal planes afloat. To this point in time there had been no need to since submarine radar still served primarily a search function. There was convincing circumstantial evidence collected over the months of drone aircraft guidance that irregular errors in the gyrocompass vertical output were occurring. There was no way of measuring these variations afloat.

Main motor problems necessitated return of *Medregal* to its home shipyard in Charleston for repairs. Scheduled as a short repair period in November 1958 when winter weather precluded operations off of Hatteras, Eckhart requested and received an open-ended period in the shipyard's primary drydock, which ended up lasting through most of January. *Medregal* was docked on a single row of blocks, braced by telephone poles to the drydock walls. Jacks were incorporated to provide vertical positioning control. Direct measurements of antenna and gyrocompass bedplates could now be made separately and precisely. Now knowing their relative misalignment, correction factors could be used to compensate in the measurement of spherical coordinates.

Even in winter, daytime heating and cooling of the exposed hull caused enough mechanical distortion between the antenna and gyrocompass bedplates to prevent reproducible measurements. All work had to be done at night, further emphasizing that missile guidance involved measurement and control accuracies several orders of magnitude greater then previously needed in the submarine force.

The drydock and surrounding industrial environment made for impossible radar measurement conditions for an antenna designed to work a few feet above an infinite water plane. Therefore, a horizontal, oval groundplane (copper on wood, 100 feet by 50 feet) was constructed and mounted on *Medregal*, a few feet under the raised radar antenna, to emulate sea conditions. Fortunately, one clear bearing through the building clutter existed on which to locate the test van as a distant target. The drydock caisson was visible in the path but was blanked out with radar absorbent material. Antenna alignment could now be completed.

Gyrocompass instability was directly measurable once drydocked but required considerable investigation to identify a penny-wise but pound-foolish cause. Much had been invested in a new gyrocompass to provide precise vertical reference for BPQ-2. It was apparently not tested sufficiently to detect its extreme sensitivity to frequency control of its 400-cycle power source. The submarine's primary 400-cycle generators had marginal voltage and frequency regulation. They were three phase generators, while most of the submarine loads were single phase. Over the years, shipyards had paid little attention to load distribution versus phase balance as they added and removed equipment during the submarine's overhaul periods. Phase imbalance was overriding voltage regulation on *Medregal*, so marked shifts in the gyrocompass vertical occurred as various equipment was switched on and off throughout the submarine. The shipyard carried out an exhaustive electric loading analysis. *Medregal's* numerous electrical switchboards were completely reworked to balance phase loading of the generators under typical operating conditions. The result was a much more stable vertical reference. When *Medregal* came out of drydock in January 1959, Eckhart had the necessary calibration criteria and measurement standards to permit aligning the bearing reference accurately. A routine guidance system calibration and checkout procedure was developed using the radar beacon transponder van as a surrogate missile pierside. Checkout of BPQ-2 control rates and typical missile autopilot responses were now part of the regular ship's routine.

The impressive results of the drydock testing allowed Eckhart to convince Washington that Regulus I could soon be made a reliable, fully mission-capable system. He had demonstrated that the systematic barriers to achieving the desired accuracy had been surmounted. Now operations capability and reliability could be confidently anticipated. The weather conditions in the Norfolk area during the winter precluded predictable submarine operations in the shoal waters off of Cape Hatteras. Eckhart received permission to move flight test operations to the Caribbean and immediately after *Medregal's* departure from the shipyard, she turned south to the Caribbean. Using the Strategic Air Command's tracking theodolite station (part of their highly instrumented practice bombing range) in Puerto Rico as the target area and the deep waters close off shore St. Maarten as the guidance area, Eckhart and the *Medregal* repeated the static and dynamic test range operations.

High intensity flight operations, final BPQ-2 modifications, and technical evaluation proceeded the request for formal OPDEVFOR evaluation in April, 1959. Eckhart and his team departed and the evaluation personnel stepped in. On 10 April 1959 two Regulus I missiles were launched, one from *Barbero* and one shore launched by GMU-51 personnel at Naval Air Station Roosevelt Roads. Both missiles were successfully and accurately guided by *Medregal* and recovered at Roosevelt Roads. Commander Eckhart signed off his final report for development program completion in early June 1959.[17]

Positive Flight Termination

Positive Flight Termination (PFT) was a range extension concept developed in 1955-56. PFT was designed to assume command of the missile if the Trounce or RCC guidance signal was lost near the target. PFT also provided a method to increase the distance from which an escort pilot could break away from the missile. PFT provided range extension only, heading guidance was provided by the missile's normal autopilot operation.

PFT used a set of sequential timers. A computerized system was rejected early in development due to the additional complexity. The PFT system installed in tactical missiles utilized four pre- set timers (pre-set according to the selected target) and three check point signals. The limiting factor for PFT accuracy, besides that of an accurate launch point, was the yaw gyro drift rate. The yaw gyro in each missile automatic stabilization system had a known drift rate. This drift rate was determined periodically as a maintenance function and then utilized to correct for drift during the PFT flight phase. PFT operation was evaluated with seven missile flights over a two and one-half year period. PFT was incorporated into all tactical as well as fleet training missiles by 1960.[18]

Regulus Internal Navigator

The Regulus Inertial Navigator (RIN) began as a Chance Vought proposal for a self-contained guidance system intended for use in Regulus I and ended up being the primary guidance system for Regulus II. RIN was never flown in Regulus I. The RIN concept was first proposed on 1 July 1954 as a new mid-course guidance system for Regulus I. Chance Vought engineers had surveyed current inertial guidance system state-of-the-art and concluded that RIN was feasible from a technological standpoint. Its advantage over the Trounce and BPN systems was that it was entirely self-contained and hence immune to electronic countermeasures. Estimated circular error probable, the circle in which 50% of the missiles launched would hit, at the missile dump point, was 3 miles for 500 nautical miles, or one hour of flight, provided the launch point was determined by accurate land fixes. Addition of a terminal guidance system to refine the missile flight path close to the target would improve the circular error probable to 1500 feet under the same flight conditions. Costs for constructing two experimental models and testing them were estimated to be $1.4 million dollars.[19]

After several conferences to discuss the RIN proposal, on 19 October 1954 BuAer's formal response was for a revised engineering and cost proposal from Chance Vought covering the development of a combination inertial and homing system. BuAer also requested a reduced accuracy inertial system that might work in conjunction with the present Trounce and BPN systems.[20] On 12 November 1954, Chance Vought responded with a proposal for the Regulus Inertial Navigator Offset Beacon (RINOB) System. This would be applicable to both Regulus I and II. The proposal also pointed out that in the not-too-distant future, navigational requirements for the missile would outstrip those available for current at-sea position determination. Thus a similar system for ships would be necessary and possibly would be an outgrowth of the RIN or RINOB systems.

RINOB was to provide the missile with inertial guidance of sufficient accuracy to direct the missile to within range of a radio beacon. Without the offset beacon update, accuracy was limited to 6000 yards after a 500 mile flight, provided the launching ship had an accurate launch reference. Once in the target area, as far as 150 miles from the actual impact point, the missile would acquire the off-set beacon guidance signal, which indicated the target location relative to the beacon. The inertial navigator would then direct the missile to the target via flight path corrections determined by the computer. Chance Vought noted that "contrary to popular conception, of all the tremendous effort on inertial systems by the industry, there is no existing complete system readily adaptable to Regulus requirements...therefore system design must be considered separately from other developments."[21]

With the offset beacon accurately placed within 100-150 miles of the target, RINOB would deliver the missile with an accuracy of 600 yards.[22] The RINOB system was proposed to be ready for flight testing in aircraft by 1956, with the second system available one year later. Total cost would still be $1.5 million. At the end of 1955, BuAer authorized development of a RIN system with emphasize for use in Regulus II but compatible with Regulus I equipment.[23] In late 1956 Chance Vought proposed a comprehensive program to place RINOB systems in Regulus I but this was apparently not pursued any further since there are no missile operations reported with RINOB guidance.

The RIN program continued without the offset beacon feature since it was to be used as the Regulus II guidance system. A critical element in the system were the gyroscope heaters. A seemingly insurmountable problem with precision gyroscopes was at this time was bearing wear. Dr. C. Stark Draper of MIT devised the floating gyroscope to circumvent this problem. By floating the entire inner gimbal of a single degree of freedom gyro, virtually all of the gravitational load on the bearings as well as that induced by linear acceleration was eliminated. The key was finding a fluid that would float the heavy rotor and spin-up motor components. After much experimentation, Draper found a fluid called fluorolube that had a very low viscosity in its fluid state, a critical factor for its use. Unfortunately, it had to be maintained above 160 F or it would gel, expanding and bursting its container. The RIN design called for the gyros to never be below 170 F and, from the time each assembly was filled with fluid at the factory until it was replaced, the heaters had to be on and the system kept at 170 F. There were allowances within the device for minor expansion and contraction, and the thermal mass was such that if a power failure occurred the system would remain stable for about one hour. BuAer was not at all please with the heating of the gyroscope flotation fluid. While accepting this concept for the developmental models of RIN, BuAer made it clear to A.C. Spark Plug that a lower temperature system would need to be found for operational use.

Radar Map Matching

Goodyear Tire Corporation was contracted to develop a radar map matching (RMM) guidance system for use with Regulus II. RMM relied on the correlation of a prepared radar map of the way point or target area with the image on a standard plan position indicator radarscope. The map was on a transparency that could be rotated and shifted to indicate needed course corrections. The main difficulty with this approach was the overwhelming amount of data the image supplied; in effect, too much. RMM was discontinued with the cancellation of Regulus II.[24]

Endnotes

[1] *The Recoverable Regulus Missile, NAMTC Memorandum Report MT3-57, 11 July 1957, pages 7-8. Pacific Missile Test Center Archives.*

[2] *Personal interview with Roy Pearson.*

[3] *Anonymous. Rigel Missile, Grumman History Center, Grumman Corporation, Bethpage, New York. page 2.*

[4] *Operational Guide for the Bipolar Navigation Guidance System as Installed in the SSM-N- 8a Regulus Missile Lot XI and Utilizing AN/BPN-1 (XN-2) Radio Beacons. 23 August 1955. Chance Vought Aircraft, Incorporated. Report #9851. pages 20-24, LVSC Library.*

[5] *Ibid.*

[6] *Ibid.*

[7] *Regulus I Flight Test Index, 1961, LVSCA A50-24, Box 7*

[8] *Regulus Mark II XN-3 Bipolar Guidance System Feasibility Program 31 January through 29 April 1955. Project TED MTC GM 2201, undated. Pacific Missile Test Center Archives.*

[9] *Personal communication with Larry Thomas, Chance Vought BPN guidance engineer.*

[10] *"Evaluate the REGULUS Guided Missile for Service Use" Final Report on Project Op/S317/X11; 721:cx FF 5-7/X11 ser 00262, 16 August 1957. Naval Historical Center, Operational Archives.*

[11] *"Project TROUNCE" CNO sec ltr 00258P34 of 17 May 1950. COMSUBDIV 51, Naval Historical Center, Operational Archives*

[12] *Submarine Guided Missile Program (Project TROUNCE - Phase II); Semi-Annual Report for October 1950-March 1951. FC4-11/PES:Fry A1 Ser 001 17 April 1951. COMSUBDIV 51,Naval Historical Center, Operational Archives.*

[13] *Submarine Launched Attack Missile (Project SLAM) Semi-Annual Report for 1 April - 30 September 1953, page III-1,. COMSUBDIV-51. Naval Historical Center, Operational Archives.*

[14] *Regulus I Progress Report #18. 1 January to 31 March 1957. LVSCA, A50-24, Box 5.*

[15] *"Semi-Annual Report on Submarine Launched Attack Missile (SLAM) Project for period 1 October to 31 March 1954. ComSubDiv 51 Conf ltr XII Ser 017 of 22 April 1954, pg IV-8. COMSUBDIV 51, Naval Historical Center, Operational Archives. Uncertainty of the points on a nautical chart was realistically a total system error and not a peculiarity of a particular submarine navigation practice.*

[16] *Personal communications with James Ottobre, BPQ-1 and -2 engineer for Stavid Engineering.*

[17] *Personal interview with Captain "Max" Eckhart, USN (Ret.). A citation from Rear Admiral F.B. Warder, Commander Submarine Force, U.S. Atlantic Fleet, further illustrates the importance of Captain Eckhart's work.*

[18] *The PFT system was used in the fleet training missiles to simulate the terminal dive or airburst signal for training purposes.*

[19] *Proposal for Project REGULUS - Inertial Navigator System. Chance Vought Aircraft, Incorporated. Record Group 72, Guided Missile Division, Bureau of Aeronautics, conference general correspondence, Box 149, 1954. National Archives*

[20] *Comments on Project REGULUS Inertial Navigation System Proposal, from Chief, Bureau of Aeronautics. Bureau of Aeronautics, Record Group 72 Box 150, 1954.*

[21] *REGULUS Inertial Navigator Offset Beacon (RINOB) System; Submission of Cost and Engineering Proposal for; 12 November 1954. Guided Missile Division, Record Group 72 Box 150, National Archives.*

[22] *Ibid., page 3.*

[23] *Regulus Progress Report #14, page 16.*

[24] *Personnel communications with Bill Hallmark, RMM guidance system engineer for Chance Vought.*

Appendix II: Navy Guided Missile Unit Histories

Three types of Regulus guided missile support units were actively involved in the Regulus program: Guided Missile Training Unit FIVE (GMTU-5), Guided Missile Units (GMUs) TEN, FIFTY, FIFTY-ONE, FIFTY-TWO, FIFTY-THREE and NINETY; and Guided Missile Groups (GMGRUs) ONE and TWO. The histories of the GMGRUs are told in the Chapter Eight. The GMU histories are more limited and given here in numerical order.

Guided Missile Training Unit FIVE

GMTU-5 was established in March 1951 to provide personnel for training in Regulus missile operations. Lieutenant Commander W.E. "Pappy" Sims was Officer-in-Charge and Lieutenant Commander I.E. "Dutch" Wetmore was Assistant Officer-in-Charge. The unit consisted of the two officers and 12 enlisted men. After several training orientation classes at Chance Vought's Regulus missile production facilities in Dallas, GMTU-5 personnel were given on-the-job assignments in April 1951 and began working directly with the contractor flight test staff at EAFB and later Pt. Mugu.[1,2]

GMTU-5 and GMU-50, which was still launching Loon missiles, were housed in the same building at Pt. Mugu, so exchange of techniques and ideas was easily accomplished. GMTU-5 was disestablished in June 1954 and its personnel absorbed into the GMU-50 structure.

Guided Missile Unit FIFTY

GMU-50 was established 1 January 1953 to replace Project Derby.[3] GMU-50 was responsible for continuing to train naval personnel in the maintenance and launch of the Loon guided missile as well as work on new operational techniques. The Trounce guidance system that had begun development as part of Project Trounce under the auspices of Project Derby, continued to be refined by GMU-50 personnel under Project SLAM (Submarine Launched Assault Missile). Loon missiles continued to be launched by GMU-50 personnel until June 1953.

In 1954 GMU-50 absorbed the personnel from GMTU-5. With Loon flight operations completed, GMU-50 had the new mission of preparing, checking out and delivering Regulus I missiles to the submarine force. GMU-50 did not launch Regulus I missiles at this time. GMU-50 moved to Port Hueneme, California, in February 1955. This move placed the unit in close proximity to the Regulus launch and guidance submarines of Submarine Division FIFTY-ONE. Captain A.R. Faust was now Officer-in-Charge of GMU-50. In June 1955, Commander James Osborn relieved Captain Faust as Officer-in-Charge. Osborn came to GMU-50 directly from duty as Commanding Officer of USS *Tunny* (SSG 282).[4]

GMU-50 continued to support submarine Regulus launch operations until 1 March 1957 when support operations ceased as the unit transferred to Pearl Harbor, Hawaii. On 15 March 1957, the Office of the Chief of Naval Operations redesignated GMU-50 as GMU-90. Arriving at the Submarine Base, Pearl Harbor, in early April, GMU-90 was fully operational by 1 May 1957.[5]

Guided Missile Unit FIFTY-ONE

Commander Eugene Pridinoff was the first Officer-in-Charge of GMU-51. He assumed command on 15 October 1955, after spending three weeks on temporary duty assigned to GMU-50. Pridinoff was quite familiar with the Regulus program as he had been Commanding Officer of the USS *Cusk* (SS 348), one of two Regulus guidance submarines on the West Coast.[6] Pridinoff and his staff arrived at the Yorktown Naval Mine Depot to a mixture of new and old buildings and dilapidated support facilities. World War II ammunition magazines were converted to missile maintenance facilities. Missile checkout pads were built, where missiles ready for shipment to a submarine were given one last simulated flight check. No missile launches were conducted by GMU-51 personnel at these facilities. GMU-51 supported submarine Regulus flight operations on the East Coast, working directly with Commander Submarine Division SIXTY-THREE. The USS *Barbero* (SSG 317), a Balao-Class fleet boat of World War II vintage, had been newly converted as a launch and guidance submarine for Regulus. She had reported to her home port, Nor-

Guided Missile Unit FIFTY facilities at Port Hueneme, California. Two fleet training missiles can be seen as well as a Short Rail Mark 2 launcher on the flat bed trailer. (Courtesy of Blount Collection)

folk, Virginia, in May 1956 (see Chapter Eleven). *Barbero* launched her first missile prepared by GMU-51 on 26 June 1956; and, after a 32 minute flight under Trounce guidance, the missile was successfully recovered at Naval Air Station Chincoteague by the aviators of GMGRU-2.

1957

Two events highlighted GMU-51 operations in 1957. The first was an East Coast launch televised live to the nation (See Chapter Eleven). The second event took place in April when, at 1600 hours on Friday, 26 April 1957, it was determined that *Barbero*, operating in Florida waters as part of the 1957 Joint Civilian Orientation Cruise, needed a replacement missile immediately. The missile had to be delivered to Key West. GMU-51 selected a missile, loaded it into its special "transicontainer"[7], arranged the logistics of transportation and then drove the missile and its checkout van 1200 miles to deliver the missile at 0800 hours 30 April 1957. Once aboard *Barbero*, the missile checked out perfectly. The entire exercise was a demonstration of the capabilities of the Regulus support units and the subject of a letter of commendation from Commander Submarine Squadron SIX, Captain Slade Cutter.

1958

From 14 February to 11 March 1958, Operation SPRINGBOARD, the annual Altantic Fleet winter exercises, were held in the Caribbean. GMU-51 moved a contingent to Naval Air Station Roosevelt Roads to support Regulus operations by the submarines of Submarine Division 63. On 1 July 1958 GMU-51's homeport was officially changed from the Naval Mine Depot, Yorktown, Virginia, to Naval Air Station Roosevelt Roads, Puerto Rico. Constant weather restrictions, increasingly heavy commercial shipping and rapidly expanding commercial air traffic in and near the firing range off Virginia Capes made the move necessary.

On 3 August 1958, the advance contingent of men and equipment from GMU-51 arrived at Roosevelt Roads. On 10 September 1958 the USS *Orion* (AS-18), a submarine tender attached to Submarine Squadron SIX in Norfolk, Virginia, moored to the fueling pier at Roosevelt Roads and GMU-51 began to disembark 240 tons of equipment. Six hours later all equipment was on its way to the new GMU-51 facilities. Lieutenant Commander Ralph "Snuffy" Jackson was Officer-in-Charge with a staff of 11 officers and 65 enlisted personnel.[8]

Jackson recalls that GMU-51 facilities were brand new, prefabricated buildings inside a fenced compound. Within the fenced area was a checkout pad with engine blast deflector, a maintenance shop and separate warhead facilities with their own security fence. Six weeks later GMU-51 personnel prepared to launch their first Regulus missile under the watchful eyes of a large crowd of spectators. The Missile Officer, Lieutenant Roy Hoffman and eight enlisted men from GMU-51, had gone to GMGRU-2, the surface Regulus support unit that launched missiles, for a two week period of hands on instruction including one launch. The launch was perfect and the missile recovered successfully.

GMU-51 launched six missiles, all successfully recovered, prior to the end of 1958. All launches were training flights. In late 1958 preparations were begun to bring Regulus II to GMU-51 at Roosevelt Roads. No facilities were constructed and when the program was terminated in December 1958, GMU-51 returned to supporting Regulus I launch operations for submarines in the Caribbean area of operations.

1959

Ten missiles were launched in 1959. One launch in particular stands out in Hoffman's memory. On 23 April 1959, FTM-1346's starboard booster did not fire and the missile crashed in a ball of fire four seconds later.[9] Amazingly enough, the standard photography sequence for the launch clearly recorded the failure of the booster to fire. The last launch of a Regulus missile by GMU-51 was on 28 May 1959. It was the 12th flight for FTM-1248 and was recovered successfully. On 30 June 1959, GMU-51 was disestablished and the missiles and support equipment transferred to GMU-90 at Pearl Harbor.

Guided Missile Unit FIFTY-TWO

Lieutenant Commander Robert Munroe reported to GMTU-5 in June 1953, at Pt. Mugu, as prospective Officer-in-Charge of the first Regulus surface unit launch team. After a year of training, GMU-52 was formed on 20 May 1954, with Munroe as Officer-in-Charge and Lieutenant Commander George Bailey as Assistant Officer-in-Charge. On 1 June 1954, GMU-52 moved from Naval Air Station, North Island to Brown Field, San Diego. No missile flight operations were conducted at Brown Field but training was conducted with manned drone aircraft flown by pilots from the Utility Squadron THREE (VU-3) Regulus Assault Missile (RAM) detachment. Several months after the move to Brown Field, the decision was made to relocate the unit back to North Island since the missiles were stored there and the warhead teams operated from the special weapons facilities located on the base.[10]

Flight operations began with a successful launch from USS *Norton Sound* (AV-11) on 20 August 1954, the first of a long series of tests for the Operational Development Force (OPDEVFOR) evaluation of Regulus I. On 18 October 1954, GMU-52 successfully launched an fleet training missile from USS *Hancock* (CV 19). While the RAM detachment naval aviators were enthusiastic about the RAM program, their counterparts on board the carriers and in command of the air groups were exactly the opposite, always looking for a reason not to have the missiles and their associated launch equipment on board (see Chapter Eight). Initial carrier operations utilized the Short Rail Mark 2 (SR MK-2) launcher, a twelve caster, lumbering monstrosity built with little attention to operational utility.

On 28 October 1954, a launch team from GMU-52 successfully launched the first guided missile from a cruiser, the USS *Los Angeles* (CA-135). Three additional launches from *Los Angeles* and four launches in support of training for the first aircraft carrier detachment, VC-61 RAM Detachment "George", closed out GMU-52 operations for 1954.

1955

GMU-52 launch operations continued in support of the OPDEVFOR program with five launches from *Hancock* before she deployed to WestPac in August 1955 (See Chapter Eight). Four cruiser launch operations were conducted by GMU-52 teams, two with *Los Angeles* before she deployed to WestPac in February 1955 (See Chapter Nine) and two for USS *Helena* (CA 75) during OPDEVFOR testing of the new SR MK-3 launcher.

In May 1955 Lieutenant Commander George E. Kaufman relieved Lieutenant Commander Munroe as Officer-in-Charge of GMU-52. In June 1955, a U.S. Army guided missile team, commanded by Captain R.M. White, successfully launched two Regulus missiles from the Marine Corp Auxiliary Air Station at Mojave. White and his team had been stationed on temporary duty at North Island with GMU-52 for 18 months, learning the guided missile business so that the Army would not be left behind in this new technology. While these were the only missiles that the Army team had complete responsibility for, they had participated in 11 previous launches in to some degree.[11] In September 1955 GMU-52 was disestablished and its personnel transferred to GMGRU-1.

Guided Missile Unit FIFTY-THREE

Lieutenant Roby A. Beal was the first Officer-in-Charge of GMU-53. Established in mid-1954 at Naval Air Station Chincoteague, Virginia, the unit had trained at Edwards Air Force Base and Pt. Mugu prior to moving to the East Coast with the pilots of Utility Squadron FOUR (VU-4) RAM Detachment in October 1954.[12] The first missile was launched on 24 March 1955 (see Chapter Eight) and was recovered after a 28 minute flight. This launch marked the first of eighteen consecutive successful launches for the GMU-53 team as they trained launch crews for deployment on soon to be converted cruisers and aircraft carriers. All of the launches were KDU-1 target drones that served as high speed, medium altitude targets for the USS *Mississippi* (AG 128) Terrier air defense missile evaluation program. Usually the flight operation would end with a practice RAM mission by the pilots of VU-4 RAM detachment prior to recovery of the missile at Chincoteague. GMU-53 was disestablished on 26 September 1955 and combined with VU-4 RAM Detachment to form GMGRU-2. GMGRU-2 provided Regulus launch services for target drone presentations as well as launch crews for the East Coast aircraft carrier and cruiser training and deployments with Regulus on board.

Guided Missile Unit FIFTY-FIVE

GMU-55 was established on 22 March 1957 at the U.S. Naval Construction Battalion Center, Port Hueneme, California. Lieutenant

GMU-51 facilities at Roosevelt Roads, Puerto Rico, during construction in 1958. The two large buildings contained missile maintenance, booster alignment and warhead facilities. The engine runup shed and missile storage area can also be seen. Two transicontainers can be seen to the left outside the fenced in area. To the right can be seen the A-frame for missile center of gravity determination. (Courtesy of Jackson Collection)

Peter L. Fullinwider was assigned as Officer-in-Charge. GMU-55's mission was to assist in the Navy test and evaluation phase of the Regulus II program. On 17 June 1957 this mission was expanded to include providing maintenance, handling and recovery services for all surface-to-surface missile operations at Pt. Mugu. Lieutenant Joseph L. Zelibor relieved Lieutenant Fullinwider as Officer-in-Charge on 17 July 1957. On 25 September 1957, GMU-55 began a string of eleven successful Regulus I launches with four of these taking place in the Fall of 1957. On 22 November 1957, GMU-55's base of operations was moved to the Naval Missile Center, Pt. Mugu, in anticipation of the start of Regulus II flight operations. In October 1957, Commander I.E. "Dutch" Wetmore relieved Lieutenant Commander Zelibor as Officer-in-Charge with Zelibor assuming duties as Assistant Officer-in-Charge. Commander Wetmore had many years of experience with Regulus, having been part of the original training unit, GMTU-5.

1958

The year began on 20 January 1958 with the televised launch of a Regulus I. After a successful flight, the missile was slightly damaged on recovery. After two additional flights in February 1958, GMU-55 focused its training efforts on the Regulus II missile being flown at Edwards.. On 28 March 1958 GMU-55 participated in the sixth boosted launch of the Regulus II program and three months later, on 5 June 1958, GMU-55 conducted the first all-Navy launch of a Regulus II, also the first launch from Point Mugu. The launch and flight were successful and missile recovery took place at Tonopah, Nevada. GMU-55 conducted two additional Regulus I launches in 1958 and participated in the only Regulus II launch from a submarine, USS *Grayback* (SSG 574), on 16 September 1958 and a surface vessel, USS *King County* (LST 157) on 10 December 1958.

1959

With the cancellation of the Regulus II program, GMU-55 continued launching Regulus I fleet training missiles and on 9 July 1959 was given the additional task of supporting KD2U-1 Regulus II target drone operations at Pt. Mugu as well as supporting all Regulus I operations on the West Coast. Commander Wetmore was relieved by Lieutenant Commander Zelibor on 30 April 1959. In July 1960, Lieutenant Commander Carl Davis, Jr., relieved Lieutenant Commander Zelibor. GMU-55's duties were expanded into a myriad of other projects of which one was continued support of KD2U-1 Regulus II target drone services.[13]

Guided Missile Unit NINETY and Guided Missile Unit TEN

GMU-50's designation was changed to GMU-90 on 15 March 1957, shortly before moving to Pearl Harbor from Port Hueneme. Commander James Osborn continued to serve as Officer-in-Charge. GMU-90 was the first guided missile unit to deploy outside the continental United States. The unit's mobile equipment made the deployment rapid and economical, demonstrating that Regulus assets could be easily deployed to forward areas.

In August 1957, Lieutenant Commander Samuel T. Bussey assumed the duties of Officer-in-Charge for GMU-90. Commander Osborn had been assigned as commanding officer of the first Polaris submarine, USS *George Washington* (SSBN 598). Bussey was intimately familiar with the Regulus program, having served in GMTU-5 in the special weapons department, as commanding officer of *Barbero*, the East Coast Regulus launch submarine and executive officer, and missile officer, on board *Tunny*, the West Coast Regulus launch submarine. As with the other submarine GMUs, GMU-90 did not launch missiles but rather serviced and repaired those launched by *Tunny*.

Tunny, Cusk and *Carbonero*, the submarines of Submarine Division FIFTY-THREE transferred their home port to Submarine Base Pearl Harbor on 7 May 1957 and Regulus training operations could be more easily and safely conducted in the wide-open operating areas around the Hawaiian Islands.[14]

1958

In October 1958, GMU-90 acquired missile launch expertise and equipment when the

Bonham Detachment, located at Bonham AFB, Barking Sands, Kauai, and previously attached to Guided Missile Group ONE, was reassigned to GMU-90. The Bonham Detachment had serviced and launched missiles in support of carrier and cruiser Regulus assault missile operations which had been phased out.[15]

With the cancellation of the Regulus II program in December 1958, the Commander Submarine Forces, Pacific Fleet, requested that all Regulus I units; the submarines, launch as well as guidance; supply and maintenance assets, be consolidated at Pearl Harbor. After several months, the decision was made to move all of the missiles and equipment from GMU-51 at Roosevelt Roads to Pearl Harbor. *Barbero*, *Growler* and *Medregal*, a guidance submarine, would also change homeport from Norfolk to Pearl Harbor, while the remaining East Coast Submarine Division SIXTY-THREE submarines had their missile guidance centers removed and were assigned other duties.

On 1 July 1959, GMU-90 and GMU-90 Bonham Detachment were redesignated GMU-10 and GMU-10 Bonham Detachment respectively. Simultaneously, GMU-51 at Roosevelt Roads was disestablished and much of its personnel and equipment sent to GMU-10.

1960

On 30 March 1960, Lieutenant Commander George P. Peed relieved Lieutenant Commander Bussey as Officer-in-Charge, GMU-10. Peed had previously served as Assistant Officer-in-Charge at GMU-51 on the East Coast. On 11 July 1960, GMU-10 began its support of the Regulus Missile School which had transferred to Hawaii from Dam Neck, Virginia, several months earlier. All Regulus support operations were now in Hawaii at GMU-10. Launch and recovery operations in support of the Regulus submarine deterrent patrols had begun in late 1959. Each of the five submarines would undergo several days of refresher training in missile launch operations, culminating in either a tactical missile Operational Suitability Test flight or a fleet training missile flight simulating the terminal dive maneuver. The 1000th launch of a Regulus I missile was made from *Barbero* on 4 August 1960 (see Chapter Eleven).[16]

1961

In 1961 the mission of GMU-10 was revised by the Chief of Naval Operations. Rather than deploy the entire unit to a remote base as needed, mobile teams were developed to carry out support operations. In addition, GMU-10 assumed responsibility for staffing the Regulus Guided Missile School. In February 1961 GMU-10 deployed a mobile missile team to Naval Air Station Naha, Okinawa with three tactical missiles. The USS *Halibut* (SSGN 587) had just completed a demonstration launch for the SEATO countries and was to leave from Naha on her first deterrent strike patrol. She had launched one of two tactical missiles with dummy warheads and needed to exchange the one remaining missile and replace the expended one (see Chapter Fourteen).[17]

As an illustration of the activity of GMU-10 during 1961, 166 missiles were processed for issue to submarines for training and deployment operations. Seven tactical missile OST flights were conducted. A Regulus program milestone was reached on 20 October 1961 when GMU-10 launched Fleet Training Missile 1246 for the twenty-first time.[18]

1962-66

On 1 August 1962, Lieutenant Commander Roy L. Hoffman relieved Commander Peed as Officer-in-Charge of GMU-10. Hoffman had been Assistant Officer-in-Charge of GMU-10, as well as Assistant Officer-in-Charge of GMU-51 before to its disestablishment on 30 June 1959.

In the last full year of operation, GMU-10 processed 99 Regulus missiles in support of Submarine Squadron ONE activities. Thirty-three missiles were launched including six OST evaluations.[19] In June 1963, Lieutenant Commander Roy Hoffman was relieved by Commander Albert H. Thomas. On 31 August 1963, Lieutenant Leonard Leo relieved Commander Thomas as Officer-in-Charge, GMU-10. On 26 October 1963, Lieutenant Commander Eugene Wells, Jr., relieved Lieutenant Leo as Officer-in-Charge, GMU-10.

With the end of the Regulus program looming on the horizon, a review of the missile inventory at Naval Air Station, Barbers Point and Bonham Auxilary Landing Field revealed a substantial quantity of missiles remaining. Since the target drone capability of the missile had been amply demonstrated earlier in the program, the Office of the Chief of Naval Operations made this surplus of missiles available for use as aerial targets for surface and air units. Commander Fleet Air, Hawaii, was delegated as the coordinating agency for the assignment of missiles to the various "customers" and Commander Submarine Squadron ONE was given launch responsibility. GMU-10, Bonham Detachment, made all of the launches while Utility Squadron ONE (VU-1) was responsible for providing the safety chase and recovery aircraft.[20]

The missiles were given new designations. RGM-6A was the fleet training missile and RGM-6B was the tactical missile with the warhead removed. Two guidance systems were available, Trounce and radio command control. Trounce control was via an AN/BPQ-2 equipped van stationed at Barbers Point. Trounce was used only with the RGM-6B missiles since right, left and dump commands were all that were available. Out-of-sight-control using radio command control was used with the RGM-6A missiles since it provided much greater flexibility in flight profile with proportional control for pitch, rate of turn and throttle setting.[21]

Four flight profiles were available, covering a wide range of intercept opportunities. While the missiles were to be "used up" in this target service, the maximum amount of training was extracted on each flight. Thus, as air-to-air or surface-to-air missiles were being launched against the missiles, telemetered or inert warheads were used on the first passes and then the target presentation ended with live warhead ordnance being used. A limited number of miss distance indicators were still available and these could be used on a selective basis.[22] Since Regulus target operations usually involved more than one ship or aircraft as a firing platform and range time and safety clearances were tightly scheduled, GMU-10 prepared two missiles for firing in case the first missile had a launch pad failure. The last Regulus I flight operation for GMU-10 Bonham Detachment took place on 6 June 1966.

Endnotes

[1] *RFTPWAM #43, February, 1951, LVSCA, A40-18, Box 1.*

[2] *RFTPWAM #44, March 1951, LVSCA, A50-18, Box 1.*

[3] *Official History, Guided Missile Unit TEN, date unknown. Provided by Commander Samuel T. Bussey, USN (Ret.)*

[4] *Ibid.*

[5] *Ibid.*

[6] *Personal interview with Captain Eugene Pridinoff, USN (Ret.).*

[7] *The transicontainer was a special canister to house the Regulus I missile with the wings and tail folded. The entire container could then be rotated 45 degrees to permit transport on the highway.*

[8] *Personal interview with Captain Ralph Jackson, USN (Ret.).*

[9] *Ibid.*

[10] *Personal interview with Captain Robert Munroe, USN (Ret.).*

[11] *Personal interview with Brigadier General Jack Koehler, USA (Ret.).*

[12] *Jet Assist Drone to Be Launched Soon, 16 March 1955, The Virginia-Pilot.*

[13] *Guided Missile Unit FIFTY-FIVE, mimeographed history, from personal records of Bill Karr, GMGC(SS) USN, (Ret.).*

[14] *Personal interview with Captain Marvin Blair, USN (Ret.).*

[15] *Guided Missile Group TEN official history, undated. Provided by Commander Samuel T. Bussey, USN (Ret.).*

[16] *Guided Missile Unit TEN Unit History, 1 January 1960 to 31 December 1960. From Captain George Pullen Peed, USN (Ret.).*

[17] *Personal communication with Captain George P. Peed, USN (Ret.).*

[18] *Guided Missile Unit TEN Unit History 1 January 1961 to 31 December 1961. From personal papers of Captain George P. Peed, USN (Ret.).*

[19] *Ibid.*

[20] *Personal communications with Lieutenant Commander L.A. "Pat" Kilpatrick, USN (Ret.).*

[21] *Ibid.*

[22] *Ibid.*

Appendix III: Regulus I Guidance Submarines

The concept of missile guidance relay between submarines was first evaluated in the Loon program in an effort to maximize effective range. This concept carried over to the Regulus program, not only for the extension of range but also as part of the BiPolar Navigation guidance system that would require two "picket" submarines. With the cancellation of the BiPolar Navigation System in December, 1955, guidance submarines continued to be involved in Trounce evaluation and training operations.

Through 1961 dual submarine Trounce operations were still being conducted. None of the officers interviewed have mentioned the use of a second guidance submarine in their launch orders, but that does not preclude their possible role. Five nuclear powered submarines carried Trounce as part of their air search radar equipment. Details of guidance submarine operations are unavailable. I can at best list the submarines equipped at one time with Trounce guidance equipment:

USS *Argonaut* (SS 475)
USS *Carbonero* (SS 337)
USS *Cusk* (SS 348)
USS *Medregal* (SS 480)
USS *Runner* (SS 476)
USS *Sculpin* (SSN 590)
USS *Seadragon* (SSN 584)
USS *Snook* (SSN 584)
USS *Swordfish* (SSN 579)
USS *Sargo* (SSN 583)
USS *Torsk* (SS 423)

Appendix IV: Nuclear Warheads for the Regulus Program

Three nuclear weapons were designed and tested for use as warheads with Regulus I: the W-5, W-8 and W-27. The W-8 was considered briefly in 1953-54 as a penetrating weapon but was dropped when accurate television guidance was not achieved. The W-27 and W-28 were to have also been used with Regulus II.

Regulus I

W-5

In January of 1950 the Secretary of Defense gave approval for the adaptation of the Mk-5 implosion-type fission weapon for use as a warhead for Regulus I. The first design meeting took place in September 1951. Two years later, in August 1953, the MK 5 Mod 0 (W-5-0) warhead design was released for production. Initial production began in April 1954.[1]

The W-5 weighed 3,000 lbs and had a varying yield depending on which nuclear components were used, giving a nominal yield range of 40-50 KT.[2,3] Regulus W-5 warheads were not stored as such because of small demand. Instead, the W-5 was stored as the Mk-5 bomb and conversion kits stockpiled. A Submarine Weapons Assembly Team converted the Mk 5 Mod O gravity bomb into the W-5 warhead using a special adaptation kit that permitted selection of the yield options.

XW-8/Regulus

On 9 April 1948, the Military Liaison Committee of the Atomic Energy Commission (AEC) recommended that the Navy begin development of a gun-type, earth-penetrating, fission warhead.[4] The Mk-8 was chosen and was a derivative of the earlier "Little Boy" gravity bomb. While originally intended for use as an aircraft bomb, by March 1950, the system was approved for study as a possible guided missile warhead. On 14 February 1951, authorization was given by AEC Santa Fe Operations to develop the XW-8/Regulus warhead.[5] One flight test was conducted (See Chapter Six). XW-8/Regulus warhead work was suspended in May 1955, when the Navy decided to investigate thermonuclear warheads for Regulus.[6]

W-27

The W-27 was the warhead derivative of the first lightweight thermonuclear bomb, which began development on 21 May 1955.[7] Production engineering began in September, 1956 and production of the W-27/Regulus adaption kit began on 27 March 1957. Production of the W-27/Regulus warhead started in September, 1958. Twenty warheads were produced. The W-27 weighed 2,800 lbs and had a yield of several megatons.[8]

The decision to utilize the W-27 warhead necessitated the only major structural change in Regulus I design. The nose section of the missile was redesigned with a "bulged chin" to accommodate either the long and narrow W-27 or the spherical W-5. Retirement of the W-27 began in 1962 and ended in July, 1965.

Regulus II

W-27 and W/28

The W-27 was also the warhead for Regulus II.[9] The W-28 lightweight (1,700 lbs) thermonuclear weapon, with a yield of approximately 1 MT was to have been a second warhead for the Regulus II missile.[10]

Submarine Warhead Assembly Team Training

In 1951 the submarine force started sending personnel to the Armed Forces Special Weapons Project at Sandia Base, Albuquerque, New Mexico, for nuclear weapons training. Personnel were ordered in groups, or teams, consisting of two officers and four enlisted men. All were qualified in submarines. These teams were assigned to one of three Naval Special Weapons Units (NSWU's) located at Sandia Base. The early role of these NSWU's was to support the aircraft carrier bomb assembly teams. On a periodic basis, teams would deploy from Sandia Base to individual carriers for a tour of sea duty, returning to Sandia Base when the tour was completed.

The early submarine warhead assembly teams did not have a Regulus warhead to train with, as the warhead had not entered operational status. Instead, the teams concentrated on training on available weapons to the maximum extent possible, augmenting that training with a variety of other courses in radar, heavy equipment operation and the like. After a year of training, team members were individually ordered to report to the West Coast, usually to the Naval Air Missile Test Center, to GMTU-5 or later, GMU-50. With no nuclear warheads available at the GMU's for training, these individuals were assigned to other duties associated with the Regulus program.

In mid-1953, two new teams, commanded by Lieutenant Marvin Blair and Lieutenant Bernie Heesacker, reported to Sandia Base for training and both were assigned to the 1233rd NSWU (mid-way through the submarine team training, their team designations were changed to SWAT). By this time, a W-5 warhead was available, so these teams undertook training on the W-5, developing the Mk 5 bomb conversion procedure. Courses on bomb assembly were attended for the Mk 5, Mk 6, Mk 7 and M7 8 bombs as well as the Army's 280 mm artillery shell. Bomb radar courses were attended by the team commander, electrical officer, electrician's mate and electronics technician. The torpedomen, gunner's mate and other team members attended courses in heavy equipment operation.

The team commanders were designated as the team's Nuclear Supervisors and the electronics technicians were designated as Nuclear Technicians. Their training varied from several weeks for the Nuclear Technicians to three months for the Nuclear Supervisors. In addition, the team commanders were ordered to Rocky Flats, Colorado, for a three week period in order to work with live nuclear components such as Uranium-238 and Plutonium-239. After the trip to Rocky Flats, the team commmanders were certified by the Atomic Energy Commission to handle, have custody of, and be personally responsible for the active nuclear components associated with nuclear weapons. To maintain certification, these team members had to return to Sandia Base each year and requalify for their certification.[11]

Endnotes

1 *U.S. Nuclear Weapons: The Secret History, by Chuck Hansen, 1988, Orion Books, page 191.*
2 *Nuclear Weapons Databook, Volume I. Tom Cochran, Wayne Nail and Stan Norris, 1987, Ballinger Publishing, page 11.*
3 *Nuclear weapon yields are referred to in terms of the tons of equivalent conventional explosive such as TNT. KT refers to thousands of tons of TNT; MT refers to millions of tons of TNT.*
4 *Nuclear weapons at this time were known as either implosion-type or gun-type.*
5 *U.S. Nuclear Weapons: The Secret History. Chuck Hansen, 1988, Orion Books; page 191.*
6 *Ibid.*
7 *Class D referred to the fact that the weapon was "maintenance" free compared to the earlier fission weapons.*
8 *Ibid., pages 192-193*
9 *Request for XW-27/Regulus II Adaption Kit Proposal, from Chief, Bureau of Aeronautics to Chance Vought Aircraft, Inc. 19 March 1956. National Archives RG 72 box 125.*
10 *Hansen, page 106.*
11 *Personal communications with Captain Marvin S. Blair, USN (Ret.).*

Appendix V: Selected Regulus Submarine Characteristics, Construction and Deck Plans

While *Tunny* and *Barbero* were conversions of "fleet type" submarines, *Grayback, Growler*, and *Halibut* were new construction. Several specific aspects of their construction are described here as well as deck plans for *Tunny, Barbero, Grayback, Growler* and *Halibut*.

Engine Installation on *Grayback* and *Growler*

New design diesel-electric engines were installed in *Grayback* and *Growler*. Recent diesel-electric submarines such *Albacore, Tang, Trigger, Trout* and *Wahoo* had received General Motors "Pancake" engines where the cylinders were vertically stacked above the generator. These engines had a very high horsepower-to-weight ratio, ran at high speed and took up much less space then the World War II diesel engines. *Harder, Darter, Gudgeon, Grayback* and *Growler* got the Fairbanks Morse aluminum block, high speed engines which also took less space and had a higher rated horsepower to weight ratio then the General Motors engines.

In the case of *Grayback* and *Growler* these engines were also shock and sound isolated, previously done only in the three "K" class hunter-killer submarines of the early 1950's. Since these engines were operating at about twice the speed of other diesels, they generated noise at a higher frequency. Since higher frequencies are attenuated underwater much faster then low frequencies, in theory, the sound detection range was much shorter.

The major drawbacks in both of the new engine installations were discovered only after operational testing. Both engine types were unsatisfactory, with broken crankshafts, cracked liners, piston failure, fatigue failures of injectors and injector fuel tubing, burned out exhaust expansion joints and much more. The joke amongst the crew and the designers was that both General Motors and Fairbanks Morse had been about 100 pounds too successful in their weight reduction program. Very small increases in strength and rigidity would have made the engines highly successful.[1]

Halibut Construction Detail

Halibut displaced 4900 tons and was 350 feet long with a main deck higher above the waterline then usual to permit a dry deck in rough weather for launching her missiles. From the forward reactor compartment bulkhead aft, *Halibut* was identical to *Sargo*, the previous and first nuclear- powered submarine built at Mare Island. The control room and sail were nearly in the same place forward of this bulkhead as on *Sargo* but the sail had been elongated to accommodate the large guidance radar antenna. The total length of the compartments forward of the reactor compartment were almost as long as the *Sargo's* overall length. This space contained the usual submarine compartments plus the missile guidance centers, missile checkout and launch consoles and the equipment for operating the various items of missile hangar machinery.

Lying above and forward of this section was the missile hangar, an integral part of the hull of the boat on a rather unusual sloping angle for a submarine. Constructed as an independent pressure compartment, it was designed sloping down to the extended keel and then continuing forward to provide sufficient housing for the missiles and torpedo tubes. The hangar itself was larger in diameter then the main hull and had two 14 foot diameter hatches to the launch area on deck.

The distinctive feature of *Halibut* was the missile hangar door hump 100 feet forward of her sail. The hangar door was really two hatches fitted as protection against rough water. Hard tanks that could be pumped or blown, of capacity matching the floodable volume between the two doors, were provided. The hangar volume was roughly 35,000 cubic feet, virtually identical to the volume of a World War II fleet-type submarine. Missile storage was aft of the torpedo tubes with centrally positioned launcher rails to move the missile through the double hatch and out on to the deck. *Halibut* was originally designed to carry four Regulus II's in storage racks. Since a Regulus I was essentially half the length of a Regulus II, in theory, eight Regulus I's could then be carried as an alternative missile loadout.

The redesign of the hangar space after Regulus II cancellation resulted in four Regulus I's being carried in shock mitigation mounts, as protection during depth charge attacks, accessible by hydraulic crane. The aft missile on each side had to be removed before the forward missile was accessible. While the full pre-launch checkout could only be done on the launcher, minor checkouts could be accomplished in the stored position. Lieutenant Charles Baron, *Halibut's* first Missile Officer, realized that a fifth missile could be stored on the launcher. While it would not have the security of the shock mitigation system, if, once on deck it was determined to be damaged, it could be jettisoned and another missile quickly brought into position. He remembers that the shock mitigation system, was a nightmare. It consisted of two large, heavy segmented metal rings fitted around each missile. The rings, in turn, were supported by large shock absorbers. These were in turn supported by heavy metal foundation structures which were welded to the hull framing. This system was to a large extent the critical item in determining the missile load. While unwieldy and a monument to Rube Goldberg, Baron recalls that while loading the missiles was arduous, the unloading sequence was straight forward and did not impede the pre-launch operations to any great extent.

The Regulus II launcher was built before the program cancellation but the foundations and hydraulic cylinder were never installed in the hull. This would have been a trainable launcher with hydraulic controls located in the missile launch control center. Once Regulus II was canceled, a concerted effort was made to eliminate any associated gear to permit larger crew spaces. In this case, elimination of the hydraulics panels permitted the officers quarters, originally for 12 officers but with only 8 berths and 8 seats at the wardroom table, to be expanded to accommodate everyone. Elimination of the associated spare parts areas permitted additional crew spaces and food storage.

The Regulus I launcher elevated but did not train to port or starboard and was virtually the same as the short rail launchers on *Tunny* and *Barbero* . Heat resistant paint was applied to the forward part of the sail as protection against the missile's engine exhaust and the blast from the booster rockets.[2]

Endnotes

[1] *Personal communication with Rear Admiral Joseph Ekelund, USN (Ret.).*

[2] *Personal communications with Rear Admiral Paul J. Early, USN (Ret.).*

USS Tunny (SSG-282)

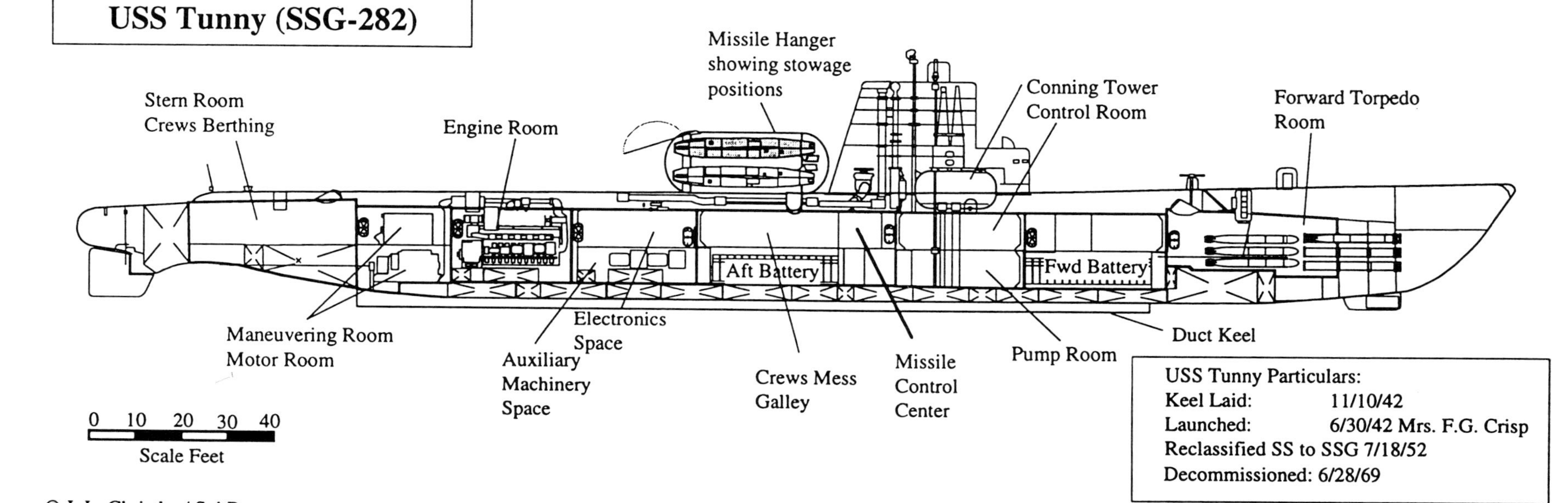

USS Tunny Particulars:
Keel Laid: 11/10/42
Launched: 6/30/42 Mrs. F.G. Crisp
Reclassified SS to SSG 7/18/52
Decommissioned: 6/28/69

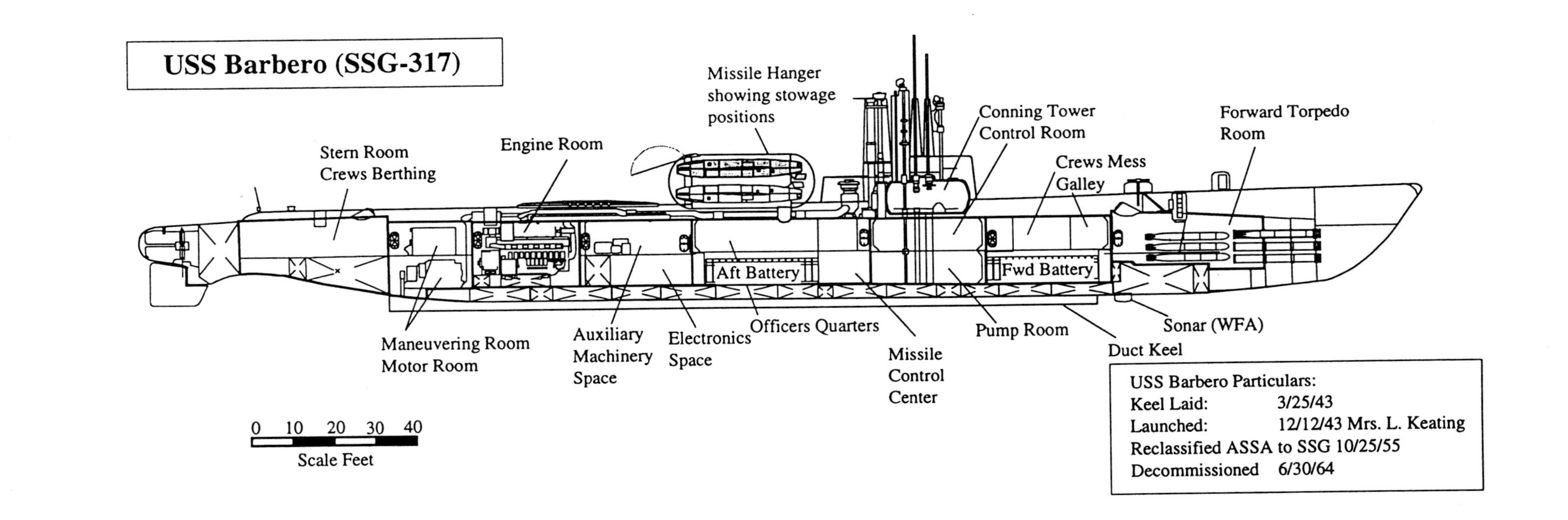
USS Barbero (SSG-317)
Missile Hanger
showing stowage
positions
Conning Tower
Control Room
Forward Torpedo
Room
Crews Mess
Galley
Stern Room
Crews Berthing
Engine Room
Aft Battery
Fwd Battery
Maneuvering Room
Motor Room
Auxiliary
Machinery
Space
Electronics
Space
Officers Quarters
Missile
Control
Center
Pump Room
Sonar (WFA)
Duct Keel
0 10 20 30 40
Scale Feet
USS Barbero Particulars:
Keel Laid: 3/25/43
Launched: 12/12/43 Mrs. L. Keating
Reclassified ASSA to SSG 10/25/55
Decommissioned 6/30/64

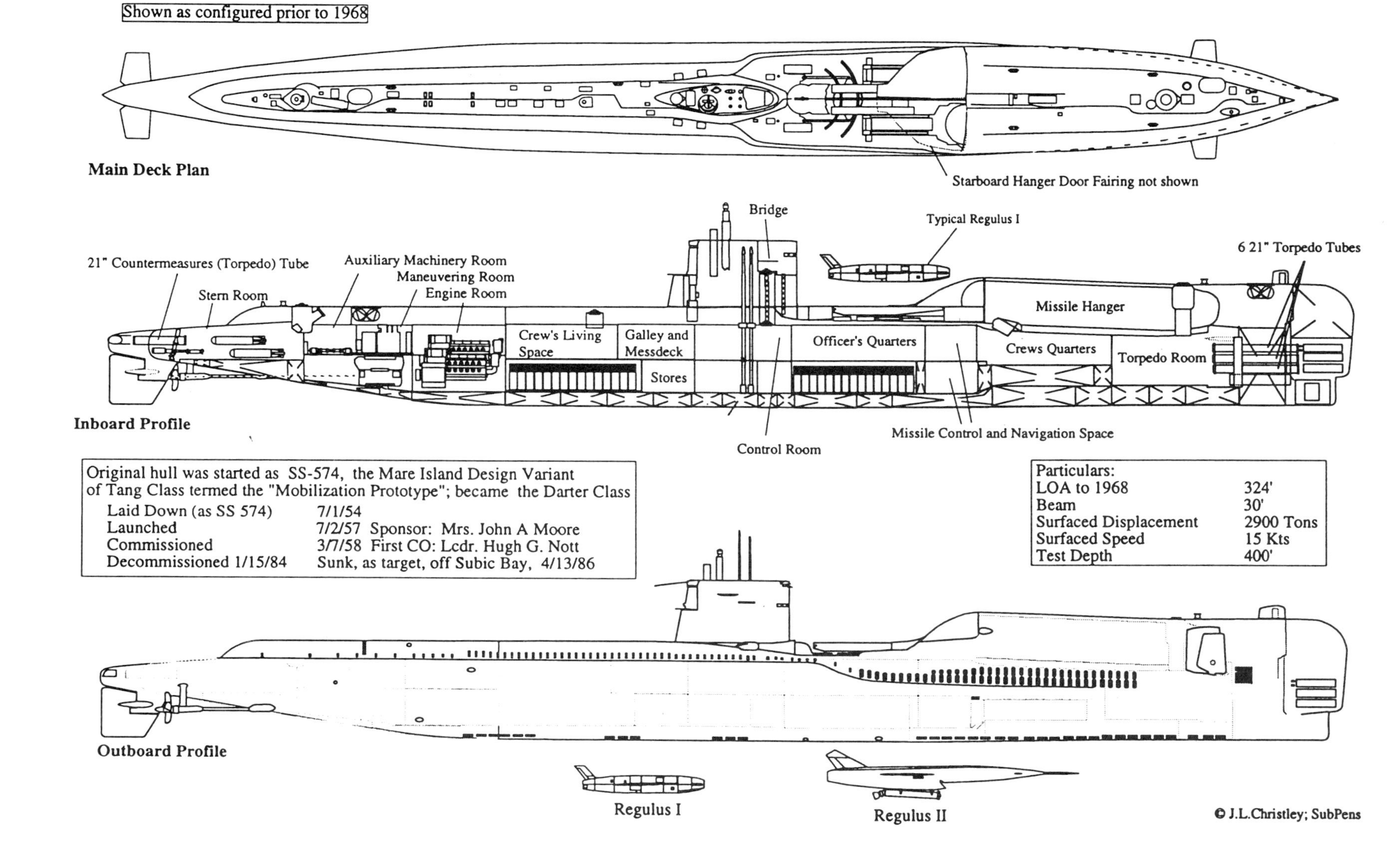

USS Grayback SSG-574
Shown as configured prior to 1968
Main Deck Plan
Starboard Hanger Door Fairing not shown
Bridge
Typical Regulus I
6 21" Torpedo Tubes
21" Countermeasures (Torpedo) Tube
Auxiliary Machinery Room
Maneuvering Room
Engine Room
Stern Room
Crew's Living Space
Galley and Messdeck
Stores
Officer's Quarters
Missile Hanger
Crews Quarters
Torpedo Room
Inboard Profile
Control Room
Missile Control and Navigation Space
Original hull was started as SS-574, the Mare Island Design Variant of Tang Class termed the "Mobilization Prototype"; became the Darter Class
Laid Down (as SS 574) 7/1/54
Launched 7/2/57 Sponsor: Mrs. John A Moore
Commissioned 3/7/58 First CO: Lcdr. Hugh G. Nott
Decommissioned 1/15/84 Sunk, as target, off Subic Bay, 4/13/86
Particulars:
LOA to 1968 324'
Beam 30'
Surfaced Displacement 2900 Tons
Surfaced Speed 15 Kts
Test Depth 400'
Outboard Profile
Regulus I
Regulus II
© J.L.Christley; SubPens

USS Grayback LPSS-574

Shown as configured after 1968 Conversion

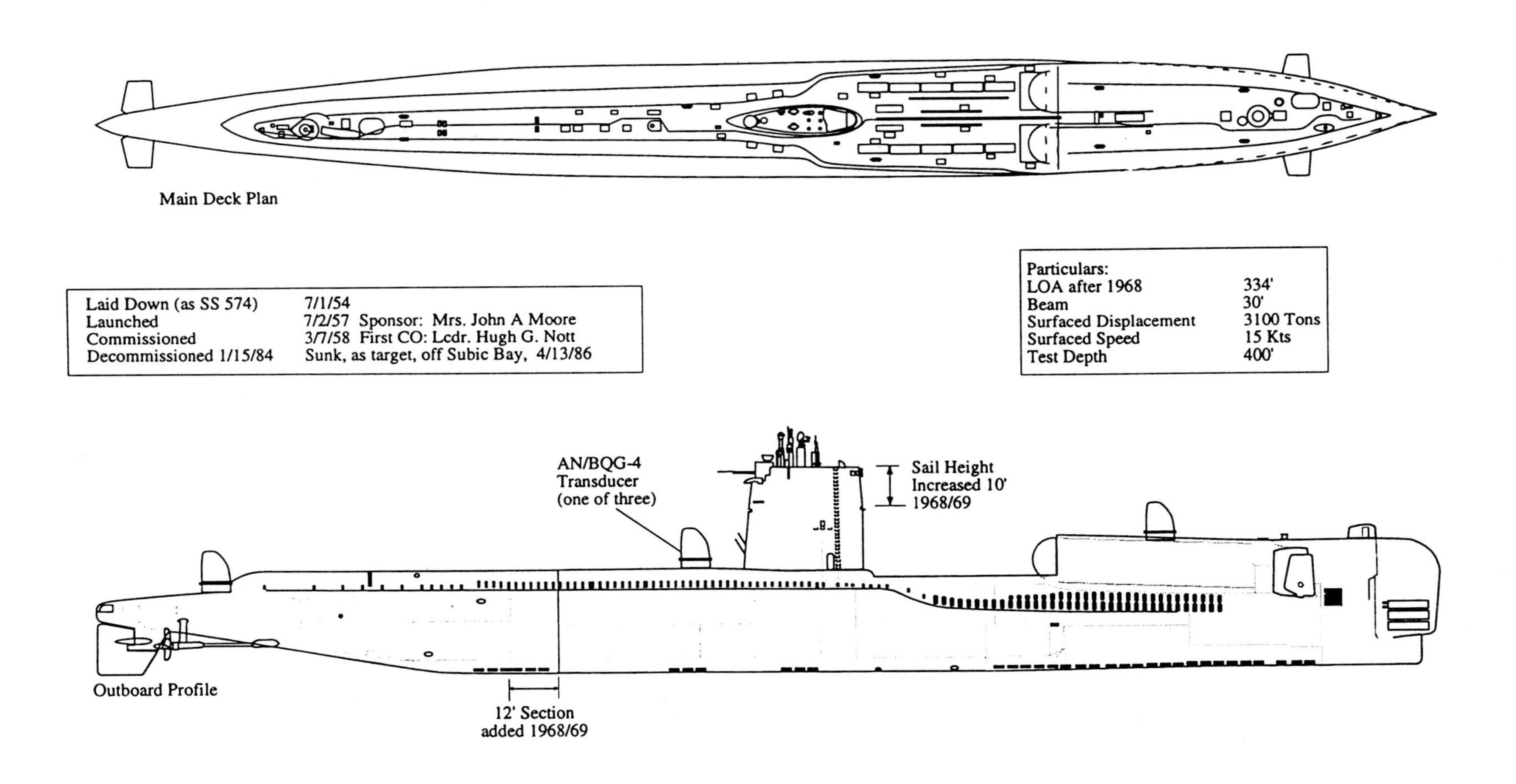

Laid Down (as SS 574)	7/1/54
Launched	7/2/57 Sponsor: Mrs. John A Moore
Commissioned	3/7/58 First CO: Lcdr. Hugh G. Nott
Decommissioned 1/15/84	Sunk, as target, off Subic Bay, 4/13/86

Particulars:	
LOA after 1968	334'
Beam	30'
Surfaced Displacement	3100 Tons
Surfaced Speed	15 Kts
Test Depth	400'

USS Growler SSG-577
(as configured prior to 1961)

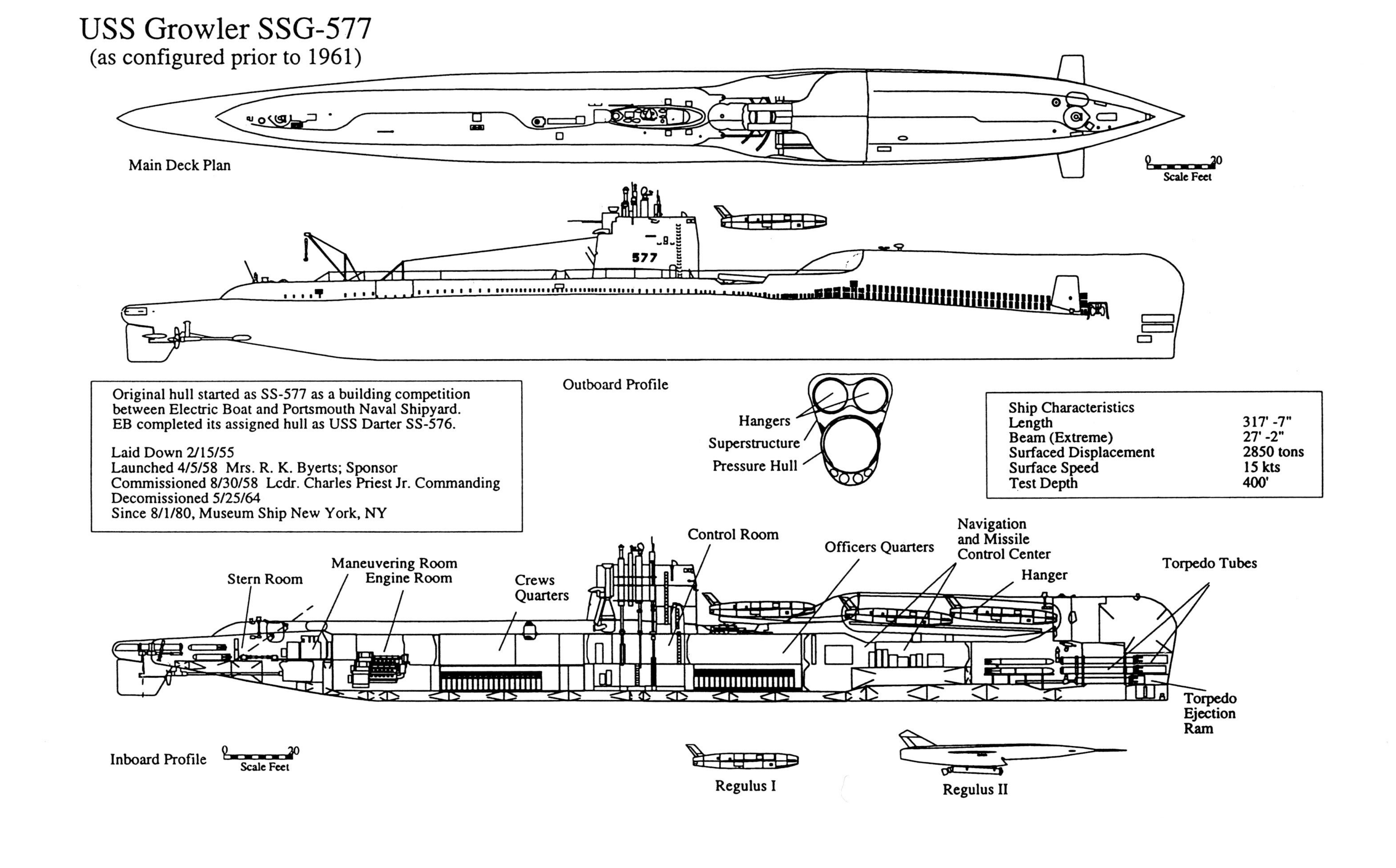

Original hull started as SS-577 as a building competition between Electric Boat and Portsmouth Naval Shipyard. EB completed its assigned hull as USS Darter SS-576.

Laid Down 2/15/55
Launched 4/5/58 Mrs. R. K. Byerts; Sponsor
Commissioned 8/30/58 Lcdr. Charles Priest Jr. Commanding
Decomissioned 5/25/64
Since 8/1/80, Museum Ship New York, NY

Ship Characteristics	
Length	317' -7"
Beam (Extreme)	27' -2"
Surfaced Displacement	2850 tons
Surface Speed	15 kts
Test Depth	400'

USS Growler SSG-577 (after 1961)

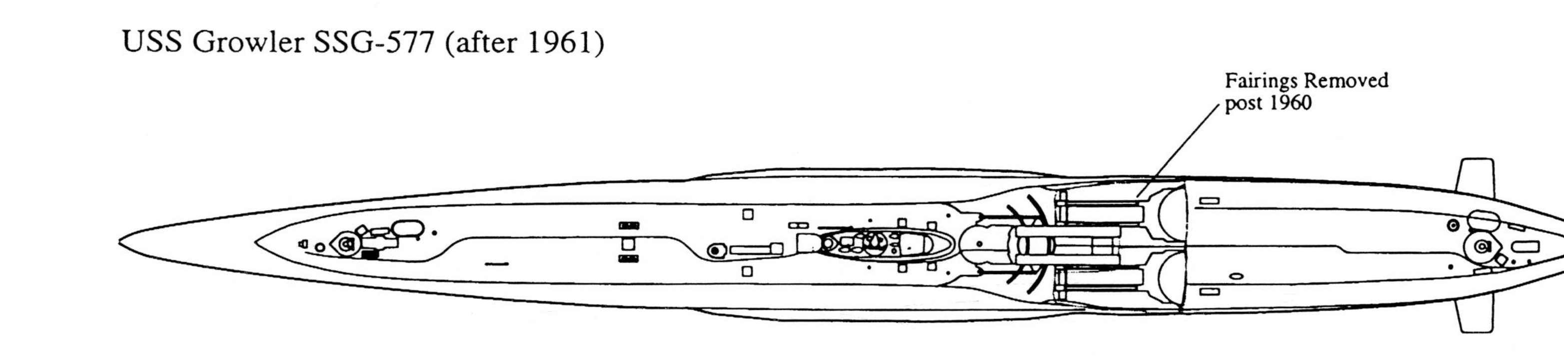

Laid Down 2/15/55
Launched 4/5/58 Mrs. R. K. Byerts; Sponsor
Commissioned 8/30/58 Lcdr. Charles Priest Jr. Commanding
Decomissioned 5/25/64
Since 8/1/80, Museum Ship New York, NY

Ship Characteristics	
Length	317' -7"
Beam (Extreme)	29' -2"
Surfaced Displacement	2850 tons
Surface Speed	15 kts
Test Depth	400'

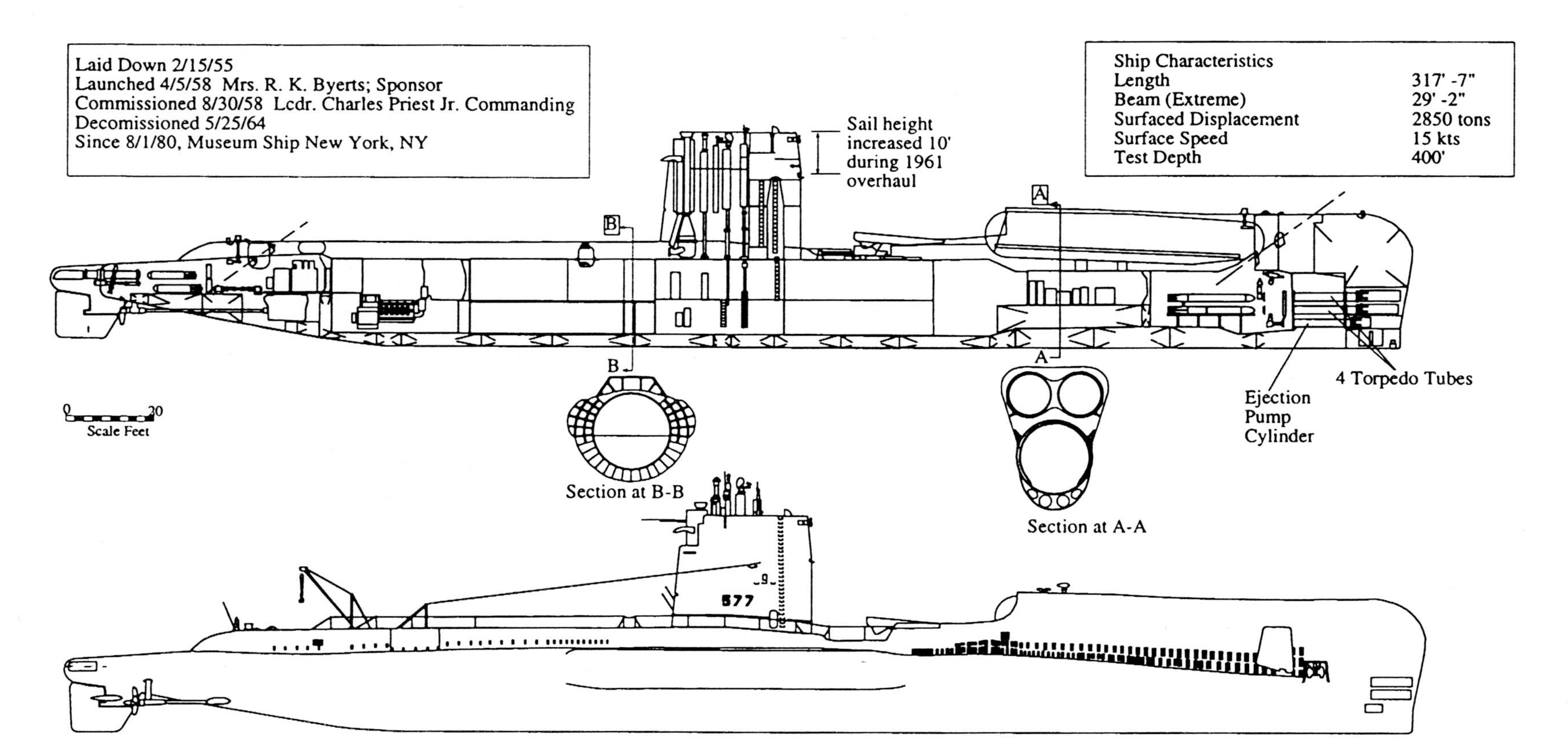

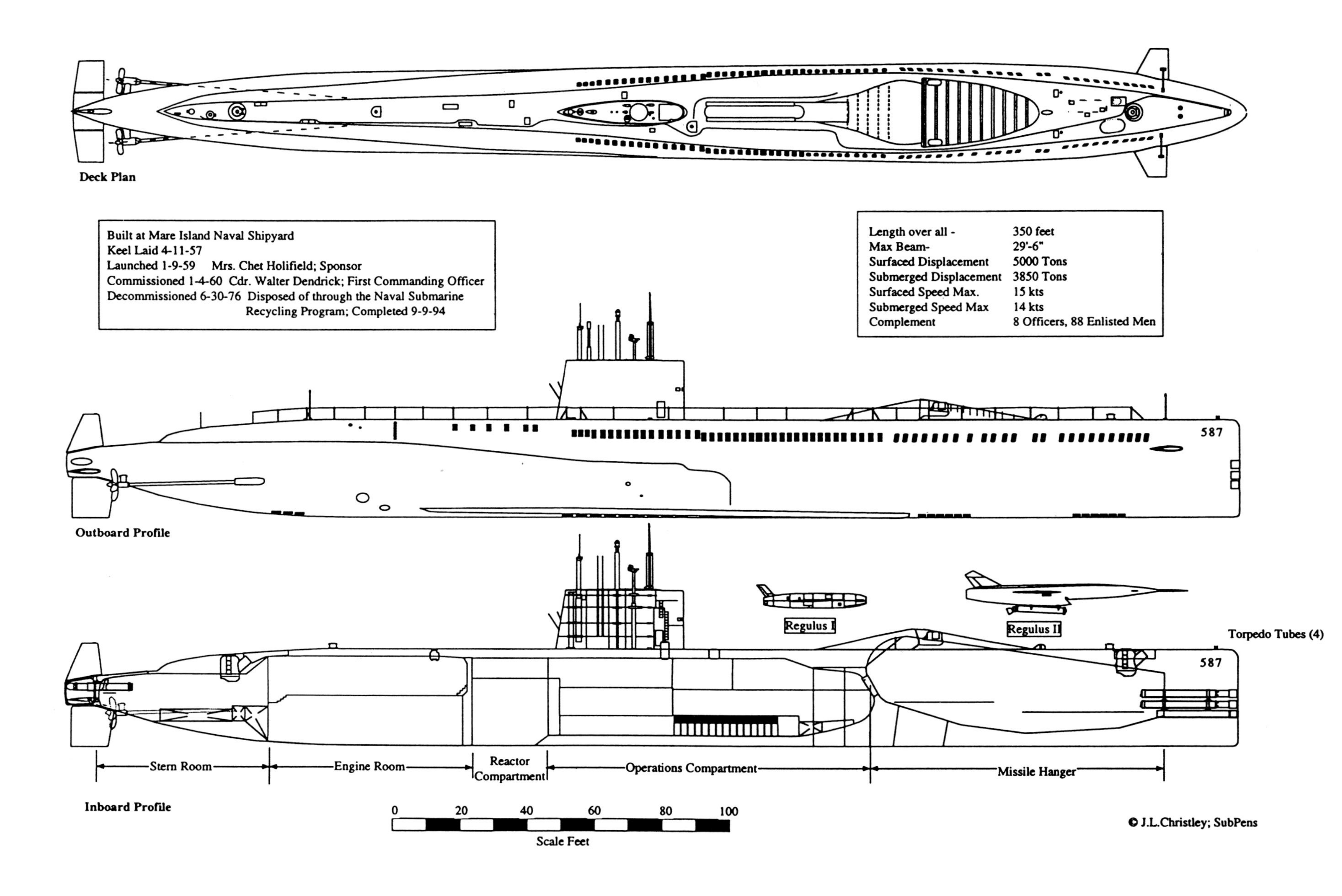
USS Halibut SSGN-587
Deck Plan
Built at Mare Island Naval Shipyard
Keel Laid 4-11-57
Launched 1-9-59 Mrs. Chet Holifield; Sponsor
Commissioned 1-4-60 Cdr. Walter Dendrick; First Commanding Officer
Decommissioned 6-30-76 Disposed of through the Naval Submarine Recycling Program; Completed 9-9-94
Length over all - 350 feet
Max Beam- 29'-6"
Surfaced Displacement 5000 Tons
Submerged Displacement 3850 Tons
Surfaced Speed Max. 15 kts
Submerged Speed Max 14 kts
Complement 8 Officers, 88 Enlisted Men
587
Outboard Profile
Regulus I
Regulus II
Torpedo Tubes (4)
587
Stern Room
Engine Room
Reactor Compartment
Operations Compartment
Missile Hanger
Inboard Profile
0
20
40
60
80
100
Scale Feet
© J.L.Christley; SubPens

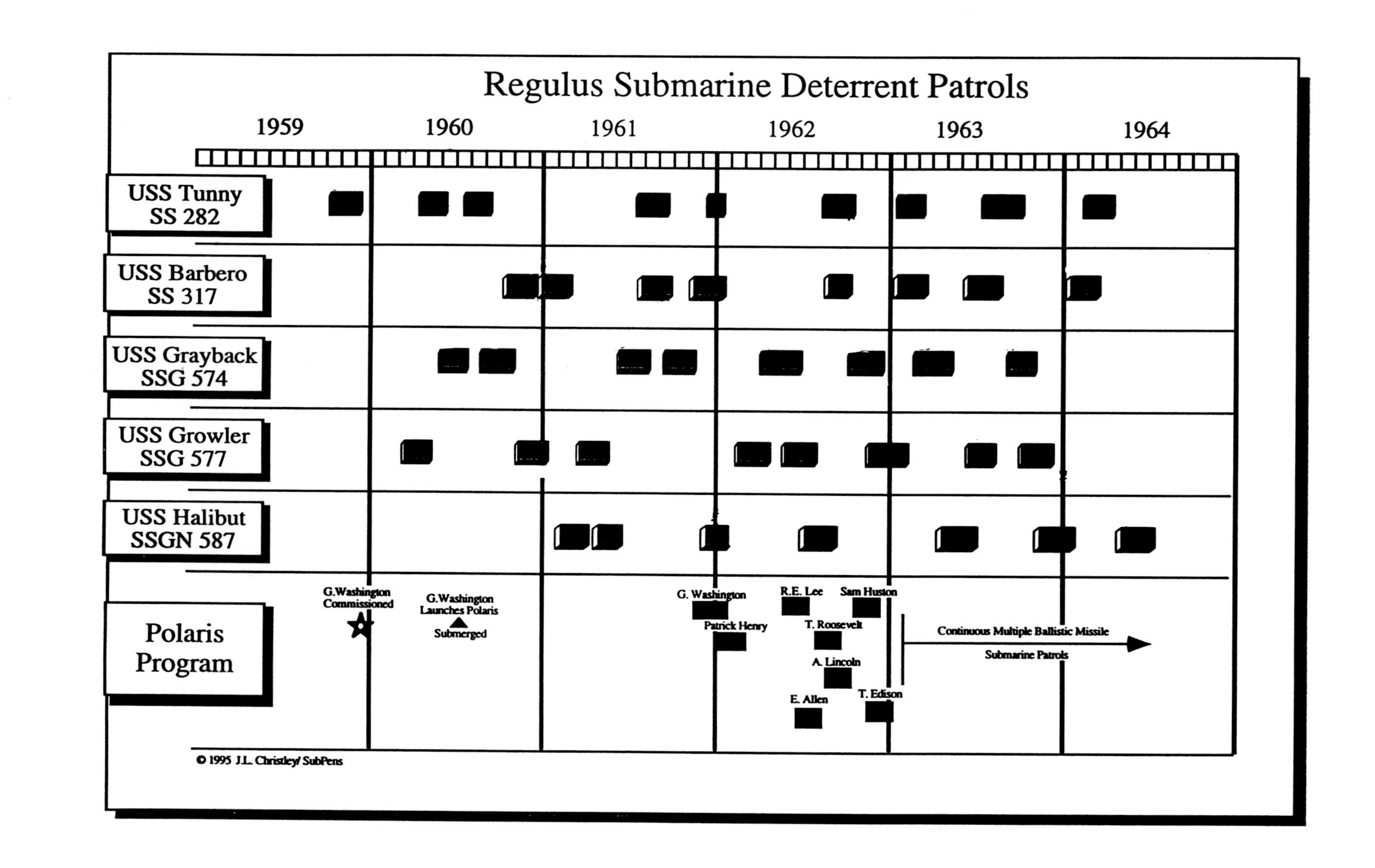
Regulus Submarine Deterrent Patrols
1959
1960
1961
1962
1963
1964
USS Tunny SS 282
USS Barbero SS 317
USS Grayback SSG 574
USS Growler SSG 577
USS Halibut SSGN 587
Polaris Program
G.Washington Commissioned
G.Washington Launches Polaris Submerged
G. Washington
Patrick Henry
R.E. Lee
Sam Huston
T. Roosevelt
A. Lincoln
E. Allen
T. Edison
Continuous Multiple Ballistic Missile Submarine Patrols
© 1995 J.L. Christley/ SubPens

Appendix VI: Regulus Submarine Strategic Deterrent Patrol Dates

Submarine	Depart	Return
Tunny	23 Oct 59	16 Dec 59
	22 Apr 60	17 Jun 60
	14 Jul 60	12 Sep 60
	23 Jul 61	28 Sep 61
	4 Nov 61	12 Jan 62
	24 Aug 62	29 Oct 62
	12 Jan 63	15 Mar 63
	13 Jul 63	3 Oct 63
	10 Feb 64	11 Apr 64
Barbero	30 Sep 60	2 Dec 60
	23 Dec 60	4 Mar 61
	23 Jul 61	28 Sep 61
	4 Nov 61	12 Jan 62
	24 Aug 62	29 Oct 62
	12 Jan 63	15 Mar 63
	10 Jul 63	28 Sep 63
	4 Jan 64	13 Mar 64
Growler	12 Mar 60	12 May 60
	10 Nov 60	18 Jan 61
	18 Mar 61	12 May 61
	11 Feb 62	24 Apr 62
	24 May 62	1 Aug 62
	24 Nov 62	11 Feb 63
	14 Jun 63	12 Aug 63
	4 Oct 63	13 Dec 63
Grayback	21 Sep 59	Dec 59
	31 May 60	30 Jul 60
	24 Aug 60	29 Oct 60
	5 Jun 61	13 Aug 61
	12 Sep 61	13 Nov 61
	2 Apr 62	3 Jun 62
	7 Oct 62	22 Dec 62
	20 Feb 63	11 May 63
	7 Sep 63	2 Nov 63
Halibut	9 Feb 61	10 Apr 61
	1 May 61	28 Jun 61
	20 Dec 61	31 Mar 62
	9 Jul 62	15 Sep 62
	29 Apr 63	6 Jul 63
	19 Nov 63	20 Jan 64
	7 May 64	14 Jul 64

Appendix VII: Regulus I Flight Operations and Production Summary

The number of Regulus I and II flights precludes the listing of all flights and their results. Flight operations are summarized by year for both Chance Vought and the Navy, in the case of Regulus I; and for the combination of the two for Regulus II. Missile production and cost information for Regulus I is complete and discussed. Missile production and costs for Regulus II is not complete but the data available is discussed.

Regulus I Flight Operations

Table I summarizes the Chance Vought flight operations from 1950 to 1957. Since an important aspect of the Regulus program was the recoverability and reuse of all but the tactical missiles, the number of missiles used each year was tabulated from the flight test record. For all of the years reviewed, several missiles were used in more then one year, explaining the discrepancy between the sum of "Number of Missiles Used" and the footnote of actual different airframes used. Chance Vought averaged 4.5 flights per recoverable missile, well above the goal of three stated in the first contract. Table II summarines the Navy flight operations from 1953 to 1961, the end of detailed flight records. Navy flight operations averaged 3.5 flights per missile, just slightly above the goal of the first contract. Navy missile teams launched nearly four times as many missiles as did the Chance Vought teams.

By the end of Chance Vought flight operations, Chance Vought personnel were averaging an 86% recovery rate while Navy personnel averaged 78% recovery over the first 7 years of missile deployment. Further review of the flight test record through 1961 indicates that 75% of the missiles flew three or more flights.

Regulus I Production and Costs

Regulus I missile contracts by year, with numbers of missiles and Bureau of Aeronautics airframe numbers are summarized in Table III.[1] While the Lot XXI missiles were ordered under the October 1956 contract, final delivery was made in December 1958. The final estimated cost of Regulus I missile production, as well as support equipment, flight testing and spare parts, was $158,000,000. Given that 514 missiles were produced, each missile cost an average of $307,392. *(See Tables I and II.)*

Regulus II Flight Operations Summary

Unlike the Regulus I program, a complete flight test history for Regulus II is unavailable. Through interviews and government documents, the flight test summary shown in Table IV was compiled.

Table I. Chance Vought Phase A Flight Operations from 1950 to 1957

Year	Flights	Recoveries	TM's	Number of Missiles Used
1950	1	0	0	1
1951	14	13	1	3
1952	20	14	3	6
1953	36	23	7	7
1954	31	22	3	8
1955	56	39	10	21
1956	29	19	7	7
1957	16	12	2	4
Totals:	203	142	33	57[a]

[a] The actual number of airframes used was 45.

Table II. Navy Phase B Flight Operations From 1953 to 1961

Year	Flights	Recoveries	TM's	Number of Missiles Used
1950				
1951				
1952				
1953	17	13	0	6
1954	30	18	0	15
1955	79	45	10	31
1956	191	140	19	77
1957	170	121	13	72
1958	158	118	15	72
1959	115	84	8	41
1960	51	38	6	17
1961[a]	19	13	3	12
Totals:	830	590	74	331[b]

[a] 1 January 1961 to 16 March 1961
[b] The actual number of airframes used was 234.

Table III. Regulus II Flight Operations Summary from 1956-1958

Year	Flights	Recoveries	TM's	Number of Missiles Used
1956	11	10	0	3
1957	10	9	0	4
1958	24	16	4	12
Totals:	45	35	4	19[a]

[a] The actual number of airframes used was 18.

Regulus II Production Summary

Regulus II missile delivery is shown in Table IV. Production information was derived from contract summaries through 1956 as shown in Table V. In early 1959, the Chance Vought inventory of Regulus II missiles included 14 FTM's in storage or ready for flight and 10 FTM's in the process of fabrication; 6 TM's in storage or ready for flight and 17 TM's in the process of fabrication.[2]

Endnotes

[1] *Regulus I Flight Test Index, 1961, LVSCA.*
[2] *Regulus II Program for NATO Application, Appendix B, page 1. Author's collection.*
[3] *Regulus II Progress Report 18, 1 January to 31 March 1957, page viii. Author's col'ection.*
[4] *A Summary of Historical and Management Data on LTV Aircraft and Missiles, Report Number 2-51130/3R-50155C, 31 October 1963. LVSC Library.*

Table IV. Regulus II Missile Delivery 1956-1962[4]

Year	Total Delivered
1956	2
1957	4
1958	27
1959	
1960	3
1961	11
1962	7
Total	54

Table V. Regulus II Contracts for Production 1953-1957[3]

Type	Model	Lot	Number	BuAer Number	Contract Number
FTM	XRSSM-N-9	I	4	GM2001-2004	NOa(s)-53-285
FTM	XRSSM-N-9	II	3	GM2005-2007	NOa(s)-55-410c
FTM	XRSSM-N-9a	III	1	GM2008	NOa(s)-55-410c
FTM	XRSSM-N-9a	IV	6	GM2009-2014	NOa(s)-56-775
TM	XSSM-N-9	V	5	GM3001-3005	NOa(s)-56-775
FTM	XRSSM-N-9a	VI	13	GM2015-2027	NOa(s)-57-199
TM	XSSM-N-9	VII	10	GM3006-3015	NOa(s)-57-199
		Total	42		

Appendix VIII: List of Interviews

Interviews were a major source of information for this history. The opportunity to put questions to the individuals that were at Edwards Air Force Base in 1950; on the cruisers and aircraft carriers; and particularly those on the submarines, made this history what it is. Their collections of photographs, newspaper articles and official documentation were of tremendous help.

Adams, Charles, *R., CDR, USN (Ret.)*
Albrecht, William
Allen, William D., CDR, USN (Ret.)
Bacon, Roger F., VADM, USN
Baker, Richard, LT, USN (Ret.)
Baron, Charles, CAPT, USN (Ret.)
Beckley, Jerry E., CWO, USN (Ret.)
Berry, Fred, CAPT, USN (Ret.) Deceased
Blair, Marvin S., CAPT, USN (Ret.)
Blount, Robert, LCDR, USN (Ret.)
Blount, Robert, RADM, USN (Ret.)
Bitter, C.J.
Brisco, Kenneth, CDR, USN (Ret.)
Bruyere, Thomas, CDR, USN (Ret.)
Burgess, James A., CDR, USN (Ret.)
Bussey, Samuel T., CDR, USN (Ret.)
Callahan, John, CDR, USN (Ret.)
Cannon, William
Christensen, Morris A., CDR, USN (Ret.)
Clarke, Richard M., CAPT, USN (Ret.)
Clegg, George, CDR, USN (Ret.)
Cobean, Warren R., CAPT, USN (Ret.)
Dagdigian, W.A., LCDR, USN (Ret.)
Dew, Carlos, CAPT, USN (Ret.)
Duke, Robert, CDR, USN (Ret.)
Early, Paul J., RADM, USN (Ret.)
Eckhart, Myron, CAPT, USN (Ret.)
Edgren, Donald, LCDR, USN (Ret.)
Ekelund, John J., RADM, USN (Ret.)
Engle, Joseph
Ferrazzano, Fred, CDR, USN (Ret.)
Fissette, Robert
Fortson, Thomas, CAPT, USN (Ret.)
Freeman, Dewitt, RADM, USN (Ret.)
Freitag, Robert, CAPT, USN (Ret.)
Fullinwider, Peter, CAPT, USN (Ret.)
Gallagher, Edward, CAPT, USN (Ret.)
Gensert, John R., CDR, USN (Ret.)
Gilchrist, R. Bruce, CAPT, USN (Ret.)
Gordon, Archer, CAPT, USN (Ret.)
Greer, James C., CDR, USN (Ret.)
Gregory, George, CDR, USN (Ret.)
Grisenti, Robert L., LT, USN (Ret.)
Gunn, William, CAPT, USN (Ret.)
Hallmark, William
Haney, John D., TMC(SS) USN (Ret.)
Hartzell, Harvey, LCDR, USN (Ret.)
Haughton, David E.
Hayward, John T., VADM, USN (Ret.)
Heckathorn, Gene, CAPT, USN, (Ret.)
Henderson, Donald, CAPT, USN (Ret.)
Heronemus, William, CAPT, USN (Ret.)
Hess, Walter, Ph.D.
Higgins, Joseph M.
Hislop, Harold
Hoffman, Roy, CDR, USN (Ret.)
Hogan, Robert, CDR, USN (Ret.)
Holcomb, Bud
Hull, Harry, RADM, USN (Ret.)
Ishii, Fredrick
Jackson, Ralph, CAPT, USN (Ret.)
Jaeger, Dolph, LCDR, USN (Ret.)
Jensen, Richard S., CDR, USN (Ret.)
Jordan, Richard, RMCS(SS), USN (Ret.)
Keeler, Norris, CAPT, USNR
Kelly, William, LCDR, USN (Ret.) Deceased
Ketner, Virgil,
Kilgore, William
Kilpatrick, Louis, LCDR, USN (Ret.) Deceased
Kintner, E.E., CAPT, USN (Ret.)
Klare, Jr., Herman, CAPT, USN (Ret.)
Kraft, Roy, CAPT, USN (Ret.)
Kurtz, Larry, CAPT, USN (Ret.)
Kutzleb, Robert, LCDR, USN (Ret.)
Leavitt, Jr., Horace, CAPT, USN (Ret.)
Lemieux, Norman, LCDR, USN (Ret.)
Leue, David, CAPT, USN (Ret.)
Lindsey, Eugene, CDR, USN (Ret.)
Lynne, Sam
Lucas, Jr., Fredrick C., CAPT, USN (Ret.)
Mangold, John F., CDR, USN (Ret.)
Mares, Ernest, CDR, USN (Ret.).
Martin, Sam, CDR, USN (Ret.)
May, Sel, CAPT, USN (Ret.)
McDonald, Carlton A.K., CAPT, USN (Ret.)
Melder, Lou
Micchelli, Bill
Miller, C.O.
Mock, Roy, CDR, USN (Ret.)
Monger, Jack, RADM, USN (Ret.)
Monthan, George, CAPT, USN (Ret.)
Morrison III, Julian K., LT, USN (Ret.)
Munroe, Robert, CAPT, USN (Ret.) deceased
Muir, Donald, MTC(SS), USN (Ret.)
Murphy, Walter P., CAPT, USN (Ret.)
Nash, Norman, CAPT, USN (Ret.)
Newbro, David
O'Connell, John F., CAPT, USN (Ret.)
Orrik, Fredrick, CAPT, USN (Ret.)
Osborn, James, ADM, USN (Ret.)
Ottobre, James,
Owens, Robert, CDR, USN (Ret.)
Palmer, Ray
Parsons, Gerald
Paulk, Glen
Pearson, Leroy, deceased
Peck, Paul, RADM, USN (Ret.)
Peed, George, CAPT, USN (Ret.)
Plog, Leonard, LCDR, USN (Ret.)
Powers, Leonard, RMC(SS), USN (Ret.)
Pridinoff, Eugene, CAPT, USN (Ret.)
Priest, Jr., Charles, CAPT, USN (Ret.)
Purkrabek, Paul, CAPT, USN (Ret.)
Rait, John, CDR, USN (Ret.)
Rader, Rex, CAPT, USN (Ret.)
Randell, Fred
Ransdell, Palmer
Reiter, Harry L., RADM, USN (Ret.)
Regnier, Donald, LCDR, USN (Ret.)
Richardson, Daniel, RADM, USN (Ret.)
Ruble, Byron C., CAPT, USN (Ret.)
Ryland, Ron, ICCS(SS), USN (Ret.)
Shanahan, Thomas L.,
Sibila, Alfred
Sims, William, CAPT, USN (Ret.)
Slonim, Charles, CAPT, USN (Ret.)
Smith, B.N.
Stahl, Douglas, CDR, USN (Ret.)
Stebbins, Tom, CWO, USN (Ret.)
Stewart, Robert
Stone, Don
Stromberg, Herman A., CDR, USN (Ret.)
Styer, Charles, CAPT, USN (Ret.)
Surman, Jr., William, CDR, USN (Ret.)
Sutherland, George
Thayer, Al, CDR, USN (Ret.)
Thomas, Larry
Tibbets, Herbert, CDR, USN (Ret.)
Travers, Paul
Trush, Steve
Welch, Bernard, LCDR, USN (Ret.)
Welch, Jack
Wetmore, I.E., CAPT, USN (Ret.)
Whitt, Jim
Wilkes, Morgan
Woods, Charles, CAPT, USN (Ret.)
Zelibor, Joseph, CDR, USN (Ret..)

Glossary

AEC: U.S. Atomic Energy Commission
ALF: Auxiliary Landing Field.
APSS: former designation for transport submarines — formerly SSP or ASSP
ATRAN: Automatic Terrain Recognition And Navigation system
BPN: Bi-Polar Navigational system
BPQ (AN/BPQ-1 & -2): designation of radar set for the Trounce guidance system
BuAer: Bureau of Aeronautics
BuOrd: Bureau of Ordnance
BuShips: Bureau of Ships
CEP: Circular Error Probable
CO: Commanding Officer
ComCruDesPac: Commander, Cruiser-Destroyer Force, U.S. Pacific Fleet.
ComSubDiv: Commander, Submarine Division (Number)
ComSubLant: Commander, Submarine Force, U.S. Atlantic Fleet.
ComSubPac: Commander, Submarine Force, U.S. Pacific Fleet.
ComSubRon: Commander, Submarine Squadron (Number)
CV: older designation for an aircraft carrier
CVA: a later designation for a conventional "attack" class aircraft carrier
DivCom: Division Commander of naval vessels
ECM: electronic countermeasures
FBM: Fleet Ballistic Missile
FTM: Regulus I "fleet training missile"
FTV: Regulus I "flight test vehicle"
GMSRON: Guided Missile Squadron
GMGRU: Guided Missile Group
GMU: Guided Missile Unit
JATO: Jet Assisted Take-Off
KDU: one of two target drone designations for Regulus I
KD2U: target drone designation for the Regulus II.
KT: kilotons or thousands of tons of TNT explosive equivalent power
Loran A,C: a radio beacon navigation system using hyperbola of constant signal time delay
MT: megatons or millions of tons of TNT explosive equivalent power
RCC: radio command control.
REAC: Reeves Electronic Analog Computer.
RIN: Regulus Inertial Navigation
SEATO: the SouthEastern Asia Treaty Organization
SHORAN: Short Range Radio Navigation System
SLAMEX: Submarine Launched Assault Missile Exercise
SLBM: Submarine Launched Ballistic Missile
SSBN: a nuclear powered submarine configured to launch ballistic missiles.
SSG: a diesel-powered submarine converted or built to launch aerodynamic guided missiles.
SSGN: a nuclear powered submarine configured to launch aerodynamic guided missiles.
SSN: a nuclear powered submarine.
TM: designation for a Regulus I or II tactical missile
TROUNCE: name given to the radar guidance system used with the Loon and Regulus I missiles.

First launch of a Loon missile from a submarine, 18 February 1947, off of Pt. Mugu, California. Missile was successfully launched, but the guidance system failed. (Courtesy of Dzikowski Collection)

Grayback *getting underway on 15 January 1962 for a training launch one month after returning from her fourth deterrent patrol. Note the missing aft port superstructure cover.*

Not all the flights ended succesfully. On 10 September 1953, FTV crashed 20 miles north of Barstow, California. L to R: "Dutch" Wetmore, Navy project officer; Olie Knightslep, inspector; Bob Grafton, field engineer; George Butler, field engineer; O. Perkins, Navy; Ray Perry, shop crew chief; Earl Young, electronics and C.O. Miller. (Courtesy of Miller Collection)

USS Cusk *(SS 348) prepares for launch. The pod-like hangar housed one missile with the wings detached. The long launch ramp was soon replaced by a shorter launcher which greatly improved the submarine's underwater preformance. (Courtesy of Berry Collection)*

Regulus Index

— H —

— I —

— J —

— K —

— L —

— M —

— N —

— O —

— P —

— R —

— S —

— T —

— U —

—V—

—W—

—Y—

—Z—

Regulus I missile, TM-1040, being readied for launch from USS Tunny *(SSG 282) on 1 June 1954.*

Regulus
CHANCE VOUGHT AIRCRAFT